Advances in Microwave Engineering

This text showcases recent advancements in the field of microwave engineering, starting from the use of innovative materials to the latest microwave applications. It also highlights safety guidelines for exposure to microwave and radio frequency energy. The book provides information on measuring circuit parameters and dielectric parameters.

- Explains microwave antennas, microwave communication, microwave propagation, microwave devices, and circuits in detail
- Covers microwave measurement techniques, radiation hazards, space communication, and safety measures
- Focuses on advanced computing technologies, wireless communication, and fiber optics
- Presents scattering matrix and microwave passive components and devices such as phase shifters and power dividers
- Showcases the importance of space communication, radio astronomy, microwave material processing, and advanced computing technologies

The text provides a comprehensive study of the foundations of microwave heating and its interactions with materials for various applications. It also addresses applications of microwave devices and technologies in diverse areas, including computational electromagnetics, remote sensing, transmission lines, radiation hazards, and safety measures. It emphasizes the impact of resonances on microwave power absorption and the effect of nonuniformity on heating rates. The text is primarily written for senior undergraduate students, graduate students, and academic researchers in the fields of electrical engineering, electronics and communication engineering, computer engineering, and materials science.

Modern Aspects of Computing, Devices, and Communication Engineering
Ankan Bhattacharya

Advances in Microwave Engineering: From Novel Materials to Novel Microwave Applications
Ankan Bhattacharya, Bappadittya Roy, Arnab De, Ujjal Chakraborty and Anup Kumar Bhattacharjee

Advances in Microwave Engineering

From Novel Materials to Novel Microwave Applications

Edited by
Ankan Bhattacharya, Bappadittya Roy,
Arnab De, Ujjal Chakraborty and
Anup Kumar Bhattacharjee

CRC Press
Taylor & Francis Group
Boca Raton London New York

CRC Press is an imprint of the
Taylor & Francis Group, an **informa** business

First edition published 2024

by CRC Press
6000 Broken Sound Parkway NW, Suite 300, Boca Raton, FL 33487–2742

and by CRC Press
4 Park Square, Milton Park, Abingdon, Oxon, OX14 4RN

CRC Press is an imprint of Taylor & Francis Group, LLC

ISBN: 978-1-032-46898-3 (hbk)
ISBN: 978-1-032-60610-1 (pbk)
ISBN: 978-1-003-45988-0 (ebk)

DOI: 10.1201/9781003459880

Typeset in Sabon
by Apex CoVantage, LLC

Contents

22 Employment of Antennas in Biomedical Applications: A Review   365

KRISHANU KUNDU AND NARENDRA NATH PATHAK

Preface

This book focuses on the recent technological breakthroughs in microwave engineering, including novel materials and novel microwave applications.

Microwave engineering and associated technologies will be examined for emerging innovations and cutting-edge research. Microwave engineering encompasses new breakthroughs and novel techniques in microwave antenna design as well as many elements of microwave propagation. Starting with single/multiband antennas, this book will demonstrate many design and performance features of wideband to super-wideband antennas, as well as multiple input/multiple outputs (MIMOs) for wireless communication, as well as various gadgets for portable wireless devices.

Reviews and research studies on microwave antennas, microwave communication, microwave devices and circuits, microwave propagation, electromagnetic interference/electromagnetic capability (EMI/EMC), radar technologies, computational electromagnetics, remote sensing, microwave measurements, microwave radiation hazards and safety measures, microwave transmission lines, space communication, radio astronomy, microwave material processing, microwave imaging, advanced computing technologies, wireless communication, microwave and fiber optics, etc., are proposed to be included in this book.

The goal of this book is to highlight various areas of microwave communication and networking in addition to highlight reviews and research on cutting-edge technologies by academicians, researchers, and scientists in the fields of microwave engineering and allied areas.

Editors

Ankan Bhattacharya earned BTech and MTech degrees in Electronics and Communication Engineering from West Bengal University of Technology, Kolkata, India. He completed his PhD at the National Institute of Technology, Durgapur, India. He is the author of several research papers which have been published in many reputed journals and conferences at the national and international levels. Dr. Bhattacharya is a life member of the Forum of Scientists, Engineers & Technologists (FOSET), a member of the Institution of Engineers India (IEI), and a member of the International Association of Engineers (IAENG). His areas of research are antenna engineering, computational electromagnetics, electronic circuits and systems, signal processing, microwave devices, and wireless communication technologies. He is also an editor or reviewer of many national and international journals of repute. He has organized and participated in many national and international conferences, seminars, workshops and webinars. Dr. Bhattacharya has been active in delivering invited talks and has also been a part of many national and international conferences in the capacity of coordinator, session chair, technical committee member, etc. Dr. Bhattacharya has been appointed as a guest editor of *SN Applied Sciences*, a multidisciplinary, peer-reviewed journal of Springer Nature (indexed in Scopus). He has also been inducted into the editorial board of the *Circuits and Systems* journal from Scientific Research Publishing (SCIRP). Dr. Bhattacharya has been appointed as an associate editor of the *Journal of Information Processing Systems* (JIPS), an official international journal of the Korea Information Processing Society (indexed in Scopus and ESCI). Presently he is associated with Hooghly Engineering & Technology College (HETC), Hooghly, India as an Associate Professor and Head of the Electronics and Communications Engineering Department.

Bappadittya Roy earned his PhD degree in electronics and communication engineering from the National Institute of Technology, Durgapur, India. He is presently associated with the Vellore Institute of Technology-AP University, Amaravati, India, as an associate professor of the School of Electronics Engineering. He is a senior member of IEEE and SLATE 2023 member from IEEE Hyderabad Section. His research areas are Millimeter Wave, Fractal Antennas, MIMO, Compact antenna for Mobile Communication and Biomedical applications. He has several indexed publications including articles in reputed journals, conferences, book chapters and books. Also, be involved in a few reputed international conferences and seminars as an organizing committee member. He has shared his experience as a Guest Editor of UGC CARE Journal, Publicity chair for IECC 2023, Osaka Japan, and Conference Chair of a series of IEEE International Conferences AISP22, AISP 23. Dr. Roy has been active in delivering invited talks and has also been a part of many national and international conferences in the capacity of coordinator, session chair, technical committee member, etc. He has received many national and international travel grants from Govt. of India and has traveled abroad for seminars, conferences and workshops. Dr. Roy received Best Research Publication (E&T), Publication Award and Raman Research awards from VIT-AP University in 2020,2021 and 2022. Dr.Roy always welcomes researchers and faculties to work in collaboration in the fields of Wireless communication and Antenna Technology.

Arnab De earned his BTech and MTech in Electronics and Communication Engineering from West Bengal University of Technology, West Bengal, India, in 2012 and 2014, respectively. He completed his PhD in Electronics and Communication Engineering in November 2021 from the National Institute of Technology, Durgapur, India, and is a member of the Institute of Electrical and Electronics Engineers (IEEE), a life member of International Association of Engineers (IAENG), a life member of Forum of Scientists, Engineers and Technologists (FOSET), and an associate member of Institution of Engineers, India (IEI). He is currently working as an Assistant Professor in the Department of Advanced Computer Science Engineering under School of Computing and Informatics at Vignan's Foundation of Science, Technology and Research, Vadlamudi, Andhra Pradesh, India since March, 2023. He has also travelled abroad for international conferences and seminars. He is an active reviewer of many journals and conferences of high repute. He has participated in numerous National/International Conferences, Seminars, Workshops and Webinars. He has also organized various Workshops and Seminars. He has also been an active member as well as Technical Committee Member of various workshops and conferences. His current research areas include compact Microstrip slotted antennas, Reconfigurable antennas, Defected Ground structures (DGS), Array Antenna, UWB and SWB communication, 5G, MIMO etc. To his credit, he has many reputed journals and conference publications in the fields of antenna engineering and wireless communications.

Ujjal Chakraborty is affiliated with the Department of Electronics and Communication Engineering, National Institute of Technology, Silchar. He is currently providing services as an assistant professor. He has authored and co-authored multiple peer-reviewed scientific papers and presented works at many national and international conferences. His contributions have garnered recognition from honorable subject experts around the world. He is actively associated with different societies and academies. His academic career is decorated with several reputed awards and funding. His research interests include compact multiband antennas, phased array antennas, body-wearable conformal antennas, and reconfigurable antennas.

Anup Kumar Bhattacharjee earned his BE degree in Electronics and Telecommunication Engineering from B.E. College, Shibpur (presently the Indian Institute of Engineering Science and Technology, Shibpur), Howrah, India, in 1983. He earned his ME and PhD degrees from Jadavpur University, Kolkata, India, in 1985 and 1989, respectively. Presently he is associated with the National Institute of Technology, Durgapur, West Bengal, India, as a senior professor with the Electronics and Communication Engineering Department. His research areas are electronic devices and circuits, microwave devices, microstrip antennas, embedded systems, VLSI systems, mobile communications, satellite communications, etc. He has successfully supervised or co-supervised 25 PhD students and has published over 150 articles in national and international journals, books, and conferences.

*Dr. Ankan Bhattacharya dedicates this book
to the loving memory of his father, the late
Amit Kumar Bhattacharya*

Contributors

Takashiro Akitsu
Tokyo University of Science
Tokyo, Japan

K. Anusha
Kumaraguru College of Technology
Coimbatore, Tamil Nadu, India

Yatish Beria
Dibrugarh University
Dibrugarh, Assam, India

Shipra Bhatia
CSIR-CEERI Pilani
Rajasthan, India

Anup Kumar Bhattacharjee
National Institute of Technology
Durgapur
Durgapur, West Bengal, India

Koyndrik Bhattacharjee
Dr. B. C Roy Engineering College
Durgapur, West Bengal, India

Ankan Bhattacharya
Hooghly Engineering & Technology
College
Hooghly, West Bengal, India

Ashim Kumar Biswas
JIS College of Engineering
Kalyani, West Bengal, India

Moumita Bose
National Institute of Technology
Silchar
Silchar, Assam, India

Akash Buragohain
Dibrugarh University
Dibrugarh, Assam, India

Ujjal Chakraborty
National Institute of Technology Silchar
Silchar, Assam, India

C. Jisha Chandra
Gnanamani College of Technology
Namakkal, Tamilnadu, India

Anirban Chatterjee
National Institute of Technology Goa
Ponda, Goa, India

Himadri Sekhar Das
Haldia Institute of Technology
Haldia, West Bengal, India

Gouree Shankar Das
Dibrugarh University
Dibrugarh, Assam, India

Arnab De
Vignan's Foundation for Science,
Technology, and Research
Guntur, Andhra Pradesh, India

Ashis De
Gargi Memorial Institute of
   Technology
Kolkata, West Bengal, India

D. Mohana Geetha
Sri Krishna College of Engineering
   and Technology
Coimbatore, Tamil Nadu, India

Anumoy Ghosh
National Institute of Technology
   Mizoram
Aizawl, Mizoram, India

Chandan Kumar Ghosh
Dr. B. C. Roy Engineering College
Durgapur, West Bengal, India

Debalina Ghosh
Indian Institute of Technology
   Bhubaneswar
Bhubaneswar, Odisha, India

J. Rajeshwar Goud
Hyderabad Institute of Technology
   and Management
Hyderabad, Telangana, India

Ayushi Jain
Symbiosis Institute of Technology
Pune, Maharashtra, India

Sweety Jain
Samrat Ashok Technological
   Institute
Vidisha, Madhya Pradesh, India

Partha Protim Kalita
Dibrugarh University
Assam, India

S. Kannadhasan
Study World College of Engineering
Coimbatore, Tamil Nadu, India

K. Karthika
Kumaraguru College of Technology
Coimbatore, Tamil Nadu, India

K. Kavitha
Velammal College of Engineering
   and Technology
Viraganur, Tamil Nadu, India

Kalyan Sundar Kola
GMR Institute of Technology
Rajam, Andhra Pradesh, India

Praveen Kumar
National Institute of Technology
   Jamshedpur
Jamshedpur, Jharkhand, India

Ranjeet Kumar
National Institute of Technology
   Jamshedpur
Jamshedpur, Jharkhand, India

Vicky Kumar
JIS College of Engineering
Kalyani, West Bengal, India

Aparna Kundu
Durgapur Institute of Advanced
   Technology and Management
Durgapur, West Bengal, India

Krishanu Kundu
GL Bajaj Institute of Technology &
   Management
Greater Noida, India

Santosh Kumar Mahto
Indian Institute of Information Tech-
   nology Ranchi
Ranchi, Jharkhand, India

Santimoy Mandal
R.V.S College of Engineering &
   Technology
Jharkhand, India

Tapan Mandal
Govt. College of Engineering and
    Textile Technology
Serampore, West Bengal, India

Barun Mazumdar
Gargi Memorial Institute of
    Technology
Kolkata, West Bengal, India

Santanu Mishra
Haldia Institute of Technology
Haldia, West Bengal, India

Pratik Mondal
Gayatri Vidya Parishad College of
    Engineering
Madhurawada, Visakhapatnam,
India

R. Nagarajan
Gnanamani College of Technology
Tamilnadu, India

Biswarup Neogi
JIS College of Engineering
Kalyani, West Bengal, India

Pravesh Pal
National Institute of Technology
    Jamshedpur
Jamshedpur, Jharkhand, India

Aparna Panja
National Institute of Technology
    Durgapur
Durgapur, West Bengal, India

Narendra Nath Pathak
Dr. B. C. Roy Engineering College
Durgapur, West Bengal, India

Kranti D. Patil
Trinity Academy of Engineering
Pune, Maharashtra, India

A. Mallikarjuna Prasad
University College of Engineering
Kakinada, Andhra Pradesh, India

S. Raghavan
National Institute of Technology Trichy
Trichy, Tamil Nadu, India

Praful Ranjan
Government Girls Polytechnic Varanasi
Varanasi, Uttar Pradesh, India

N. V. Koteswara Rao
Chaitanya Bharathi Institute of
    Technology
Hyderabad, Telangana, India

Vanitha Rani Rentapalli
VIT AP University
Amaravati, Andhra Pradesh, India

Bappadittya Roy
VIT AP University
Amaravati, Andhra Pradesh, India

Koushik Roy
United International University
Dhaka, Bangladesh

Gourisankar Roy Mahapatra
Haldia Institute of Technology
Haldia, West Bengal, India

Vaibhav Saini
Symbiosis Institute of Technology
Pune, Maharashtra, India

Amit Sarkar
Indian Institute of Technology
    Bhubaneswar
Bhubaneswar, Odisha, India

Manoj Sarkar
CSIR-CEERI Pilani
Rajasthan, India

Partha Pratim Sarkar
University of Kalyani
Kalyani, West Bengal, India

Abhishek Sarkhel
National Institute of Technology
 Meghalaya
Shillong, Meghalaya, India

M. Shobana
JJ College of Engineering and
 Technology
Tiruchirappalli, Tamil Nadu, India

Rashmi Sinha
National Institute of Technology
 Jamshedpur
Jamshedpur, Jharkhand, India

G. Srividhya
Gnanamani College of Technology
Namakkal, Tamilnadu, India

D.M. Yadav
SND College of Engineering &
 Research Center
Yeola, Maharashtra, India

Chapter 1

Microstrip Antennas

Theory, Principles and Review
of Literature

Ankan Bhattacharya

1.1 INTRODUCTION

The most popular and revolutionary wireless technology available to date
are microstrip antennas. A radiating patch element and ground plane, both
made of conducting materials (copper) on either side of a dielectric substrate
material, are the basic constituents of a microstrip antenna [1]. Compared
to conventional antennas, microstrip antennas have many advantageous
features. They are cost-effective, lightweight, easy to reproduce, reliable,
easy to fabricate, etc. They can have various configurations like rectangular,
square, trapezoidal, circular, elliptical, etc. [2, 3]. Microstrip antennas can
be applied from low- to high-frequency ranges, approximately from the
MHz range to GHz. With the advancement of wireless technology, multiple
band systems have come into the picture. However, this type of antenna
has drawbacks that reduce their performance. The quality factor, Q, is very
high in the case of microstrip antennas [4]. Q represents the various associated antenna losses. A narrow bandwidth results if the value of Q is very
high. The substrate thickness can be reduced to reduce the Q value. But
with the increment in substrate thickness, a major portion of the power
from the source is lost as surface waves. Antenna efficiency, η, is also an
important point of consideration. As pointed out by H. A. Wheeler in [5],
the efficiency is limited due to coupling closeness of the antenna and its
tuning element. The effect of antenna size on efficiency is discussed in [6]
by R. F. Harrington. Radiation efficiency is related to the dissipation factor, D, which in turn depends on the electrical size of an antenna. For an
antenna to be reasonably efficient, D should be close to unity. The maximum radiation efficiency of a small antenna, in fact, depends on its electrical size, frequency of operation and the conductivity of the metal. For a
broadband antenna, at the low frequencies, the antenna might well become
electrically small. Therefore, the size of an antenna should be optimized
to achieve the maximum radiation efficiency. Ideally a microstrip antenna
should be able to transmit all the power i.e. it is fully efficient in technical

DOI: 10.1201/9781003459880-1

terms. However, it fails to do so because of some unwanted power losses like metallic losses, dielectric losses, mismatch of impedance, etc. Furthermore, L. J. Chu discussed the certain physical limitations of omnidirectional antennas in [7]. Band-gap structures can be employed to minimize surface waves. Antenna arrays can be implemented as an alternative to overcome other drawbacks like low gain and low power. In wireless communication systems a rapid evolution has taken place. So the researchers are developing new radiating devices. Therefore, use of compact microstrip antennas is one of the solutions. In spite of a few disadvantages, microstrip antennas have other advantages. In the modern world, due to the tremendous technological advancements, the demand for compact antennas with a wide bandwidth is increasing constantly. The two most popular wireless technologies prevailing today are wide local area network (WLAN) and WiMAX [8]. Antenna researchers are engaged in inventing novel techniques and formulas to achieve wide-bandwidth microstrip antennas to fulfill the demand of advancing communication technology. In earlier times, for a particular wireless system, a particular antenna type with a certain radiation pattern was implemented. If the antenna bandwidth is increased, it will help to fulfill the bandwidth requirement of other wireless communication standards. Microstrip antennas can be applied for G.S.M. 1800 (01.71–01.88 GHz), P.C.S. 1900 (10.93–01.99 GHz), G.P.S. (01.57–1.58 GHz), WiMAX (03.3–03.7 GHz) and WLAN bands (05.15–05.85 GHz) and also in several other applications suited for commercial or military purposes [9]. The applications of microstrip antennas can also be extended to satellite communications and remote sensing [10].

1.2 MICROSTRIP PATCH ANTENNA DESIGN

Researchers in antenna design have proposed various formulas and numerical methods to ease the computational means for designing microstrip antennas [11]. A microstrip patch antenna is basically made of a ground plane and patch element, both composed of conducting materials (generally Cu). In between the patch element and the ground plane, there is a substrate material, which possesses definite dielectric properties. The patch can be of various shapes and sizes. Photo-etching can be performed to design the patch and the feed. Patches may be of various shapes and sizes like square, circular, rectangular, elliptical, fractal, etc. The thickness of the patch should be, $t \ll \lambda_o$, where λ_o is the wavelength (of free space). The dielectric substrate height, h, should be usually between $00.003\lambda_o \leq h \leq 00.05\lambda_o$. The substrate's relative permittivity (ε_r) is generally between $02.2 \leq \varepsilon_r \leq 12$. Different feeding techniques and basic formulas for obtaining patch length and width are discussed in the following sections.

1.3 FEEDING TECHNIQUES

Microstrip antennas have various feeding techniques. The analysis and design of different microstrip antennas using different feeding techniques are discussed here. The method of feeding is either of a contacting type or non-contacting type. In the contacting method, for feeding power to the antenna directly, a microstrip line is applied, but in the non-contacting scheme, power is transferred to the patch by means of electromagnetic field coupling. The coaxial probe, microstrip line, proximity and aperture coupling are the four commonly used feeding techniques [12].

A strip of conducting material is fixed at the edge of the patch directly in the microstrip line feeding technique. The width of the strip should be smaller than the patch width. In this kind of feeding technique, the feed is placed on the same substrate as a planar structure could be obtained. Thus this type of feeding technique is easy and also provides good impedance matching.

Another popular feeding technique is the coaxial or probe feeding technique. The conductor originating from the inner portion of the connector finds its way all within the substrate material and is affixed to the patch element, whereas the conductor at the outer side is attached to the ground plane only. In this method of feeding, the feed point can be selected anywhere on the patch for matching input impedance.

In the aperture coupled feeding method, the separation between the patch and feed line is made by the ground plane. A slot is made through the ground plane for coupling of the feed and the patch element. The aperture is generally centered below the patch to minimize cross-polarization, which may occur due to the structure symmetry. The percentage of mutual coupling between feed and patch is decided by the aperture's dimension and position.

The proximity coupled feeding technique is a type of coupling scheme that is electromagnetic in nature. Here, in between two substrate materials, the feed line exists. On the top of the upper substrate material lies the patch element. Spurious radiation can be minimized by applying this type of feeding technique. Due to the increase in overall thickness, high bandwidth can be obtained.

1.4 MODELS OF ANALYSIS OF MICROSTRIP ANTENNAS

Figure 1.1 displays the structure of a microstrip transmission line. The line's width is represented by W and h represents the substrate height. It basically consists of a strip line made of some conducting material and is isolated from the ground plane by means of some dielectric substrate, which is characterized by its dielectric permittivity.

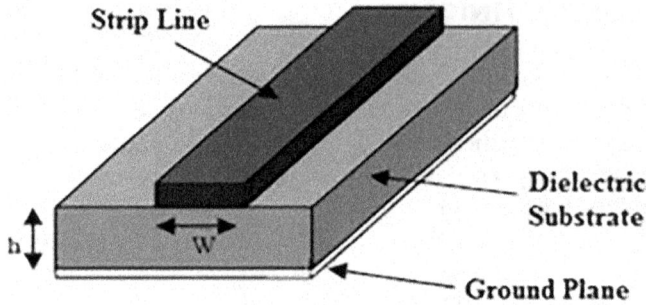

Figure 1.1 Microstrip line with dielectric substrate and ground plane.

Transverse electromagnetic (TEM) mode is not supported by the transmission line, since the phase velocities in the substrate and air are different. However, it supports quasi-TEM mode. Therefore, to account for the effects of fringing and wave propagation, it is essential to obtain the effective electric permittivity (ε_{eff}). Since fringing fields are spread in the dielectric substrate as well as in the air, the magnitude of ε_{eff} is less than ε_r. The effective dielectric constant ε_{eff} can be represented in [13] as,

$$\varepsilon_{eff} = \frac{\varepsilon_r + 1}{2} + \frac{\varepsilon_r - 1}{2} \left[1 + \frac{12h}{W} \right]^{\frac{-1}{2}} \tag{1.1}$$

where,
ε_r = Substrate's permittivity
ε_{eff} = Substrate's effective permittivity
h = Substrate's height
W = Patch's width

Along the x direction is considered the patch's length, the patch width has been considered along the direction of the *y-axis* and the height of substrate has been considered along the direction of the *z-axis*. The patch's length should be less than half of the wavelength $(\lambda / 2)$, and the antenna supports the fundamental transverse magnetic (TM) mode of operation, where λ denotes the wavelength of the electromagnetic (EM) wave and is almost equal to $\lambda_o / \sqrt{\varepsilon_{eff}}$ where λ_o denotes the wavelength of free space. The TM mode signifies that variation of the field is $\lambda / 2$ cycles only along the patch length, but it does not vary along the patch width. Figure 1.2 displays the structure of a microstrip patch antenna along with a feed line. The voltage is maximum along the patch width, and due to the open-ended structure, the current is minimum. Near the edges, the fields are resolved into tangential and normal components with the ground plane as the reference. Along the

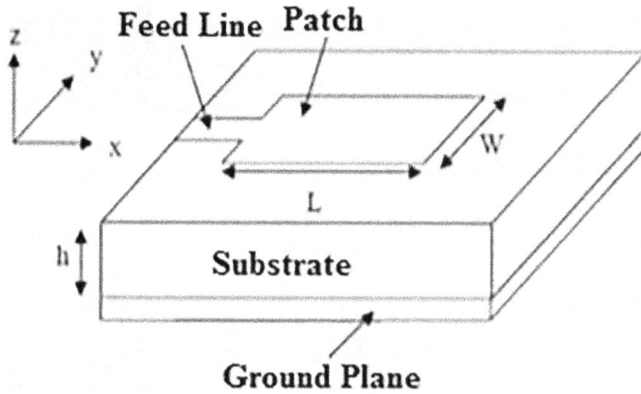

Figure 1.2 3D view of a microstrip patch antenna.

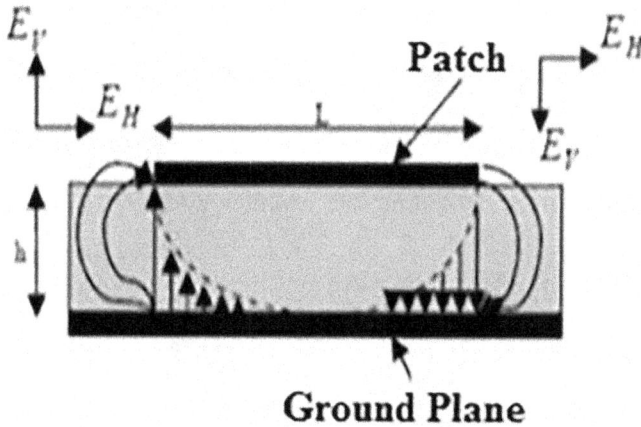

Figure 1.3 Side view of antenna, displaying the orientation of the radiating field components.

width, near the edges, the resolved normal components of the electric field are oppositely directed and therefore they are not in phase. As the patch length is $\lambda/2$, in the broadside, they nullify one another. In-phase tangential components, as shown in Figure 1.3, generate a resulting field, which is directed orthogonally to the structure's surface. Along the patch width, the edge behaves as a radiating slot, which is separated by $\lambda/2$ distance, and they radiate above the ground plane.

The electrical dimension of the patch is larger than its original physical dimension. If along the patch length the dimension has been extended by ΔL, it can be formulated [14] as,

$$dL = 00.412 \frac{h(\varepsilon_{re} + 0.30)\left(\dfrac{W}{h} + 0.264\right)}{(\varepsilon_{re} - 0.258)\left(\dfrac{W}{h} + 0.80\right)} \tag{1.2}$$

The patch's effective length, L_{eff}, can be formulated as,

$$L_{eff} + 2\,\Delta L \tag{1.3}$$

Considering the resonant frequency, f_r, the effective length is given as,

$$L_{eff} = \frac{c}{2f_r\sqrt{\varepsilon_{eff}}} \tag{1.4}$$

The resonant frequency of the TM mode of the rectangular patch is given as,

$$f_r = \frac{c}{2\sqrt{\varepsilon_{eff}}}\left(\left(\frac{m}{L}\right)^2 + \left(\frac{n}{W}\right)^2\right)^{1/2} \tag{1.5}$$

where n and m are modes along the width W and length L, respectively.

The patch's width, W, is now given as,

$$W = \frac{c}{2f_r\sqrt{\dfrac{\varepsilon_r + 1}{2}}} \tag{1.6}$$

where c denotes the light wave's velocity.

The transmission line model has some disadvantages. It is somewhat helpful for rectangular patch designs since it neglects the variations of the field. The cavity model can be considered to overcome the disadvantages of the primitive model of the transmission line. Though the cavity model gives accurate results and provides better physical insight, it is somewhat complex in nature. The dielectric substrate behaves as a cavity with metallic walls on both sides. The height of the substrate considered should be $h \ll \lambda$. The substrate thickness is very less, as there is not much variation of the interior fields along the z-axis. The electric field, or E field, is directed along the z-axis. The magnetic field, or the H field, is decomposed into transverse components H_x and H_y [15].

1.5 OTHER POPULAR ANALYTICAL MODELS

Apart from the transmission line model and cavity model presented earlier, antenna full wave analysis can also be done using the method known as the method of moment (MoM). Here, the patch can be modeled using the surface current, and the dielectric field can be modeled using the volume polarization current. In [16], Newman and Tulyathan have shown how these unknown currents can be analyzed using integral equations. They have also shown how the MoM can be applied for converting these electric field equations to matrix form. Different algebraic techniques can be applied for obtaining the solution of the equations. Other methods of analysis include the finite element method, finite difference time domain (FDTD) method, etc. [17]

1.6 REVIEW OF THE LITERATURE

The literature review is systematically presented in the following sections.

1.6.1 Introduction

Rapid growth has been observed in the wireless communication field since the last decade. Today, huge demand exists for portable wireless devices for fast and reliable communication purposes. The antenna is a major component of all these wireless devices. An antenna is considered a device for transmitting or receiving radio frequencies. A wireless system's performance is hugely antenna dependent. Microstrip antennas are used everywhere, from personal area networks to core sectors like defense and military applications. The first antenna chapter was presented by Deschamps and Sichak in 1953 [18]. The basic theory and design facts for circular and rectangular microstrip patch antennas were presented in 1972 by Howell [19]. Conformal microstrip antennas and phased arrays were published in work 1974 by R.E. Munson [20]. For the purpose of mathematical modeling of microstrip antennas, the basic transmission line analogy was applied to a simple rectangular patch fed by a microstrip line [21]. The pattern of radiation of a microstrip disc antenna was reported by Carver in [22]. Here, the author reported on the various design methodologies of circular patches. The effects of the dielectric substrate on the antenna's resonant frequency were discussed. Discussion was made on the basic design techniques of rectangular and circular patches. The basic design concepts of circularly polarized patch antennas were also discussed. Bandwidth, efficiency and quality factor were also taken into consideration.

1.6.2 Mathematical Approaches in Antenna Design

Experimental and analytical design techniques on microstrip antennas were discussed in [23, 24]. In 1977, Lo et al. presented a detailed mathematical analysis on microstrip patch antennas [25], where the technique of modal expansion was applied to analyze rectangular, triangular, circular and semi-circular shaped patches. Lo et al. provided theoretical explanations for analyzing microstrip patches based on the cavity model [26]. the formulas for various shapes and sizes were provided. Various experiments and theories on microstrip antennas were presented in [27]. In general, antennas having various shapes and sizes were investigated, and the theoretical radiation pattern and impedance closely agreed with measured values. In the Smith chart plot, the locus of input admittance usually follows a circular path, and the conductance remains somewhat constant, but the center gradually shifts toward the inductive portion. In microstrip antennas, a dielectric substrate separates the planar radiating element or the patch element from the ground plane. The thin and planar design makes it easy to be mounted. Microstrip antennas may be fed either from the edge by depositing microstrip lines on the dielectric substrate or from the back through the ground plane.

1.6.3 Efforts in Resonant Frequency Determination

In articles [28–31], various useful methods are proposed for to determine an antenna's resonant frequency. Mythili and Das presented a simple approach of regular geometry microstrip antennas to determine the accurate resonant frequencies [32]. In this article a generalized formula is introduced for determining the effective dielectric constant, which is given by the ratio of the area of fringing to the total patch area. A modification of the effective dimension of an elliptical microstrip antenna is also presented. The obtained resonant frequencies of several geometries like circular, elliptical, rectangular and triangular shaped antennas well agreed with other available results. This approach in resonant frequency determination in [33], however, is simpler and more efficient than previous approaches.

To control the operating frequency and polarization of microstrip antennas, a new technique is introduced in [34]. Because of the fringing effect, the analysis of microstrip antennas becomes very complex, and the antenna's resonant frequency shifts from the desired resonant frequency point(s).

1.6.4 Bandwidth Enlargement Techniques

In [35], K.L. Wong focused on the techniques for compact design of microstrip antennas using a different patch structure. In [36], several techniques for the design of wideband antennas are explained along with suitable examples. Several theoretical simulations and parametric studies were carried out, which provided a physical insight into microstrip antennas. In [37], J. Sze and K. Wong demonstrated a new technique for bandwidth enhancement using a U-shaped rectangular microstrip patch. Various dimensions of right-angular and U-shaped slots are employed to obtain a larger bandwidth and desired radiating characteristics in contrast to unslotted versions. In [38], radiation and impedance matching characteristics of linearly polarized, slotted antennas within the frequency band of 01.9–02.4 GHz are discussed. Investigations on a printed microstrip slot antenna with wideband characteristics are presented in [39]. Here, a slot having dimension of $\lambda / 4$ is etched from the ground plane, which is electromagnetically fed by a microstrip transmission line. This resulted in a wide bandwidth, which is adjustable by means of parametric variation. An impedance bandwidth of approximately 60% is achieved by parametric optimization, and the simulation results tallied well with the results of measurement. A wideband WLAN application with an E-shaped patch antenna is presented in [40]. To achieve a wideband frequency by exciting two resonant modes, a zig-zag shaped slot is employed. Simulation results and experimental measurements showed good agreement. A U-slotted, dual-band dipole antenna is presented in [41]. In [42], a U-slotted antenna with a frequency selective surface along with a modified cross-shaped element is presented. Improved return loss, gain and bandwidth are obtained at 02.45 and 05.8 GHz and are generally applicable for Bluetooth and WLAN frequency bands.

1.6.5 Monopole Antennas with Improved Frequency Response

Previously, for a specific wireless system, an antenna with fixed radiation characteristics was used. Another disadvantage was the narrow bandwidth of a traditional microstrip antenna preventing its applications in many practical microwave applications. With the necessity to develop a single antenna that fulfils the bandwidth requirements of several wireless communications, many efforts are given by researchers for improving the frequency response of microstrip antennas, a printed disc monopole antenna with ultra-wideband (UWB) applications is presented in [43] by J. Liang et al. The various parameters affecting the antenna performance are studied. A triangular monopole antenna in a planar configuration is presented in [44] for UWB communication purposes. The FR4 substrate is used to design a printed tapered microstrip antenna (PTMA). Here, the voltage standing wave ratio (VSWR) measured is less than 3 from 04.0 to 10.0 GHz. A printed monopole antenna with a ring geometry applicable for UWB operation is proposed in [45]. A circular ring and a multistep ground structure are the basic components of the antenna proposed. The VSWR measured remains less than 2 from 03.0 to 12.0 GHz. Numerical parametric studies are performed in this article. In [46], a compact printed monopole antenna having a compact dimension of 30 × 30 mm² for application in UWB is presented. FR4 is used as substrate material. The proposed structure consists of an asymmetric ring-shaped patch and is fed by a 50.0 Ω line. The antenna proposed exhibited a wide impedance bandwidth from 02.6 to 14.3 GHz covering the entire bandwidth prescribed by the Federal Communications Commission (FCC) for UWB communication.

1.6.6 Novel Feeding Techniques for Bandwidth Improvement

An experimental investigation on the frequency response of common microstrip patch antennas is carried out in [47], which clearly shows that the impedance variation largely governs the impedance bandwidth of microstrip antennas. Using the modal expansion theory, this behavior is explained in the theory of cavities [48] and applied in cavity analysis models of microstrip antenna [49]. From these models it can be interpreted that the total input impedance is a collective sum of model impedances. Previously, microstrip antennas consisted of a feed line or coaxial probe. In [50] a novel feeding technique is proposed for designing a multilayer, coaxial-fed, wideband microstrip antenna structure.

A narrow-frequency bandwidth is the most serious disadvantage of traditional microstrip antennas and limits its applicability in microwave communications. So, efforts are made to develop new antenna designs for bandwidth improvement. The microstrip antenna's bandwidth can be increased by the use of a thick substrate material possessing low permittivity or by slot etching within the patch. The antenna bandwidth can also be improved using the proximity feeding technique. In [51], Deshmukh and Ray used a combination of thick substrate and the proximity feeding technique to develop different E-shaped antennas and achieved an enlarged bandwidth.

1.6.7 Coplanar Waveguide Technique for Bandwidth Improvement

The coplanar waveguide (CPW) technique can be applied for improvement in impedance bandwidth. Microstrip antennas using a coplanar feed line have been presented in articles [52–62]. A CPW-fed, slotted antenna with circular polarization features is presented in [63]. A reflector is incorporated for improvement in antenna gain. The bandwidth can also be enhanced by using the reconfigurability technique. Researchers are designing a reconfigurable antenna by using a different approach [64–66]. In [67] novel reconfigurable antenna structures for modern wireless communication application are presented. Positive-intrinsic-negative (PIN) diodes that are biased with on-board biasing circuitry are used to achieve reconfigurability in these antennas. Full EM and transient analysis is carried out for radio frequency (RF) and time domain modeling. These include frequency reconfigurability, radiation pattern configurability and polarization configurability. Reconfigurable antennas are applicable for RF and have unique switching characteristics.

1.6.8 Application of Fractal Geometries in Designing Antennas

Fractal geometry can be applied to antenna elements for achieving unique characteristics [68]. By incorporating fractal structures, the size and area of the antenna can be minimized. Multiband resonant frequencies can be achieved, the gain level can be optimized and a wider-frequency bandwidth can be achieved. Several fractal geometries have been found beneficial for various low-profile antenna designs like the Sierpinski Carpet [69], Koch Snowflake Curves [70], Cantor Set [71], Sierpinski Gasket [72], etc. Among all these mentioned fractal designs, the Koch Snowflake fractal geometry is advantageous since it generates comparatively more resonant frequencies and also helps to achieve the size miniaturization requirements of the wireless communication industries [73, 74].

1.6.9 Selection of a Suitable Substrate for Frequency Response Optimization

The choice of a suitable substrate is a prime task of microstrip antenna manufacturing. Dielectric material plays a vital part in governing the frequency response characteristics of microstrip patches. Various dielectric substrates are used by researchers to fabricate microstrip patch antenna. Therefore, among the common substrates available in the market which dielectric substrate is chosen to obtain a better performance and what are the properties that affect antenna performances are important considerations. Electrical permittivity and loss tangent are the basic properties of a substrate material. Homogeneity, adhesion of metal-foil cladding and moisture absorption factor are the prime criteria for determining a substrate's quality. The effect of substrate on radiation is studied in articles [75–79]. Experimental studies on a rectangular microstrip patch antenna operating in the X-band using various dielectric substrates are presented in [80, 81]. The results indicate that the resonant frequency does not shift much for a thick substrate, but the radiation pattern splits into multiple lobes, depending on substrate thickness. The state-of-the-art techniques of compact microstrip antennas with UWB responses have been reflected in [82, 83].

REFERENCES

1. J. Howell, "Microstrip antennas," IEEE Transactions on Antennas and Propagation, vol. 23, no. 1, pp. 90–93, 1975.
2. K. Carver, J. Mink, "Microstrip antenna technology," IEEE Transactions on Antennas and Propagation, vol. 29, no. 1, pp. 2–24, 1981.
3. M. Gustafsson, S. Nordebo, "Bandwidth, Q factor, and resonance models of antennas," PIER, vol. 62, pp. 1–20, 2006.
4. H. A. Wheeler, "Fundamental limitations of small antennas," Proceedings of the IRE, vol. 35, pp. 1479–1484, 1947.
5. R. F. Harrington, "Effect of antenna size on gain, bandwidth and efficiency," Journal of Research of the National Bureau of Standards-D. Radio Propagation, vol. 64D, no. 1, 1960.
6. L. J. Chu, "Physical limitations of omni-directional antennas," Journal of Applied Physics, vol. 19, pp. 1163–1175, 1948.
7. S. Song, B. Issac, "Analysis of WiFi and WiMAX and wireless network coexistence," International Journal of Computer Networks and Communications, vol. 6, no. 6, 2014.
8. I. Chen, C. Peng, "Printed broadband monopole antenna for WLAN/WiMAX applications," IEEE Antennas and Wireless Propagation Letters, vol. 8, pp. 472–474, 2009.
9. V. Dongre, A. K. Mishra, "Design of single band and dual band microstrip antennas for GSM, GPS and Bluetooth applications," International Conference & Workshop on Electronics & Telecommunication Engineering, Mumbai, 2016, pp. 207–212.
10. Z. Qiurong, Lu Wanzheng, Liu Feng, "A microstrip antenna element for satellite communication," 6th International Symposium on Antennas, Propagation & EM Theory, Proceedings, Beijing, China, 2003, pp. 81–85. doi: 10.1109/ISAPE.2003.1276633.
11. M. I. Nawaz, Z. Huiling, M. S. Sultan Nawaz, K. Zakim, S. Zamin, A. Khan, "A review on wideband microstrip patch antenna design techniques," 2013 International Conference on Aerospace Science & Engineering (ICASE), Islamabad, Pakistan, 2013, pp. 1–8. doi: 10.1109/ICASE.2013.6785554.
12. D. M. Pozar, D. H. Schaubert, "Basic microstrip antenna elements and feeding techniques," in Microstrip Antennas: The Analysis and Design of Microstrip Antennas and Arrays, 1995, pp. 57–104. IEEE.
13. H. Pues, A. van de Capelle, "Accurate transmission-line model for the rectangular microstrip antenna," IEE Proceedings H-Microwaves, Optics & Antennas, vol. 131, no. 6, pp. 334–340, 1984.
14. E. Newman, P. Tulyathan, "Analysis of microstrip antennas using moment methods," IEEE Transactions on Antennas & Propagation, vol. 29, no. 1, pp. 47–53, 1981.
15. F. L. Teixeira, "Time-domain finite-difference and finite-element methods for Maxwell equations in complex media," IEEE Transactions on Antennas and Propagation, vol. 56, no. 8, pp. 2150–2166, 2008.
16. G. Deschamps, G. Sichak, "Microstrip microwave antenna," Proceedings of the Third Symposium USAF Antenna Research Development Program, 18–22 October 1953.

17. J. Q. Howell, "Microstrip antennas," Antennas and Propagation Society International Symposium, Williamsburg, VA, Dec. 1972, pp. 177–180. doi: 10.1109/APS.1972.1146932.
18. R. E. Munson, "Conformal microstrip antennas and microstrip phased arrays," IEEE Transactions on Antennas and Propagation, vol. 1, pp. 74–78, 1974.
19. A. G. Derneryd, "Linear microstrip array antennas," Chalmers University of Technology, Goteborge, Technical Report, TR 7505, October 1975.
20. K. R. Carver, "The radiation pattern of a microstrip disc antenna," Physics and Sci. Lab., New Mexico State University, Las Cruces, Technical Memo, 29 November 1976.
21. D. M. Pozar, D. H. Schaubert, "Microstrip antennas: The analysis and design of microstrip antennas and arrays," John Wiley & Sons, Inc, 1995.
22. J. Anguera et al., "A systematic method to design single-patch broadband microstrip patch antennas," Microwave and Optical Technology Letters, vol. 31, pp. 185–188, 2001.
23. M. Bailey, M. Deshpande, "Integral equation formulation of microstrip antennas," IEEE Transactions on Antennas and Propagation, vol. 30, pp. 651–656, 1982.
24. R. Mosig, F. E. Gardiol, "General integral equation formulation for microstrip antennas and scatterers," IEEE Transactions on Antennas and Propagation, vol. 132. pp. 424–432, 1985.
25. J. Ashkenazy, S. Shtrikman, D. Treves, "Electric surface current model for the analysis of microstrip antennas on cylindrical bodies," IEEE Transactions on Antennas and Propagation, vol. 33, pp. 295–300, 1985.
26. D. R. Jackson, N. G. Alexopoulos, "Simple approximate formulas for input resistance, bandwidth, and efficiency of a resonant rectangular patch," IEEE Transactions on Antennas and Propagation, vol. 39, pp. 407–410, 1991.
27. Y. T. Lo, D. Solomon, F. R. Ore, D. D. Harrison, G. A. Deschamps, "Study of microstrip antennas, microstrip phased arrays and microstrip feed networks," Rome Air Development Center, Technical Report TR-77-406, 21 October 1977.
28. S. Sagiroglu, K. Güney, "Calculation of resonant frequency for an equilateral triangular microstrip antenna with the use of artificial neural networks," Microwave and Optical Technology Letters, vol. 14, pp. 89–93, 1997.
29. W. C. Chew, Q. Liu "Resonance frequency of a rectangular microstrip patch," IEEE Transactions on Antennas and Propagation, vol. 36, pp. 1045–1056, 1988.
30. X. Gang, "On the resonant frequencies of microstrip antennas," IEEE Transactions on Antennas and Propagation, vol. 37, pp. 245–247, 1989.
31. K. Güney, S. Sagiroglu, M. Erler, "Generalized neural method to determine resonant frequencies of various microstrip antennas," International Journal of RF and Microwave Computer-Aided Engineering, vol. 12, pp. 131–139, 20 02. https://doi.org/10.1002/mmce.10006.
32. R. W. Dearnley, A. R. F. Barel, "A comparison of models to determine the resonant frequencies of a rectangular microstrip antenna," IEEE Transactions on Antennas and Propagation, vol. 37, pp. 114–118, 1989.
33. P. Mythili, A. Das, "Simple approach to determine resonant frequencies of microstrip antennas," IEE Proceedings – Microwaves, Antennas and Propagation, vol. 145, no. 2, 1998.

34. D. Schaubert, F. Farrar, A. Sindoris, S. Hayes, "Microstrip antennas with frequency agility and polarization diversity," IEEE Transactions on Antennas and Propagation, vol. 29, no. 1, pp. 118–123, January 1981. doi: 10.1109/TAP.1981.1142546.

35. K. L. Wong, "Compact and broadband microstrip antennas," John Wiley & Sons, Inc, 2002.

36. G. Kumar, K. P. Ray, "Broadband microstrip antennas," Artech House, 2003.

37. J. Sze, K. Wong, "Slotted rectangular microstrip antenna for bandwidth enhancement," IEEE Transactions on Antennas and Propagation, vol. 48, pp. 1149–1152, 2000.

38. A. A. Deshmukh, G. Kumar, "Compact broadband U-slot-loaded rectangular microstrip antennas," Microwave and Optical Technology Letters, vol. 46, pp. 556–559, 2005. https://doi.org/10.1002/mop.21049.

39. S. K. Sharma, L. Shafai, N. Jacob, "Investigation of wide-band microstrip slot antenna," IEEE Transactions on Antennas and Propagation, vol. 52, no. 3, pp. 865–872, March 2004. doi: 10.1109/TAP.2004.825191.

40. D. Caratelli, R. Cicchetti, G. Bit-Babik, A. Faraone, "A perturbed E-shaped patch antenna for wideband WLAN applications," IEEE Transactions on Antennas and Propagation, vol. 54, no. 6, pp. 1871–1874, 2006. doi: 10.1109/TAP.2006.874364.

41. C. M. Su, H. T. Chen, K.-L. Wong, "Printed dual-band dipole antenna with U-slotted arms for 2.4/5.2 GHz WLAN operation," IEEE Transactions on Antennas and Propagation, vol. 38, pp. 1308–1309, 2002.

42. H. Y. Chen, Y. Tao, "Performance improvement of a U-slot patch antenna using a dual-band frequency selective surface with modified Jerusalem cross elements," IEEE Transactions on Antennas and Propagation, vol. 59, pp. 3482–3486, 2011.

43. J. Liang, C. C. Chiau, Xiaodong Chen, C. G. Parini, "Study of a printed circular disc monopole antenna for UWB systems," IEEE Transactions on Antennas and Propagation, vol. 53, no. 11, pp. 3500–3504, November 2005. doi: 10.1109/TAP.2005.858598.

44. C. C. Lin Y.-C. Kan, L.-C. Kuo, H.-R. Chuang, "A planar triangular monopole antenna for UWB communication," IEEE Microwave and Wireless Components Letters, vol. 15, no. 10, pp. 624–626, October 2005. doi: 10.1109/LMWC.2005.856694.

45. J. X. Xiao, M. F. Wang, G. J. Li., "A ring monopole antenna for UWB application," Microwave and Optical Technology Letters, vol. 52, pp. 179–182, 2010.

46. W. Ren, J. Y. Deng, K. S. Chen, "Compact PCB monopole antenna for UWB applications," Journal of Electromagnetic Waves and Applications, vol. 21, pp. 1411–1420, 2007.

47. H. F. Pues, A. R. V. D. Capelle, "An impedance-matching technique for increasing the bandwidth of microstrip antennas," IEEE Transactions on Antennas and Propagation, vol. 37, pp. 1345–1354, 1989.

48. R. F. Harrington, "Time-harmonic electromagnetic fields," McGraw-Hill, pp. 431–440, 1961.

49. K. R. Carver, J. W. Mink, "Microstrip antenna technology," IEEE Transactions on Antennas and Propagation, vol. AP-29, pp. 2–24, 1981.

50. K. S. Fong, H. F. Pues, M. J. Withers, "Wideband multilayer coaxial-fed microstrip antenna element," Electronics Letters, vol. 21, pp. 497–499, 1985.

51. A. Deshmukh, K. P. Ray, "Broadband proximity-fed modified rectangular microstrip antennas," IEEE Antennas and Propagation Magazine, vol. 53, pp. 41–56, 2011.

52. W. Menzel, W. Grabherr, "A microstrip patch antenna with coplanar feed line," IEEE Microwave and Guided Wave Letters, vol. 1, no. 11, pp. 340–342, November 1991. doi: 10.1109/75.93905.

53. L. Giauffret, J. M. Laheurte, "Theoretical and experimental characterization of CPW-fed microstrip antennas," IEE Proceedings – Microwaves, Antennas and Propagation, vol. 143, pp. 13–17, 1996.

54. R. L. Smith et al., "Coplanar waveguide feed for microstrip patch antennas," Electronics Letters, vol. 28, pp. 2272–2274, 1992.

55. T. G. Ma, C. H. Tseng, "An ultra wide band coplanar waveguide-fed tapered ring slot antenna," IEEE Transactions on Antennas and Propagation, vol. 54, pp. 1105–1110, 2006.

56. N. Behdad, K. Sarabandi, "A multiresonant single element wide-band slot Antenna," IEEE Transactions on Antennas and Propagation, vol. 53, pp. 994–1003, 2005.

57. K. Siwiak, D. McKeown, "Ultra-wide band radio technology," 2005, pp. 97–111. Wiley Online Boo ks. https://onlinelibrary.wiley.com/doi/book/10.1002/0470859334.

58. V. A. Shameena, S. Mridula, Anju Pradeep, Sarah Jacob, A. O. Lindo, P. Mohanan, "A compact CPW fed slot antenna for ultra wide band applications," International Journal of Electronics and Communications, vol. 66, pp. 189–194, 2012.

59. H. Liu, Y. Liu, S. Gong, "Broadband microstrip-CPW fed circularly polarised slot antenna with inverted configuration for L-band applications," IET Microwaves, Antennas & Propagation, vol. 11, pp. 880–885, 2016.

60. R. L. Haupt, M. Lanagan, "Reconfigurable antennas," IEEE Antennas and Propagation Magazine, vol. 55, pp. 49–61, 2013.

61. T. Ali, R. C. Biradar, "A compact hexagonal slot dual band frequency reconfigurable antenna for WLAN applications," Microwave and Optical Technology Letters, vol. 59, pp. 958–964, 2017.

62. M. I. Hossain, M. R. I. Faruque, M. T. Islam, M. T. Ali, "Design and analysis of coupled-resonator reconfigurable antenna," Applied Physics A, vol. 122, 2, pp. 2–4, 201 6. https://doi.org/10.1007/s00339-015-9520-6.

63. M. S. Parihar, A. Basu, S. K. Koul, "Reconfigurable printed antennas," IETE Journal of Research, vol. 59, pp. 383–391, 2013.

64. J. Gianvittorio, Y. Rahmat-Samii, "Fractal geometry in antenna system design: Miniaturized-multiband element, phased array and frequency selective surface design," International Conference on Microwave and Millimeter Wave Technology, 2002, pp. 508–511.

65. M. F. A. Kadir, A. S. Jaafar, M. Z. A. A. Aziz, "Sierpinski carpet fractal antenna," Asia-Pacific Conference on Applied Electromagnetics, Melaka, 2007, pp. 1–4.

66. Ankan Bhattacharya, Bappadittya Roy, Santosh K. Chowdhury, Anup K. Bhattacharjee, "Design and analysis of a koch snowflake fractal monopole antenna for wideband communication," Applied Computational Electromagnetics Society Journal, vol. 32, no. 06, pp. 548–554, July 2021.

67. B. Manimegalai, S. Raju, V. Abhaikumar, "A multifractal cantor antenna for multiband wireless applications," IEEE Antennas and Wireless Propagation Letters, vol. 8, pp. 359–362, 2009.

68. N. Kaur, J. S. Sivia, M. Kaur, "Design of modified sierpinski gasket fractal antenna for C and X-band applications," International Conference on MOOCs, Innovation and Technology in Education, Amritsar, 2015, pp. 248–250.

69. K. J. Vinoy et al., "Fractal dimension and frequency response of fractal shaped antennas," IEEE Antennas and Propagation Society International Symposium, 2003 Digest Held in conjunction with: USNC/CNC/URSI North American Radio Sci. Meeting (Cat. No. 03CH37450), Columbus, OH, vol. 4, 2003, pp. 222–225.

70. A. Ismahayati, P. J. Soh, R. Hadibah, G. A. E. Vandenbosch, "Design and analysis of a multiband koch fractal monopole antenna," 2011 IEEE International RF & Microwave Conference, Seremban, Negeri Sembilan, Malaysia, 2011, pp. 58–62. doi: 10.1109/RFM.2011.6168695.

71. P. B. Katehi, N. G. Alexopoulos, "On the effect of substrate thickness and permittivity on printed circuit antennas," IEEE Transactions on Antennas and Propagation, vol. AP-31, pp. 34–39, 1983.

72. N. G. Alexopoulos, D. R. Jackson, "Fundamental superstrate (cover) effects on printed circuit antennas," IEEE Transactions on Antennas and Propagation, vol. AP-32, pp. 807–816, 1984.

73. N. G. Alexopoulos, P. B. Katehi, D. B. Rutledge, "Substrate optimization for integrated circuit antennas," IEEE Transactions on Microwave Theory and Techniques, vol. 31, no. 7, pp. 550–557, July 1983. doi: 10.1109/TMTT.1983.1131544.

74. D. H. Schaubert, D. M. Pozar, A. Adrian, "Effect of microstrip antenna substrate thickness and permittivity: Comparison of theories with experiment," IEEE Transactions on Antennas and Propagation, vol. 37, no. 6, pp. 677–682, June 1889. doi: 10.1109/8.29353.

75. Huang Chang-Hsiu, Hsu Powen, "Superstrate effects on slot-coupled microstrip antennas," IEEE Transactions on Magnetics, vol. 27, no. 5, pp. 3868–3871, September 1991. doi: 10.1109/20.104945.

76. R. Afzalzadeh, R. N. Karekar, "Effect of dielectric protecting superstrate on radiation pattern of microstrip patch antenna," Electronics Letters, vol. 27, pp. 1218–1219, 1991.

77. X. H. Shen, P. Delmotte, G. A. E. Vandenbosch, "Effect of superstrate on radiated field of probe fed microstrip patch antenna," IEEE Microwaves, Antennas and Propagation, vol. 148, pp. 141–146, 2001.

78. A. De, N. S. Raghava, S. Malhotra, P. Arora, R. Bazaz, "Effect of different substrates on compact stacked square microstrip antenna," Journal of Telecommunications, vol. 1, no. 1, pp. 63–65, February 2010.

79. J. Deepika, M. Mathivanan, A. Muruganandham, R. Vivek, "Parametrical variation and its effects on characteristics of microstrip rectangular patch antenna," 2017 Second International Conference on Electrical, Computer and Communication Technologies (ICECCT), Coimbatore, India, 2017, pp. 1–6. doi: 10.1109/ICECCT.2017.8117913.

80. A. Esmaeilkhah, C. Ghobadi, J. Nourinia, M. Majidzadeh, "Effect of substrate scaling on microstrip patch antenna performance," Advanced Electromagnetics, vol. 7, no. 5, pp. 82–86, 2018.

81. H. Fallahi, Z. Atlasbaf, "Study of a class of UWB CPW-fed monopole antenna with fractal elements," IEEE Antennas and Wireless Propagation Letters, vol. 12, pp. 1484–1487, 2013.

82. Ankan Bhattacharya, Bappadittya Roy, Santosh K. Chowdhury, Anup K. Bhattacharjee, "Computational and experimental analysis of a low-profile, isolation-enhanced, band-notch UWB-MIMO antenna," Journal of Computational Electronics, vol. 18, pp. 680–688, 2019.

83. A. Bhattacharya, B. Roy, R. Caldeirinha, A. Bhattacharjee, "Low-profile, extremely wideband, dual-band-notched MIMO antenna for UWB applications," International Journal of Microwave and Wireless Technologies, vol. 11, no. 7, pp. 719–728, 2019.

Chapter 2

The Role of Microstrip Antennas in Microwave Engineering and Research

Barun Mazumdar and Ashis De

2.1 INTRODUCTION

In recent years, a compact antenna with multiband characteristics is a topic of interest for research work for application in wireless communication systems [1] as well as microwave engineering. There are various techniques of feeding a microstrip antenna (MSA). The most popular methods are 1. microstrip line, 2. coaxial probe (coplanar feed), 3. proximity coupling and 4. aperture coupling. Here we use the coaxial probe feeding technique. To design a compact MSA—2–9] we use cutting slits on the radiating patch to increase the length of the patch of the surface current. MSAs are used in a wide range of applications, from communication systems to biomedical systems, due to several attractive properties such as light weight, low profile, low production cost, reproducibility, conformability, reliability and ease in fabrication and integration with solid-state devices. The work to be presented in this chapter is also a compact MSA made by cutting slits on the patch [10–12]. A microstrip device is like a sandwich of two parallel conducting layers separated by a single thin dielectric substrate. The lower conductor is called the ground plane, and the upper conductor is a simple resonant circular/rectangular patch. Our aim in this chapter is to reduce the size of the antenna as well as increase the operating bandwidth.

2.2 LITERATURE SURVEY

The antenna [13–16] is one of the most important electronics elements in the radio frequency (RF) [2] system for receiving and transmitting signals from and into the air as a medium. We know that without proper design of the antenna, the signal generated by the RF system will not be transmitted and received. Antenna design is one of the active fields in communication for future development. Many types of antennas have been designed to suit most devices. The concept of the MSA, which contains a conducting patch on a ground plane separated by a dielectric substrate, was undeveloped until

DOI: 10.1201/9781003459880-2

the revolution in electronic circuit miniaturization and large scale integration (LSI) in 1970. In this section, the design of MSA literature survey is discussed. Various papers on the design of microstrip patch antennas with slots are analyzed. Summarized details of the research papers are discussed as well. The study shows that the microstrip patch antenna exhibits many advantages like light weight, low cost, low profile, planar configuration, easy conformability, superior portability, suitable for arrays, ease of fabrication, easy integration with external circuitries, etc. Though the typical MSAs have so many advantages, they also suffer from three basic disadvantages like narrow bandwidth, low gain and relatively large size, and many times these factors create an adverse impact in the antenna's efficiencies. We use various types of slots [1–2] (L, H, T, U) on both sides of the patch to design the proposed antenna. Our main aim is to reduce the size of the antenna as well as to operate the antenna in multiple frequencies. We compare various slots and their simulated results. The most common possible shapes are square, circular, rectangular and elliptical, but any other continuous shape is possible.

2.3 ANTENNA DESIGN FORMULA

When designing a microstrip patch [2–4] antenna, we must select the resonant frequency and a dielectric medium for which the antenna is to be designed. The various parameters are calculated as shown:

Width (W):

The width of the patch is calculated using the following equation 2.1:

$$W = C_0 / 2f_r \sqrt{2 / \varepsilon_r + 1} \qquad (2.1)$$

Where,
W = Width of the patch
C_0 = Speed of light
ε_r = Value of the dielectric substrate

Effective refractive index:

The effective refractive index value of a patch is one of the important parameters in the design of an MSA. The radiation that travels from the patch towards the ground passes through air and through the substrate, which is known as fringing. We know that air and the substrates both have different dielectric values; therefore we find the value of the effective dielectric constant. The value of the effective dielectric constant (ε_{reff}) is calculated using the following equation 2.2:

$$\varepsilon_{reff} = \varepsilon_r + 1/2 + \varepsilon_r - \tfrac{1}{2}[1+12h/w]^{-1/2}, w/h > 1 \qquad (2.2)$$

Length:

Due to fringing, the size of the antenna is increased by an amount of (ΔL). Therefore, the actual increase in length (ΔL) of the patch is calculated using the following equation 2.3:

$$\Delta L/h = 0.412(\varepsilon_{reff} + 0.3)(w/h + 0.264)/(\varepsilon_{reff} - 0.258)\ (w/h + 0.8) \quad (2.3)$$

Where $h=$ height of the substrate
The length (L) of the patch is calculated using equation 2.4:

$$L = C_0 / 2f_r\sqrt{\varepsilon_{reff}} - 2\Delta L \qquad (2.4)$$

Length (L_g) and width (W_g) of the ground plane:
Now the dimensions of a patch are known. The length and width of a substrate are equal to the ground plane. The length of the ground plane $(L_g$) and the width of the ground plane (W_g) are calculated using the following equations 2.5 and 2.6:

$$L_g = 6h + L \qquad (2.5)$$

$$W_g = 6h + W \qquad (2.6)$$

2.4 IE3D (INTEGRAL EQUATION THREE-DIMENSIONAL) SOFTWARE

IE3D [17] is the first scalable electromagnetic (EM) design and verification platform that delivers the modeling accuracy for the combined needs of high-frequency circuit design and signal integrity engineers across multiple design domains. For various companies, there is no longer just one EM problem at hand, but several different ones, each presenting a unique bottleneck and delaying overall design closure. IE3D's multithreaded and distributed simulation architecture and high-design capacity are the most cost-effective EM simulation and modeling solution for component-level and circuit-level applications. IE3D software offers the highest simulation capacities and fastest turnaround times for a great number of applications, making it the best choice for improving your design team's productivity and meeting design schedules on time.

2.4.1 HFSS Software (High-Frequency Structure Simulator)

Ansys HFSS [18] is a 3D EM simulation software that is used for designing and simulating high-frequency electronic products such as antennas, RF or microwave components, antenna arrays, high-speed interconnects, filters, connectors, integrated circuit (IC) packages and printed circuit boards. We use Ansys HFSS software to design high-frequency, high-speed electronics found in communications systems, advanced driver assistance systems (ADAS), satellites, and internet-of-things (IoT) products. HFSS Mesh Fusion's patented technology enables far more complex designs to be simulated with the same rigor, accuracy and reliability of Ansys HFSS. It offers this by applying targeted meshing technologies within the same design, appropriate to the local geometry.

2.5 ANTENNA CONFIGURATION

2.5.1 Patch Antenna with L-slot

The shape of the conventional antenna is shown in Figure 2.1. The antenna configuration is a 24 mm × 18 mm rectangular patch. The FR4 epoxy substrate is selected for this design with the dielectric constant (ε_r) = 4.4 and height of the substrate (h) = 1.5875 mm. We use the coaxial probe feed technique whose radius is 0.5 mm, with a simple ground plane arrangement used at the point (0,-3) where the center of the patch is considered at point (0,0). In Figure 2.2

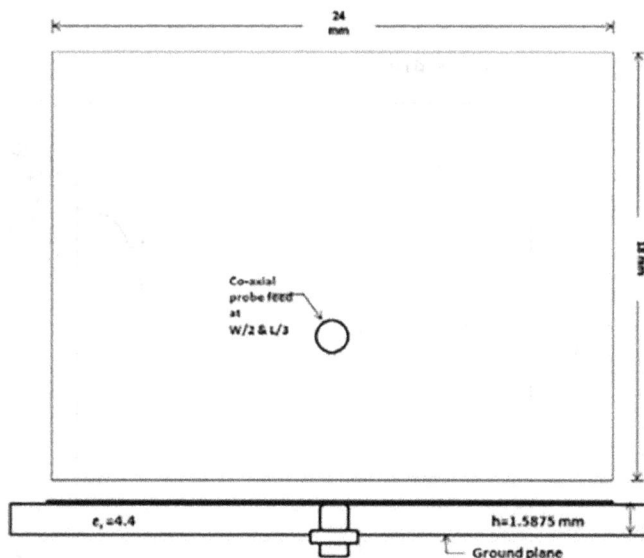

Figure 2.1 Antenna 1 configuration.

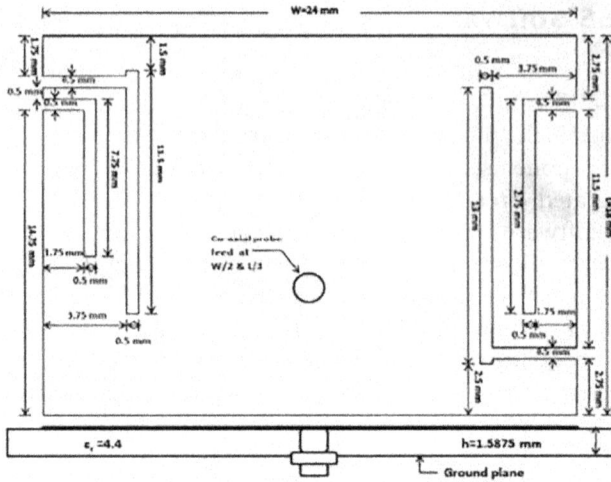

Figure 2.2 Antenna 2 configuration.

a configuration of the proposed antenna which is designed with the similar substrate is shown. The antenna size is also a 24 mm × 18 mm rectangular patch. Four L-shaped slots are created on the rectangular patch (as shown in Figure 2.2). The location of the coaxial probe-feed (radius = 0.5 mm) is also same as shown in antenna 1. The optimal values of the parameter of the L-slots and their positions are given in Figure 2.2. The simulated return losses of antenna 1 and antenna 2 are shown in Figure 2.3 and 2.4 respectively.

2.5.1.1 Simulated Results and Discussion

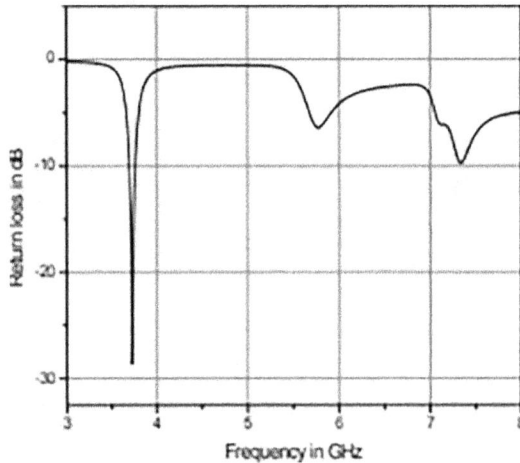

Figure 2.3 Simulated return loss of antenna 1.

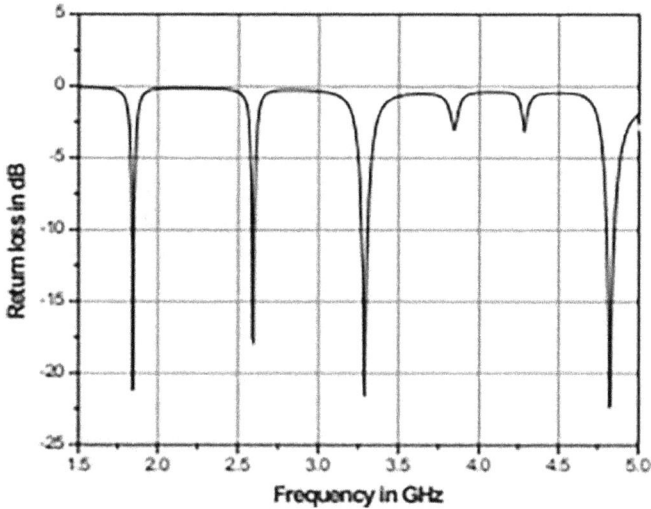

Figure 2.4 Simulated return loss of antenna 2.

Following is the comparison of the antenna without the slot and antenna with the L-slot using IE3D [17]:

Antenna without Slot	Antenna with L-slot
Works with a single frequency	Works with quad frequency
At 3.725 GHz frequency, return loss is −28.59 dB with 68 MHz bandwidth	At 1.845 GHz frequency, return loss is −28.59 dB with 16 MHz bandwidth
	At 2.59 GHz frequency, return loss is −21.19 dB with 16 MHz bandwidth
	At 3.29 GHz frequency, return loss is −−21.56 dB with 40 MHz bandwidth
	At 4.825 GHz frequency, return loss is −22.31 dB with 49 MHz bandwidth

2.5.2 Patch Antenna with T-slot

The shape of the conventional antenna is shown in Figure 2.5. The antenna is a 20 mm × 16 mm rectangular patch. The dielectric material selected for the antenna design is $\varepsilon_r = 2.4$ and the substrate height = 1.5875 mm.

Figure 2.6 shows the configuration of antenna 2, which is designed with the same substrate. Two unequal T-slots having lengths l_1 and l_2 are created. The dimensions of l_1 and l_2 and the location of coaxial probe-feed (radius = 0.5 mm) are shown in Figure 2.6. The optimal parameter values of the T-slots are listed in Table 2.1.

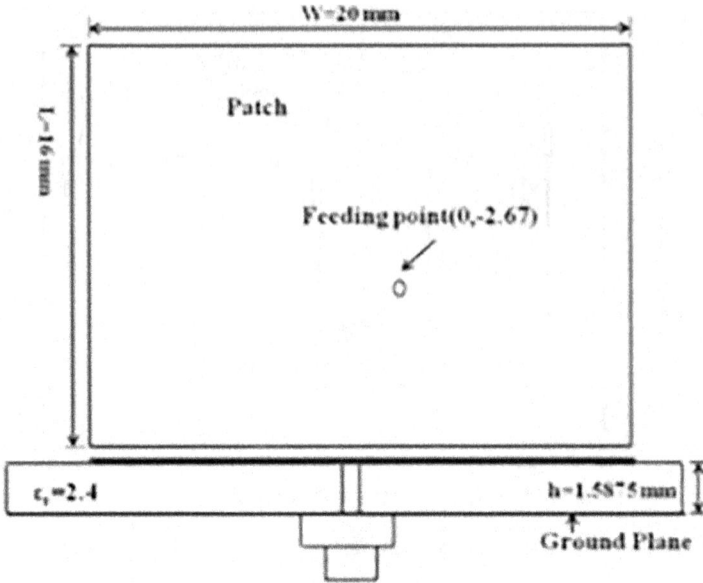

Figure 2.5 Antenna 1 configuration.

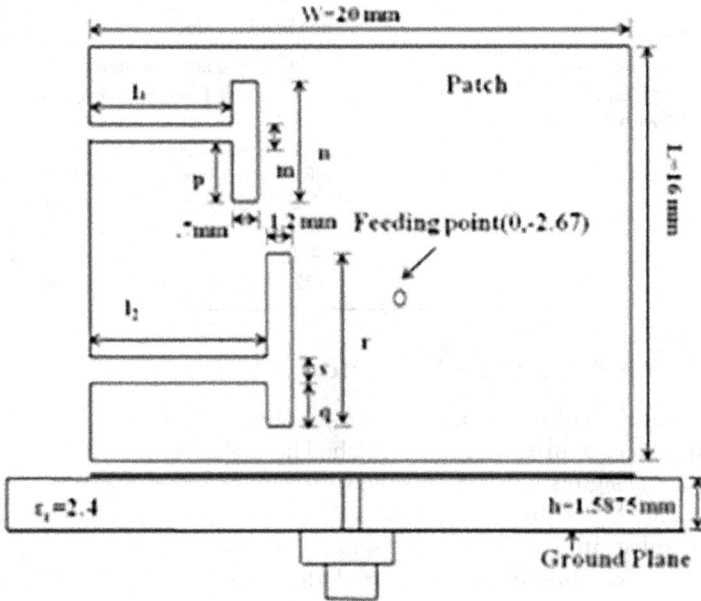

Figure 2.6 Antenna 2 configuration.

Table 2.1 Optimal Parameter Values of the T-slots of Antenna 2

Parameters	m	n	p	l_1	q	r	s	l_2
Values (mm)	1	6	1.5	3.65	2.3	5.8	.5	6.75

2.5.2.1 Simulated Results and Discussion

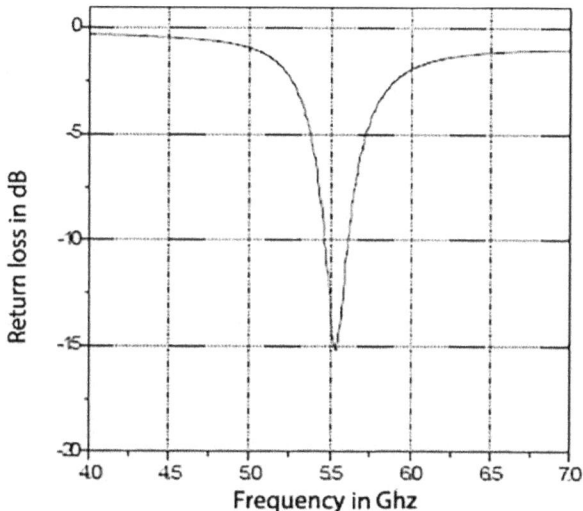

Figure 2.7 Simulated return loss of antenna 1.

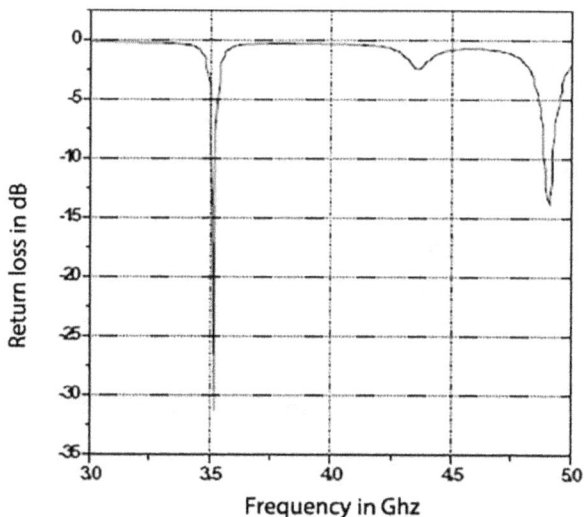

Figure 2.8 Simulated return loss of antenna 2.

Following is the comparison of the antenna without the T-slot and antenna with the T-slot using IE3D [17]:

Antenna without Slot	Antenna with T-slot
Works with a single frequency	Works with dual frequency
At 5.54 GHz frequency, return loss is −15.1 dB with 144.42 MHz bandwidth	At 3.51 GHz frequency, return loss is −31 dB with 15.8 MHz bandwidth
	At 4.9 GHz frequency, return loss is −13.5 dB with 35.19 MHz bandwidth

2.5.3 Patch Antenna with H-slot

Reference antenna with no slots is shown in Figure 2.9. The configuration of the proposed antenna is shown in Figure 2.10. The size of the antenna is a 18 mm × 14 mm rectangular patch. The dielectric material which is selected for the design is an FR4 epoxy with dielectric constant (ε_r) = 4.4 and substrate height (h) = 1.6 mm.

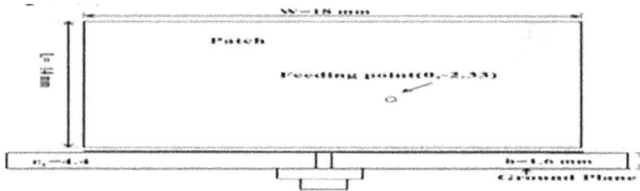

Figure 2.9 Antenna 1 configuration.

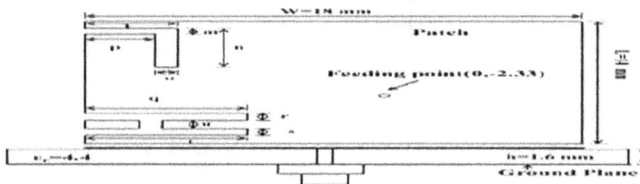

Figure 2.10 Antenna 2 configuration.

The optimal parameter values of the L-slots and H-lots are listed in following table:

Parameters	l	m	n	o	P	q	r	s	t	u
Values (mm)	4.45	.45	9.85	1.2	3.25	7.5	1	.8	7.5	1

2.5.3.1 Simulated Results and Discussion

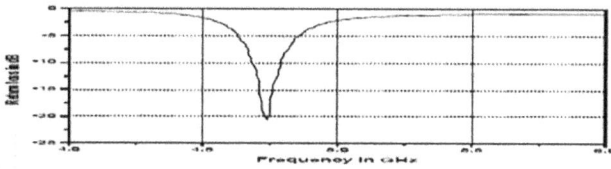

Figure 2.11 Simulated return loss of antenna 1.

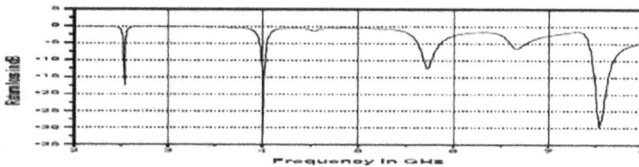

Figure 2.12 Simulated return loss of antenna 2.

Following is the comparison of the antenna without slots and antenna with the L-slot using IE3D [17]:

Antenna without Slot	Antenna with L-slot
Works with a single frequency	Works with quad frequency
At 4.73 GHz frequency, return loss is −20.43 dB with 103.09 MHz bandwidth	At 2.53 GHz frequency, return loss is −17.4 dB with 12.48 MHz bandwidth
	At 4 GHz frequency, return loss is −32.5 dB with 37.97 MHz bandwidth
	At 5.73 GHz frequency, return loss is −12.4 dB with 80.68 MHz bandwidth
	At 7.54 GHz frequency, return loss is −29.7 dB with 230.67 MHz bandwidth

2.5.4 Patch Antenna with U-slot

The U-slot rectangular microstrip patch antenna design is described in Figure 2.13. Here, L is the length, W is the width of the patch, F is the feed point, Ls is the vertical slot length, Ws is the horizontal slot length and t_L and t_w are slot widths in the vertical and horizontal, respectively.

The dielectric constant $\varepsilon_r = 4.4$ is used as the substrate material with a thickness of 1.6 mm. The length and width of the patch are 39 mm and 28.2 mm, respectively. The feeding point is 7.5 mm from the center of the patch.

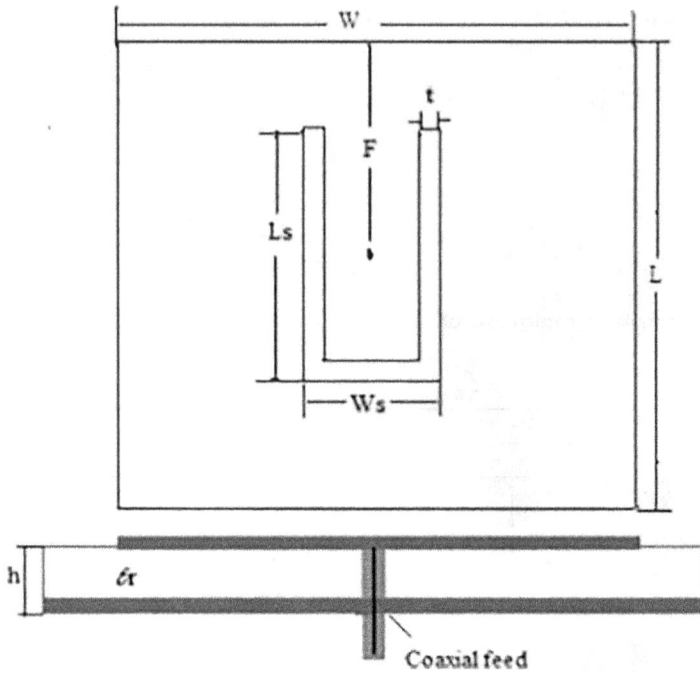

Figure 2.13 U-slot loaded rectangular microstrip antenna.

2.5.4.1 Simulated Results and Discussion

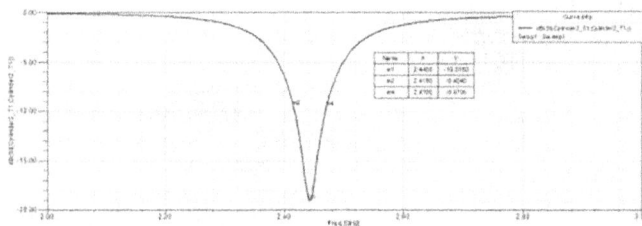

Figure 2.14 Simulated return loss of antenna 2.

Following is the comparison of antenna without the slot and antenna with the U-slot using HFSS [18]:

Antenna without Slot	Antenna with L-slot
Works with a single frequency	Works with quad frequency
At 2.43 GHZ frequency, return loss is −16.21 dB with 68 MHz bandwidth	At 2.44 GHz frequency, return loss is −19.01 dB with 51 MHz bandwidth

2.6 CONCLUSION

Due to its inherent advantages like small size and weight, low cost, printed directly on the circuit board, low profile and easy fabrication, the microstrip patch antennas are the most preferable antennas. The slotted antennas are used in GSM, WiMAX, Wi-Fi, HIPERLAN, etc., which are very important nowadays. This chapter describes the different slots on the antenna that have resulted in an improvement of various performance parameters of the antenna like gain, bandwidth, return loss, etc.

REFERENCES

1. U. Chakraborty, B. Mazumdar, S. K. Chowdhury, A. K. Bhattacharjee, "A compact L-slot microstrip antenna for quad band applications in wireless communication," Global Journal of Researches in Engineering Electrical and Electronics Engineering, vol. 12, no. 2, 2012.
2. B. Mazumdar, A. Kumar, "A compact dual band printed antenna for WiMAX & HIPERLAN applications," International Journal of Electronics Communication and Computer Engineering, vol. 3, no. 3, 2012.
3. B. Mazumdar, "A compact printed antenna for Wimax, Wlan & C band applications," International Journal of Computational Engineering Research (ijceronline.com), vol. 2, no. 4, 2012.
4. M. Chakraborty, B. Rana, P. P. Sarkar, A. Das, "Design and analysis of a compact rectangular microstrip antenna with slots using defective ground structure," Procedia Technology, vol. 4, pp. 411–416, 2012.
5. U. Chakraborty, "Compact dual-band microstrip antenna for IEEE802.11a WLAN application," IEEE Transactions on Antennas and Propagation, vol. 13, no. 3, pp. 407–410, 2014.
6. A. De, B. Roy, A. Bhattacharya, A. K. Bhattacharjee, "Bandwidth-enhanced ultra-wide band wearable textile antenna for various WBAN and Internet of Things (IoT) applications," Radio Science, vol. 56, p. e2021RS007315, 20 21. https://doi.org/10.1029/2021RS007315.
7. A. Bhattacharya, B. Roy, R. Caldeirinha, A. Bhattacharjee, "Low-profile, extremely wideband, dual-band-notched MIMO antenna for UWB applications," International Journal of Microwave and Wireless Technologies, vol. 11, no. 7, pp. 719–728, 20 19. https://doi.org/10.1017/S1759078719000266.
8. Ankan Bhattacharya, Bappadittya Roy, Santosh K. Chowdhury, Anup K. Bhattacharjee, "Computational and experimental analysis of a low-profile, isolation-enhanced, band-notch UWB-MIMO antenna," Journal of Computational Electronics, vol. 18, pp. 680–688, 20 19. https://doi.org/10.1007/s10825-019-01309-3.
9. E. O. Hammerstad, "Equations for microstrip circuit design," Proceeding of the Fifth European Microwave Conference, pp. 268–272, September 1975.
10. H. R. Kharat, M. D. Khetmalis, S. H. Pimpalgaonkar, R. Shamalik, "Design and analysis of compact U slot microstrip patch antenna for wireless applications,"

International Journal of Wireless Networks and Communications, vol. 8, no. 1, pp. 7–14, 2016.

11. N. Kaur, N. Sharma, "Designing of slotted microstrip patch antenna using inset cut line feed for S, C and X band applications," International Journal of Electronics Engineering Research, vol. 9, no. 7, pp. 957–969, 2017.

12. S. Singh, J. Kumar, "A review paper on rectangular microstrip patch antenna," National Conference on Industry 4.0 (NCI-4.0), 2020.

13. C. A. Balanis, "Advanced engineering electromagnetic," John Wiley & Sons, 1989.

14. I. J. Bahl, P. Bhartia, "Microstrip antennas," Artech House, 1980.

15. J. R. James, P. S. Hall, C. Wood, "IET digital library: Microstrip antenna theory and design," 19 81. https://digital-library.theiet.org/content/books/ew/pbew012e.

16. G. Ramesh, P. Bhartia, I. Bahl, A. Ittipiboon, "Microstrip antenna design handbook," Artech House Inc., 2001.

17. www.zeland.com.

18. www.ansys.com.

APPENDIX
List of Symbols and Abbreviations

Serial No.	Symbols/Abbreviations	Significance
1	ε_r	Dielectric constant
2	ε_{reff}	Effective dielectric constant
3	ΔL	Increase in length
4	C_0	Speed of light
5	L_g	Length of a ground plane
6	W_g	Width of a ground plane
7	h	Height of the substrate
8	W	Width of the patch
9	Ls	Vertical slot length
10	Ws	Horizontal slot length
11	t_L	Slot widths in the vertical
12	t_W	Widths in the horizontal
13	MSA	Microstrip patch antenna
14	FR4	Flame retardant
15	RF	Radio frequency
16	MoM	Method of moments
17	IE3D	Integral Equation Three-Dimensional

Continued

Serial No.	Symbols/Abbreviations	Significance
18	EM	Electromagnetic
19	RFIC	Radio frequency integrated circuits
20	HFSS	High-frequency structure simulator

Chapter 3

Recent Developments in Low-Cost Manufacturing Antennas and Their Challenges

S. Kannadhasan, R. Nagarajan,
C. Jisha Chandra, and G. Srividhya

3.1 INTRODUCTION

Transmission of data that is of a higher technological standard middleware is just one link in the network of components that must be present for effective data transmission. The efficiency of the cellular communication channel is yet another important factor to take into account. However, utilising optimum antennas that take into consideration the environment in which the constrained device will be deployed, this quality can potentially be improved. This is because external variables, such as protecting environments and other signals that produce interference, have a significant impact on this quality. Industry 4.0 is one of the businesses that relies significantly on reliable cellular communication in order to function effectively. A well-optimised data transmission for Industry 4.0 is now one of the most essential cornerstones that must be present for a successful implementation of Industry 4.0. The optimisation of data communication for Industry 4.0 is a challenging task because it requires antennas that are both low cost (in terms of construction cost and power utilisation) and effective (e.g., total radiation efficiency). The most prevalent wireless technologies required for such devices are the Internet of Things frequencies and 4G technology. (e.g., Long-Term Evolution [LTE], WiMAX). Due to the closeness of carrier frequencies, it is extremely difficult to construct an antenna that is compatible with each of those devices and can function properly. As a consequence of this, new media devices require additional technologies such as Universal Mobile Telecommunications System (UMTS), LTE, and WiMAX in order to be capable of transmitting industrial data together with other administration data to central computers. In recent years, there has been considerable progress made in the scientific study of Internet of Things antennas. A comprehensive analysis of the differences and similarities between the microstrip patch, helical, and conical spiral antennas is presented. An outstanding multiband antenna that is functional for global system for mobile communication (GSM), LTE, and UMTS has been developed; however, the global positioning system (GPS) band is not supported, and the antenna is too massive to be integrated into devices with a

DOI: 10.1201/9781003459880-3

low profile. Both the standard implanted array antenna and the shortened implantable array antenna are capable of providing faultless coverage of the GPS bands; however, neither design is capable of covering the LTE bands and is rather bulky. An extremely small antenna was proposed by Liu and colleagues for use in GPS devices; however, it is challenging to manufacture and does not encompass the UMTS or LTE frequencies. Although a respectable antenna that is capable of covering LTE bands is incorporated into the device in the form of a miniaturised low-profile antenna, this antenna does not cover the GPS band. An integrated antenna performs faultlessly in both the GPS and industrial, scientific, and medical (ISM) bands, and the antenna itself is constructed from high-tech components to ensure its reliability. It is suitable for use in a variety of portable products, such as smart timepieces. Our goal is to design an inexpensive incorporated multiband antenna that is capable of operating in the GPS1600, UMTS1800, LTE2300, ISM2400, LTE2600, WIMAX3500, WIMAX5800, and ISM5800 frequency bands. The antenna, which will be installed inside industrial machines (metallic construction), must function well in terms of return loss, gain, and transmission pattern within the machine at all frequencies necessary [1–5].

The ability to receive adequate medical attention is not a privilege that should be reserved for a limited group of people; rather, it is a fundamental human right. Nevertheless, reductions in spending and evaluations of expenditures are having a substantial influence on public services in a variety of domains, including health care. In these kinds of situations, the elderly are frequently the demographic that is most adversely affected by the circumstance; in point of fact, institutions are frequently unwilling to spend their money on the treatment of senior citizens. On the other hand, an increase in life expectancy, along with the fundamental right of the elderly to enjoy their golden years, pushes the demand for innovative technological solutions that are both low-cost and inventive in order to make health care more cost-effective and efficient. In this context, telemedicine is regarded as one of the most important methods for substantially reducing communal expenditures related to health care, while at the same time maintaining the provision of necessary assistance to the elderly and guaranteeing a high standard of life for all individuals. Wireless technology is an invaluable partner for performing distant surveillance of patients' health. The elderly may be able to receive the necessary assistance without having to move into unpleasant and expensive nursing facilities if information technology is combined with unobtrusive external devices. This scenario is possible thanks to advances in medical science. As a consequence of this, there are many solutions available at the current state of the art that are dedicated to activity identification. These solutions are able to collect information about the behaviour of habitats and identify potential irregularities in health indicators. Among these was a description of a system for distant fall detection in an indoor environment.

This particular system included a microwave radar sensor and a base station for data processing that was electronically connected.

In addition to that, it was proposed to implement a sensing network system for in-home care. It is composed of biosensors that are attached to the patient's body and transmit measured signals to a remote wireless monitor in order to acquire witnessed human physiological signals. The monitor's purpose is to observe human physiological signals. Both ZigBee and GSM technologies were utilised in the process of developing the surveillance platform. In conclusion, a framework for textile detection was demonstrated. This portable monitoring device has sensors that are stitched into the clothing. This reduces the amount of pressure that is placed on the patient and ensures a higher level of sensitivity than printed sensors. Put simply, users should find wearing the gadget to be beneficial, pleasurable, non-intrusive, and unobtrusive at the very least. Memory loss accelerates the ageing process, which is why Alzheimer's disease is one of the most common causes of dementia in older people. It is extremely unusual for people who have Alzheimer's disease to wander around their homes for no apparent reason. Although people with Alzheimer's disease may forget their phones or identity papers, it is much less likely that they will forget to wear their clothing, which is an essential part of a person's daily routine [6–10]. Therefore, wearable electronics that are incorporated into garments become the ideal method for implementing, on the one hand, pervasive and continuous health monitoring and, on the other hand, a tracking system that permits the individual to be located even if they meander off. The components that are necessary for virtually monitoring an individual and the utilisation of the monitoring system would be included in the clothing, turning the process into an automatic one that is analogous to dressing yourself.

In this particular scenario, it is possible to use textile materials for both the electrical sections and the substrate. In addition, the electrical sections ought to be made of materials that permit tin soldering of the necessary electronics, and this ought to be possible (electronic chips and, in general, surface-mounted components). The amount of time and money required for the manufacturing procedure are also important considerations for portable devices. When considering large-scale production and cost management, it is important to select manufacturing processes that are both economically viable and easily adaptable to industrial settings. GNSS, which stands for Global Navigation Satellite System, is the official general term for worldwide navigation satellite systems. The GNSS is administered by the government of the United States. GNSS permits independent worldwide geospatial positioning. The GNSS is a system that allows users to determine the longitude, latitude, and altitude position of a desired location by using time signals that are transmitted from a constellation of satellites. This can be accomplished with a certain degree of accuracy by using a very small electrical detector.

The GPS is a space-based satellite system that provides information on the location and time of any preferred location on or near the Earth, regardless of the weather. This includes any location on or near the ocean. The GPS is now accessible to anyone who possesses a GPS receiver.

Because of how dependent all positioning systems are on GPS, there is a potential downside in the event that the United States GPS becomes inoperable. Access to global navigation satellite systems administered by foreign governments is not always guaranteed in dangerous environments, which is one of the reasons why an independent navigation system is necessary in these kinds of situations. As a consequence of this, India has decided to utilise geostationary satellite applications by means of a regional satellite system, which has been given the name Indian Regional Navigational Satellite System (IRNSS). This innovative concept is considered to be a pioneering one. When it comes to antenna construction, manufacturing, and analysis, pricing presents one of the most significant difficulties from an educational point of view. In order to conduct an accurate evaluation of a manufactured antenna design, several different types of high-priced radio frequency (RF) test and measurement instruments are required. RF power metres, spectrum analysers, vector network analysers, and RF generators are some instances of the instruments that fall under this category. Every one of these things has the potential to surpass the price of a brand-new premium automobile.

Before an antenna design can be evaluated, it must first be physically manufactured based on the theoretical design. This can only be done after the design has been finalised. Printed circuit boards (PCBs) can even have modern antennas integrated into them as an integral component of the board. The majority of device manufacturers subcontract the manufacturing of their devices to third-party PCB businesses due to the high expense of the machinery required to manufacture PCBs. This technique could end up being quite pricey, specifically for manufacturing prototypes and in small quantities. In conclusion, in order to construct antennas, one needs to have a strong mathematical foundation in addition to a significant amount of experience in RF engineering [11–18]. This is because there are numerous complicated mathematical principles connected to electromagnetic wave propagation, as well as specific design ideas pertaining to the intended antenna application.

In this document, a comprehensive discussion is had regarding the enhancement of a rectangular microstrip patch antenna's bandwidth as well as its gain when the antenna is operated at 5 GHz. The microstrip patch antenna has attracted a lot of interest over the past two decades, and it is presently serving as an essential component in the research and development of wind profile radar. A dielectric substrate is placed between a ground plane and a patch in a printed microstrip antenna. This creates a sandwich-like structure for the antenna. In order to construct an antenna that is suitable for wireless portable radio (WPR), the micro strip patch antenna technology is utilised. This is due to the fact that it is commercially viable. It finds application in a wide variety of applications, including wireless local area networks (WLAN), personal communication systems (PCS), and microwave systems, amongst others. In comparison to other types of radiators, they are preferred due to their low profile and light weight; however, their restricted frequency and low gain are two of their primary drawbacks. Academics from all corners of the world have been focusing their attention on finding solutions to problems like this one. Both the amplification and the frequency of the patch antenna improved as a direct consequence of this. It has been noticed that the bandwidth measurements as well as the amplification observations have substantially increased.

3.2 LOW-COST ANTENNA MANUFACTURING

Since the Federal Communications Commission (FCC) officially designated the 3.1–10.6 GHz frequency to ultrawide-band (UWB) communications applications in 2002, UWB technology has sparked the attention of both scholarly and industrial entities in the wireless sector. Since the 3.1–10.6 GHz frequency was opened up for commercial use, there has been a lot of interest in the development of UWB technologies for wireless applications such as radar, remote sensing, and location confirmation applications. These applications have also generated a lot of interest in short-range wireless communications. The primary purpose behind the construction of the UWB antenna is to enhance the manufacturing efficiency and gain and provide a broad frequency while still maintaining an outstanding radiation efficiency. The compact and unobtrusive antenna system serves a purpose that is different from that of traditional narrowband systems. In addition, planar UWB antennas have attracted a lot of attention because of their ability to facilitate comfortable wireless access to multimode communication systems. These antennas are compact, lightweight, and straightforward to construct. On a substance that has a large frequency and is capable of supporting UWB transmissions, a printed monopole antenna can be fabricated. An antenna in the form of a monopole cylindrical band is recommended for use in UWB applications. Numerous configurations of monopole antennas, such as rectangular, circular, elliptical, and curved monopole antennas, have been outlined as potential means of achieving UWB communication. It is possible to create microstrip patches in a variety of shapes, such as dipoles, squares, rectangles, triangles, circles, ring sectors, and disc sectors. Circular patches have a lot of advantages, such as the flexibility to create whatever you want, the greatest bandwidth in GHz, reasonable lossy characteristics, increased gain, and perfect electric and magnetic field intensity patterns. Circular patches also have a lot of advantages.

Band-40 of the 4G LTE standard, which operates over a frequency range of 2300–2400 MHz and is referred to as TD-LTE, is now live in India and is being offered by operators within Indian territory. The government of India is also holding a bidding for the other LTE band, which operates on the FDD-LTE frequency of 1800 MHz. However, there is not a single telecommunications company that provides this frequency for LTE. The 100 MHz of bandwidth that Band-40 provides, which ranges from 2300 MHz to 2400 MHz, is divided into two parts: the downlink and the uplink. The term "downlink" refers to transmissions made from the eNB (also known as a base station) to the UE (also known as a mobile user), whereas "uplink" refers to transmissions made from the UE to the eNB. The purpose of this chapter is to demonstrate a fundamental planar inverted F antenna (PIFA) layout for a mobile phone application. The fundamental frequency of the PIFA is 2300 MHz, making it an excellent candidate for 4G-LTE.

The human race has had to contend with difficulties ever since the Stone Age. Communication has come a long way from the days of smoke messages to the cellular advancements of today. An antenna is an essential component in cellular communication. There are many different types of antennas that can be used for wireless communication, including microstrip patch antennas, reflector antennas, aperture antennas, transverse wave antennas, vertical antennas, and more. Figure 3.1 illustrates how microstrip patch antennas

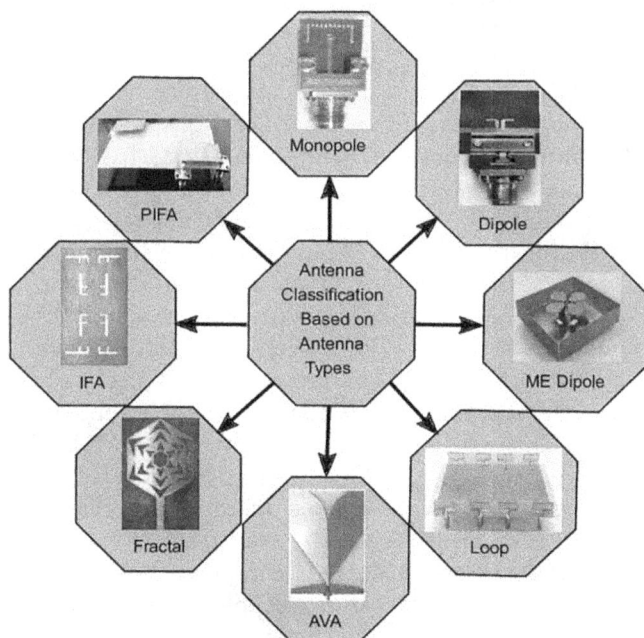

Figure 3.1 Various types of low-cost antennas.

guarantee low-profile, compact, and cost-effective manufacturing for real-time applications. These antennas are frequently observed on the exterior of aircraft and satellites, in addition to being utilised in transportable wireless communication devices. These antennas can be constructed to be lightweight, have a low volume, and have a low production cost. Additionally, they are readily able to be incorporated with microwave integrated circuits (MICs). On the other hand, microstrip patch antennas confront the substantial risk of having a restricted bandwidth and subpar gain, both of which can be alleviated by using slot techniques.

Slotting is one of the most efficient ways to improve the implementation of microstrip patch antennas, and it's also one of the simplest techniques. By utilising a particular type of slot, the implementation can be enhanced in terms of gain, bandwidth, and directivity while also experiencing a reduction in size. It is possible for the length of a particular space as well as its position to have an effect on an antenna's resistance as well as its frequency. It is possible to expand the frequency of a patch antenna by arranging appropriate spaces along the transmitting sides of the patch. The term "narrowband" refers to information transmission and broadcast communications apparatus, advancements, and services that make use of a reduced variety of frequencies in the transmission channel. Narrowband also refers to the use of a narrower bandwidth in the transmission channel. These make use of a channel frequency that is considered to be standard or a fewer number of different frequency bands. Microstrip antennas are utilised in all of these fields: satellite communications, direct broadcast television, missile defence systems, and military applications. This antenna has many advantages like ease of integration in printed circuit board, but it also has some disadvantages, such as poor efficiency, low gain, and limited frequency. These disadvantages are caused by the PCB. Because it is simply a leading strip that attaches to the patch and is therefore capable of being regarded as an expansion of the patch, the microstrip feed line is one of the techniques that is one of the easiest to produce. The offset location can be easily controlled, which makes modelling and coordinating the process much simpler.

The microwave radio region of the electromagnetic spectrum includes the X band as one of its subareas. For instance, when it comes to designing correspondence, the repetition scope of the X band is sometimes described inconclusively as ranging from approximately 7.0 to 11.2 GHz. When it comes to radar construction, the Institute of Electrical and Electronics Engineers (IEEE) guidelines recommend using a frequency spectrum of 8.0–12.0 GHz. The X band is utilised frequently for a variety of purposes, including radar, satellite communication, wireless computer networks, and conventional communications and networking. Applications in the field of radar and the military that make use of it include synthetic aperture radar, phased arrays, continuous-wave radar, pulsed radar, single-polarisation radar, and dual-polarisation radar.

Building microstrip antennas with multiband, dual band, and wideband characteristics, as well as evaluating functional parameters like return loss, gain, and bandwidth, have all been accomplished through the application of a wide variety of techniques that have been described in published works. The antenna can be found in a variety of shapes, such as a T-shaped microstrip patch antenna; a rectangular, circular, or triangular shaped antenna; a P-shaped resonator; a triangular microstrip patch antenna; a polygon patch antenna; and a W-shaped microstrip patch antenna, and it is utilised in wireless communication systems, satellite communication, and WLAN or WiMAX. Additionally, in recent years, fractal antennas, array antennas, and dielectric resonator antennas have been developed specifically for use in WLAN applications. The published studies indicate that the return loss is negligible and that the bandwidth is not only wider but also bigger in size for the frequency range and antenna construction that was chosen. This research proposes and develops a circular split ring resonator (SRR)–based microstrip patch antenna in order to solve the issues brought up by the previous sentences. In this, a microstrip patch antenna for X band applications that is built on an SRR is proposed and designed. Radiating elements are frequently employed in antenna designs, and the slit and breaks in the suggested microstrip patch antenna are no exception. This antenna is constructed using an FR4 substrate that has a thickness of 0.8 mm, a breadth of 10 mm, and a length of 7.7 mm. The piece is 8.2 millimetres in length and 6 millimetres in breadth.

3.3 CHALLENGES IN LOW-COST ANTENNAS

Both the business and scholarly spheres have demonstrated a significant amount of interest in the field of portable and malleable electronics in the most recent years. Flexible electronics, which can be crumpled, contorted, stretched, and compressed, would significantly extend the applicability of current electronic devices to a variety of non-flat circumstances, including the shape of the human body. This would be a significant advancement in the field of flexible electronics. As a consequence of this, flexible electronics combined with textile materials provide a number of benefits that make them a potential technology for improving the next generation of consumer electronics. These benefits include low-cost manufacturing, low-cost flexible substrates, light weight, and simplicity of construction, among other advantages.

Antenna sensors have recently garnered a lot of attention as a result of their user-friendly designs, bidirectional detecting capabilities, cost-effective inactive operation, and straightforward designs. Electrical devices known as antenna sensors have the ability to communicate in addition to their detecting functions, and they can be constructed using a limited amount of parts.

The functional concept of antenna sensing is demonstrated by the effect that a change in either the geometry or the intrinsic material has on the antenna's resonance frequency. This effect can be quantified by the influence it has on the reflection coefficient. In addition, antenna sensors have been developed as a technique for detecting a wide range of physical properties. It is necessary to incorporate flexible antenna devices with flexible electrical systems in order to facilitate the wireless communication that is required in today's information-oriented culture. A spherical antenna will be used to determine the amount of humidity that is contained in sewage samples. A sample of the sediment was placed inside a plastic container, which was then placed on top of the antenna sensor. Readings of the antenna sensor's dielectric effective permittivity were used in conjunction with the Bottcher model to arrive at a conclusion regarding the amount of moisture present in a sample. There is a possibility that antenna capacitance sensors are present in each of these antenna sensors. The relationship between the physical measurement and the transmission parameters of the antenna can be understood in terms of the material characteristics. This difficulty cannot be explained mathematically in the overwhelming majority of cases, and it is one of the challenges that must be overcome when developing antenna dielectric sensors. As a consequence of this, the vast majority of research on antenna dielectric sensors is concentrated on direct characterisation and experimental testing. The first antenna temperature monitor to be identified in the published research is one that is specifically designed for temperature threshold detection. In order to construct the antenna sensor, a sheet of shape memory polymer (SMP) paper was sandwiched in between a radio frequency identification (RFID) tag and a sheet of aluminium material. The relative permittivity of the SMP changed as the temperature increased beyond a certain benchmark. This change was recognised by the turn-on power of the RFID tag as the temperature continued to rise.

In the past, antenna sensors, which are typically made of solid materials, were utilised in activities such as agriculture and horticulture, structural health, biological sensing, food quality monitoring, and other applications. There have been a number of studies that have been previously published in the literature that make use of stiff materials for a variety of antenna sensors. These studies include temperature sensing, fracture detection, strain sensing, and dielectric sensing. Microstrip patch antennas are frequently utilised in detecting applications due to the numerous advantages they provide, some of which include low manufacturing costs, low weight, sturdiness, trustworthiness, and a small dimension. Patch antennas are able to perform the role of sensing thanks to the interaction of their insulating characteristics and electromagnetic radiation. For instance, it has been demonstrated that a sensor can be built on a microstrip patch antenna and utilise a flame resistant 4 (FR-4) substrate. The proposed antenna functions as a sensor that can determine, based on the electrical properties of the solution, the relative amounts of sugar and salt present in varying proportions by measuring the return loss. A microstrip patch antenna sensor was constructed using a Rogers (R03006) substrate as the base material. The temperature can be read off of an object using the specified antenna, which was created by electrically changing a patch antenna that was connected to a variety of metal bases. These substrates do not lend themselves well to use as clothing antenna devices due to their inflexibility; they cannot be twisted or stretched. When designing a portable antenna sensor, it is often necessary to include additional features such as resilience and flexibility. This requirement makes it necessary to use adaptable nonconventional materials in place of standard PCBs.

Wearable textile antenna sensors have become increasingly essential in on-body applications over the course of the past decade. This is primarily attributable to their ability to monitor and keep an eye on human health in addition to detecting microstructure deformations and human movements. Textile antenna sensors, as opposed to the more conventional types of antenna sensors, have the potential to be incorporated into garments and offer significant advantages such as comfort, low weight, and the ability to be washed. A variety of portable bendable wireless devices have been recommended in the research that has been done so far. An example of a wireless sensor system that is capable of both detecting and communication was provided, and it was in the form of a finger motion antenna sensor that was built on a dipole antenna. This dipole antenna sensor was also connected to a glove so that it could evaluate the impact of human twisting on a portable device in a real-world environment. In light of what has been discussed thus far, it is possible that in the not too distant future, antenna sensing technology will be employed in the fields of human-machine interaction, health care, automation, and virtual reality.

It is getting close to implementation time for the 5G network in a number of countries because it provides substantial enhancements in terms of cellular data throughput, connection, bandwidth, coverage, and latency while consuming less energy overall. It is hoped that in the not too distant future, it will be possible to incorporate multiple cellular devices. The Internet of Things (IoT) is a quickly developing technology with the goal of changing and connecting the world through the use of a network of various types of intelligent devices. The IoT has an effect on all spheres of human activity and commercial enterprise, but specifically on medical treatment. It is anticipated that a substantial role will be played in the IoT by a variety of different kinds of wearable technology.

In addition, because there are limits placed on the amount of continuous radiation exposure that can be received by the body, the miniaturisation of an antenna is a practical choice for the wireless body area network (WBAN) system. This is due to the fact that a smaller antenna requires less room on or off the body. It is difficult to achieve miniaturisation of a portable antenna while maintaining a satisfactory level of performance. In contemporary wireless communications systems, the capacity to maintain a fundamental antenna design with broadband or multiband functionality has made easy incorporation and adaptability into the system into one of the most significant challenges. As a consequence of this, numerous considerations need to be given attention in the process of designing portable transmitters that can be attached to the garments of the user. It is necessary to conduct research into the relationship that exists between the antenna and the human body before any performance degradation can be prevented. The second issue is dealing with the deformation of the portable antenna caused by its placement on the body, which has a significant effect on overall performance. It is also difficult to get portable antennas to function consistently under a variety of conditions, such as the dampness, temperature, and distance between the body and the antenna. This makes it one of the more challenging aspects of this technology. Essential aspects include having high electrical and mechanical characteristics, a low cost, being lightweight, having a low loss, being flexible, having user convenience, and having high resolution and precision in the manufacturing techniques used.

3.4 CONCLUSION

The goal of the fourth industrial revolution, also known as Industry 4.0, is to improve understanding of production procedures across all of an organisation's industrial locations in what is considered to be "near real-time." This will allow for increased productivity, which will be accomplished by integrating the aforementioned characteristics into each and every process and asset at all times. Given the anticipated increase in data requirements varying from mission-critical to enormous machine connections, the implementation of 5G has generated expectations that it will offer new possibilities for industrial business models. Automating industrial technology and making use of other supporting technologies like artificial intelligence (AI) and machine learning are examples of what is referred to as Sector 4.00, also known as the fourth industrial revolution. This is how the manufacturing industry plans to use the advancements made possible by 5G cellular communications. The business world anticipates that this will result in more accurate decision making, such as the automation of physical tasks based on historical data and knowledge, or improved outcomes for a wide range of vertical marketplaces, such as agriculture, supply chain logistics, health care, and energy management. Additionally, a growing number of businesses are becoming more aware of the potential of these technologies.

REFERENCES

[1] S. Majumder, T. Mondal, and M. Deen, "Wearable sensors for remote health monitoring," Sensors, vol. 17, no. 1, p. 130, 2017.

[2] Carmine Garripoli, Marco Mercuri, Peter Karsmakers, Ping Jack Soh, Giovanni Crupi, Guy A. E. Vandenbosch, Calogero Pace, Paul Leroux, and Dominique Schreurs, "Embedded DSP-based telehealth radar system for remote in-door fall detection," IEEE Journal of Biomedical and Health Informatics, vol. 19, no. 1, pp. 92–101, 2015.

[3] R. Nakamura and H. Hadama, "Target localization using multi-static UWB sensor for indoor monitoring system," in 2017 IEEE Topical Conference on Wireless Sensors and Sensor Networks (WiSNet), pp. 37–40, Phoenix, AZ, January 2017.

[4] F. Viani, M. Martinelli, L. Ioriatti, G. Oliveri, P. Rocca, and A. Massa, "WSN for real-time localization and tracking of elderly people," in OASIS 1st International Conference, Florence, November 2009.

[5] F. Viani, P. Rocca, G. Oliveri, D. Trinchero, and A. Massa, "Localization, tracking, and imaging of targets in wireless sensor networks: An invited review," Radio Science, vol. 46, 2011.

[6] Constantin Ungureanu, Vinh Bui, Wouter Roosmalen, Ronald M. Aarts, Johan B. A. M. Arends, Richard Verhoeven, and Johan J. Lukkien, "A wearable monitoring system for nocturnal epileptic seizures," in 2014 8th International Symposium on Medical Information and Communication Technology (ISMICT), pp. 1–5, Firenze, April 2014.

[7] S. Jegadeesan, Z. Mansouri, A. Veeramani, and F. B. Zarrabi, "Ultra wideband PIFA antenna with supporting GSM and WiMAX for mobile phone applications," in International Conference on Advanced Communication, Control and Computing Technologies, vol. 2015, April, pp. 15–20, 2015.

[8] F. N. M. Redzwan, M. T. Ali, M. N. M. Tan, and N. F. Miswadi, "Design of planar inverted F Antenna for LTE mobile phone application," 2014 IEEE Region 10 Symposium, pp. 19–22, 2014.

[9] K. T. Kim, J. H. Ko, K. Choi, and H. S. Kim, "Robust optimum design of PIFA for RFID mobile dongle applications," IEEE Transactions on Magnetics, vol. 47, no. 5, pp. 962–965, 2011.

[10] J. S. Kim, K. H. Shin, S. M. Park, W. K. Choi, and N. S. Seong, "Polarization and space diversity antenna using inverted-F antennas for RFID reader applications," IEEE Antennas and Wireless Propagation Letters, vol. 5, no. 1, pp. 265–268, 2006.

[11] "LTE Frequency Bands | LTE Bands 1–12, 13–25, 33–43 LTE bands." Available: www.rfwireless-world.com/Terminology/LTE-frequency-bands.html [Accessed: 17 August 2017].

[12] J. A. Ansari, S. Verma, and Ashish Singh, "Design and investigation of disk patch antenna with quad C-slots for multiband operations," Journal of Microwave Science and Technology, vol. 2014, pp. 1–6, 2014.

[13] N. Ortiz, F. Falcone, and M. Sorolla, "Gain improvement of dual band antenna based on complementary rectangular split-ring resonator," ISRN Communications and Networking, vol. 2012, pp. 1–9, 2012.

[14] J. G. Joshi, S. S. Pattnaik, and S. Devi, "Geo-textile based metamaterial loaded wearable microstrip patch antenna," International Journal of Microwave and Optical Technology, vol. 8, no. 1, pp. 25–33, 2013.

[15] D. C. Nascimento and J. C. D. S. Lacava, "Probe-fed linearly-polarized electrically-equivalent microstrip antennas on FR4 substrate," Journal of Microwaves Optoelectronics and Electromagnetic Applications, vol. 13, no. 1, pp. 55–66, 2014.

[16] A. De, B. Roy, A. Bhattacharya, and A. K. Bhattacharjee, "Bandwidth-enhanced ultra-wide band wearable textile antenna for various WBAN and Internet of Things (IoT) applications," Radio Science, vol. 56, p. e2021RS007315, 20 21. https://doi.org/10.1029/2021RS007315.

[17] A. Bhattacharya, B. Roy, R. Caldeirinha, and A. Bhattacharjee, "Low-profile, extremely wideband, dual-band-notched MIMO antenna for UWB applications," International Journal of Microwave and Wireless Technologies, vol. 11, no. 7, pp. 719–728, 20 19. https://doi.org/10.1017/S1759078719000266.

[18] Ankan Bhattacharya, Bappadittya Roy, Santosh K. Chowdhury, and Anup K. Bhattacharjee, "Computational and experimental analysis of a low-profile, isolation-enhanced, band-notch UWB-MIMO antenna," Journal of Computational Electronics, vol. 18, pp. 680–688, 20 19. https://doi.org/10.1007/s10825-019-01309-3.

Chapter 4

Design and Performance Analysis of a Miniaturized UWB Monopole Antenna with Embedded Octagonal Slot and DGS for Various Wireless Applications

Aparna Panja, Arnab De, Koyndrik Bhattacharjee, Partha Pratim Sarkar, and Anup Kumar Bhattacharjee

4.1 INTRODUCTION

Recently, wideband antennas for wireless communication systems have come under more and more attention. Many systems can now perform in more than one frequency band due to advancements in wireless technology. Its capacity to operate over various frequency bands ultimately hinges on how well its antenna performs. Many devices use multiple antennas to meet this requirement, each of which can cover one or more working bands. Unfortunately, because there are so many antennas, the system becomes more complicated and takes up a lot of space, which is usually limited in devices. Such numerous antenna deployments prohibit future system modifications that need the usage of bands that still need to be supported. Because of this, it is better to have an antenna with an operating bandwidth that can accommodate the working frequency ranges of various wireless communication networks. Across the full frequency region, such an antenna ought to exhibit steady radiation-pattern features. There is an increase in demand for smart cities that require high channel capacity and dependability, smart homes, wearables, and other applications (Internet of Things [IoT]). In most cases, the word "ultra-wideband" (UWB) is typically used for describing types of techniques meant to transmit information dispersed across a wide working bandwidth where electronic systems must cohabit with some other electronic users. UWB technology has existed for many years. The majority of its initial applications were in military systems, but this technology has recently gained much attention thanks to its wider bandwidth, simplicity of fabrication, minimal spectral power density, and minimal energy requirement, narrower pulses, less interference,

DOI: 10.1201/9781003459880-4

45

less multipath fading, narrow-band signal, better radiation, faster data rate, and closer communication range characteristics. The objective is to enable radar and imaging systems and wireless personal area networks (WPANs) with a small range yet high data rates and a wide range yet low data rates for various wireless applications. In 1960, UWB technology was first introduced by studying a linear time-invariant system using a phenomenal impulse response. The enthusiasm for UWB was revived in the late twentieth century, but there was no infrastructure to create it. UWB was explicitly designed in the United States for military uses like communications security and radar [1]. Only the military and government organizations used UWB. UWB communication systems use a 7.5 GHz spectrum from 3.10 to 10.60 GHz, adhering to the guidelines established by the Federal Communications Commission (FCC) in the United States [2]. To promote wireless multimedia communication and interoperability across devices in personal area networks, the WiMedia Alliance, Microsoft, Intel, and HP were also members of the consortium established in 2002. Wi-Fi, a competitor of UWB, introduced IEEE 802.11ac as a reaction to wireless universal serial bus in 2010, ending WiMedia's efforts to make wireless universal serial bus prosperous [3, 4]. Recently, there has been a resurgence in interest in UWB technology. Manufacturers of smartphones have started integrating location estimation technologies by inserting a UWB chip as of 2019. Also, many companies are creating UWB-focused indoor location tracking systems in various market segments, such as smart factories, smart homes, etc. [5]. Creating a suitable or ideal antenna is the primary obstacle to the application of UWB systems. From a systems perspective, the antenna's response should be encompass the whole operational bandwidth. An antenna's performance or specs will change depending on the system's needs. As a result, an antenna engineer must be knowledgeable about the system requirements before constructing an antenna. Wi-MAX (IEEE 802.16) at 3.5 GHz, WLAN (IEEE 802.11a) operating at 5.2 or 5.8 GHz, higher satellite X-band operating at 8.5 GHz, and the most recent 5G spectrum allocations for satellite communication in the C band where 5G cellular transmission takes place from 3.3 to 4.2 GHz, and narrow band for 5G use at 4.5 GHz to 5.5 GHz are just a few examples of how UWB technology is used in wireless technology today.

Consumer electronics and WPANs are among a couple of the many uses for UWB technology. Using unique smart technologies like radar, wireless systems, and medical research, UWB radio develops into a new technology [6, 7]. As depicted in Figure 4.1, an UWB antenna system has a several applications, including smart cities with smart medical care, homes, security, automobiles, industrial Internet, connectivity, and mobile/mobile accessories.

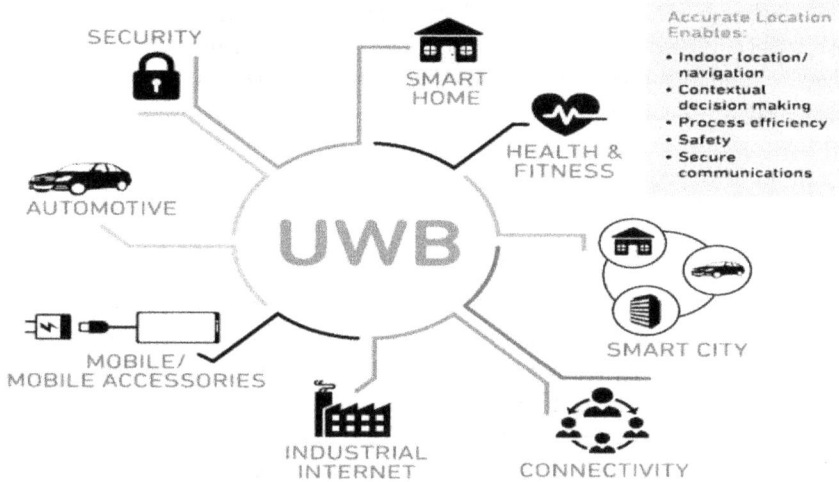

Figure 4.1 Applications of the ultra-wideband antenna.

4.2 LITERATURE SURVEY

A variety of antenna construction strategies and techniques are used to achieve UWB functioning. A thorough literature review indicates that UWB antenna-based research has recently attracted the antenna community's attention. Several researchers have recently proposed UWB antennas that, in many ways, meet the required criteria [8–27]. Several planar, monopole antennas for UWB systems were investigated for downsizing using electromagnetic bandgap [8], a meta-material [9], a partial ground plane [10], and twice triangular and single circular section slots [11]. Asymmetrical hexagon-shaped radiators [12], z-shaped radiators [13], compact tapered-shape slots [14], bow tie–shaped antennas [15], and other shapes of compact UWB planar antennas have also been proposed [16–22]. An antenna that is applied to mobile terminal devices can be created using a variety of approaches. Utilizing a type of shorted strip positioned just on top of the bottom plane for creating a planar inverted F antenna (PIFA) is one of the most often used techniques [23, 24]. Although these antennas have a narrow bandwidth and poor efficiency, their height is sensitive to resonance frequency. Use of a monopole antenna configuration is one of the simplest ways to obtain UWB characteristics. In this context, it has been revealed that some traditional UWB monopole antenna systems have simpler geometries. There are numerous shapes, such as rectangles [25], circles [26–28], ellipses [29, 30], semi-ellipses [31], and complicated shapes such as U-shaped [32, 33], cactus design [34], and so on. These antennas may be more compact and operate at lower frequencies due to the long, thin strips that are integrated into them. Monopole antennas have drawn increased interest in various wireless applications because of their desirable attributes of affordable cost, simple construction, and broad bandwidth. Most mobile terminal applications, such as Bluetooth, LTE, WiMAX, GPS, DVB-H, and UWB, are said to be covered by the bandwidth of certain antennas, which are reportedly capable of operating in the 1.575–10.6 GHz range of frequency (UWB). The operating bandwidth of the monopole antennas is being increased using two alternative strategies. A standard method to obtain the necessary characteristics is to modify the bottom plane as well as the metallic patch of a UWB antenna. Techniques include introducing slots like an inverted T shape and L to the bottom plane and metallic patch, metal strips additionally to slots in the radiating patch, and uniform stubs used inside the ground plane mainly to influence the antenna current distribution in order to null the unexpected effects of the back side of the antenna [35–38]. Etching the electromagnetic bandgap (EBG) structure forms onto an antenna is another popular technique [39, 40]. The impedance bandwidth is made better by the properties of dielectric materials.

Combining UWB properties with the antenna's small size is essential for smaller wireless devices. Therefore, it is difficult for researchers to make UWB antennas compact. So many researchers have also reported on the presence of different compact UWB antennas. Researchers from around the world have recently carried out several investigations on the design of compact UWB antennas.

Liu has described a 124×110 mm^2 elliptical monopole antenna with a modified feeding structure that operates between 1.02 GHz and 24.10 GHz and has a maximum gain of roughly 7.0 dBi [41]. Bhattacharya in [42] has proposed a small filter with dual band-notched printing supported an antenna with reduced radiation properties beyond the ultra-wide frequency range. UWB was created to function in a certain range of frequencies (03.1–10.6 GHz). Except for two frequency notches, 03.5 GHz and 05.5 GHz center frequency, it was possible to attain 03.1–10.6 GHz of impedance bandwidth (S_{11}–10.0 dB). De described a novel, small, circular-sectioned UWB antenna with dimensions of 40×40 mm^2 and gain of 6.1 dBi that functioned in frequency ranging from 2.5 to 20.25 GHz [43]. Nikolaou proposed a 20×28 mm^2 miniaturized UWB monopole antenna with a 2.85–11.85 GHz band feature and maximum gain of 3.2 dBi in [44]. De in [45] described a circular monopole antenna with a radius of 12 mm^2. It operates between 3.05 and 12.11 GHz and has a maximum gain of roughly 8.36 dBi. In refs. [46–48], other antenna types with UWB characteristics are discussed; the current issue in each case is that as an antenna gets smaller, its performance (gain, bandwidth) deteriorates. Gopikrishna describes a printed planar UWB antenna with a novel notch band–like structure that measures 34×25 mm^2 and operates in the 2.85–20 GHz frequency range [49]. In [50], J. Y. Siddiqui proposed a small, split ring resonator (SRR)–loaded, UWB circular monopole antenna of dimensions of 50×50 mm^2 and UWB behavior at 2.60–11.20 GHz. De proposed a circular monopole antenna with a 38.87×24 mm^2 footprint, an operating bandwidth of 9.42 GHz, frequency range of 3.12–2.54 GHz, and a maximum gain of approximately 11.30 dBi in [51]. D. Gaetano reports narrowband linking performances at 2.45 GHz vs dual UWB channels cantered at 3.95 GHz as well as 7.25 GHz utilizing a single-sided printed monopole antennas having a dimension of 40×50 mm^2 and optimized for on-body performances ranging from 2.4 GHz to 10 GHz in [52]. In [53], A. Wu presented a miniaturized and coplanar waveguide (CPW)–fed planar UWB with dual band-notched antenna performance. This suggested design has a voltage standing wave ratio (VSWR) of less than 2, an impedance bandwidth from 2.4 GHz to 12.5 GHz in the UWB range of frequency, and two stop bands for filtering the wireless local area network (WLAN) and WiMAX signals. The manufactured antenna is just 32 mm $\times$ 32 mm $\times$ 0.508 mm. The 2.4–12 GHz frequency band is covered by an antenna designed by Y. K. Soni, which itself is fed by a CPW and developed on a cheaper (FR-4) substrate with only a measure height of 0.76 mm in [54]. For use in WLAN and UWB bands of 2.4 GHz, Y.-B. Yang performed an investigation on a unique wideband planar monopole antenna that produces a bandwidth of 2.2–11 GHz with a VSWR value less than 2 in [55]. J. Ren presented a miniaturized multiple-input, multiple-output (MIMO) antenna encompassing the 2.4 GHz (WLAN) and UWB range of 3.1–10.6 GHz for the wireless device [56]. It has a modest dimension of 40×40 mm^2. The antenna is a strong contender for portable applications due to its impedance bandwidth, which is larger than 2.4 GHz to 10.6 GHz and has a lower mutual coupling than 20 dB in the WLAN spectrum and less than 18 decibels in the 3.1–10.6 GHz band. In [57], F.T. Zha suggests a brand-new, small microstrip slot antenna for 2.4 GHz to 2.5 GHz WLAN and UWB. The antenna takes up a space of 33 mm $\times$ 35.5 mm $\times$ 1 mm. From range of 2.4–11 GHz, the operating bandwidth (VSWR 2) is available. In [58], Y. Y. Sun describes an antenna that provides industrial, scientific, and medical (ISM) or UWB using a defective bottom plane for on-body usage. The proposed antenna comprises three rectangular slots on the bottom plane's upper edge and a feed line that is slightly off-center along the radiator side and a radiator that is very nearly square. The antenna is a possible contender for on-body UWB communications since it can produce a wide operation range from 2.38 to 14.5 GHz, with much higher omnidirectional emission characteristics in the E-plane than a traditional monopole. According to the *Journal of Electrical and Electronic*

Engineering, the antenna's measured gain peaks at 5.9 dBi at 12.7 GHz and averages 4.42 dBi from 2.38 to 14.5 GHz. Between 2.38 and 14.5 GHz, its measured efficiency varies from an average of 86.6% to a maximum of 95.3% while at 3 GHz. In [59], Bhattacharya proposes a MIMO antenna having dimensions of 36×63 mm^2 and impedance bandwidth of 3.0–11.1 GHz gained as well as a band-notch feature cantered at 3.5 GHz. De offers a small, asymmetric, CPW-fed monopole antenna suitable for various types of WBANs and UWB communication. The antenna's design comprises a circular shape with a reconstructed bottom side of the antenna and the addition of defective ground structure (DGS) to provide UWB (3.10–10.6 GHz) properties. The presented body-worn antenna provides a fractional bandwidth (FBW) of approximately 144.44%, which operates between 2.40 and 14.88 GHz. Inserting a rectangular slit with an open end into the ground plane increases bandwidth. At 13.05 GHz, the final antenna achieves omnidirectional radiating patterns with a maximum gain of about 6.57 dBi [60]. In [59], Bhattacharya describes a MIMO antenna having an incredibly broadband, isolation-enhanced, low-profile and dual band-notched characteristics. Except for two band notches located at 3.5 and 5.5 GHz, the antenna's 20 mm × 20 mm (0.08 λo × 0.08 λo) dimensions demonstrate an extensive bandwidth of frequency encompassing 1.2–19.4 GHz. According to Pandey's proposal [61], a UWB monopole antenna using gate-like structures with dimensions of 70×60 mm^2 can achieve a higher bandwidth between frequencies of 2 and 11 GHz and a gain of 8.52 dBi.

This research has looked at a miniaturized UWB monopole antenna with a surface area of 20×23 mm^2 that operates in the UWB and is suited for several wireless communication systems. The suggested design incorporates a DGS, a square patch with truncated edges, and an octagon-shaped slot at the canter point to increase radiation properties and produce a wider impedance bandwidth. The suggested antenna is easier to construct and configure than other available UWB antennas. It has more compact dimensions and many helpful qualities regarding FBW, gain, and radiation pattern.

4.3 ANTENNA DESIGN

This designed antenna with a miniaturized UWB and monopole incorporated with an octagonal slot and DGS is shown in this section. The reported UWB monopole antenna is depicted in Figure 4.2, and this is constructed on substrate FR-4 epoxy of dimensions $20 \times 23 \times 1.6$ mm^3, loss tangent (δ) value of 0.02, and dielectric constant (ε_r) value of 4.4. The arrangement has an octagonal-shaped slot at the mid-point of the metal patch and a DGS with open-ended slits on the bottom. The back plane and radiation patch are designed on the printed circuit board's (PCB's) both sides. Figure 4.2

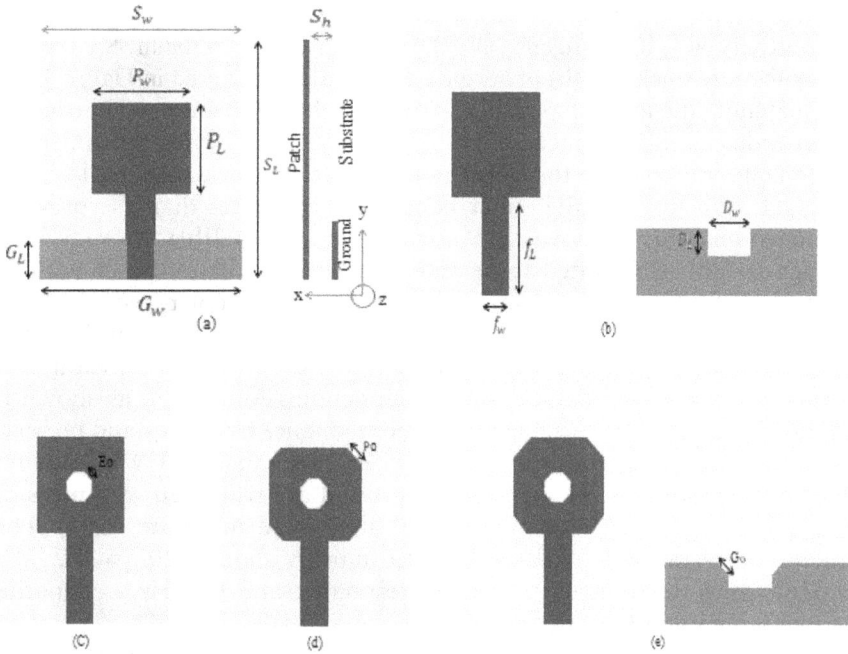

Figure 4.2 (a) Antenna 1, (b) Antenna 2 (using an optimized ground plane), (c) Antenna 3 (using octagonal slot), (d) Antenna 4 (using truncated edges along with an octagon shaped slot), and (e) Antenna 5 or suggested antenna.

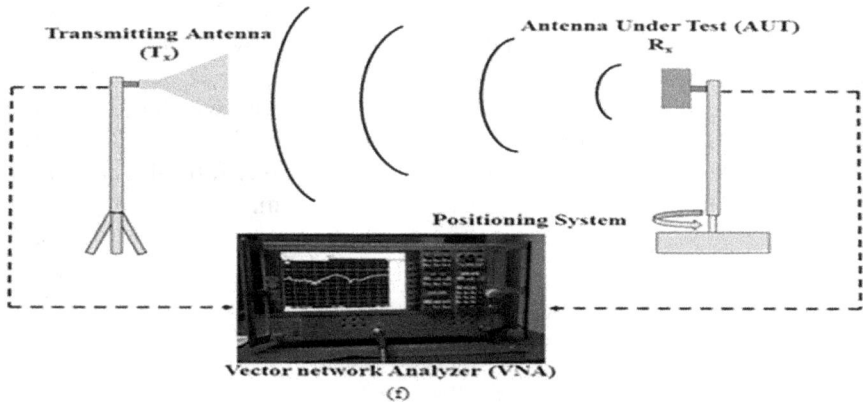

Figure 4.2 (f) Antenna measurement test setup.

depicts the many structural layout processes for the overall antenna. The suggested antenna's design phases are classified into five types, called Antenna 1, 2, 3, 4, and 5. First to design Antenna I, which is a reference antenna, a square monopole antenna of sides $P_L = P_W = 8$ mm through a strip line having width $W_f = 3mm$ is drawn on an FR-4 epoxy material of relative-permittivity 4.4, loss-tangent $(\tan \delta) = 0.02$ having $S_L = 20mm$, $S_W = 23mm$ and substrate thickness (h) of about 1.60 mm demonstrated in Figure 4.2(a) for matching the input impedance to 50 Ω of the UWB antenna. Further, an open-ended slit is implemented into the ground plane having the dimension $(D_L \times D_w)$ mm^2 to move the lower resonance frequency to the left less than 3.8 GHz and increase the wider bandwidth responses than the previous design. Thus, this is shown as Antenna 2 (Figure 4.2[b]). Moreover the octagonal slot is imprinted at the mid-point of the radiation patch with the radius E_0 of 1.5 mm to demonstrate a UWB response portrayed in form of Antenna 3, as seen in Figure 4.2(c). In the next step, the edges of the radiation patch with the octagonal-slotted area are cropped of dimension P_0 to improve the working bandwidth, which denotes Antenna 4, as shown in Figure 4.2(d). Finally, to achieve a lower-resonance frequency and preserve the impedance bandwidth across the operating range of UWB, Antenna 5 describes the edges of the grooved ground are truncated of dimension G_0, which is the suggested antenna as illustrated in Figure 4.2(e). This antenna's overall area of 20 × 23 × 1.6 mm^3 is about $0.20\lambda_0 \times 0.23\lambda_0 \times 0.016\lambda_0$ with the frequency of lowest resonance of 3.10 GHz, λ_0 denoting the wavelength in free space.

Figure 4.2(f) depicts the system setup needed to conduct measurements using the suggested antenna. For antenna measurement, the most fundamental device is the vector network analyzer (VNA). It is used to measure antenna gain, reflection coefficient, radiation pattern, etc., so that we can compare it with our simulated results. The specification of VNA which is used for measurement purposes is ZNB 20 (Vector Network Analyzer: R&S®ZNB vector network analyzer | Rohde & Schwarz (rohde-schwarz.com)); it can measure up to 18 KHz to 20 GHz, Rohde & Schwarz made. The positioning system regulates the radiation patterns, which are observed and covered later on with parts. Data were calculated by providing a DC voltage to the antenna while using a DC power source. Table 4.1 contains a list of all the optimized dimensions in summary form.

Table 4.1 Optimized Values of Various Antenna Parameters

Parameters	Value (mm)
S_L	23
S_W	20
S_h	1.6
P_L	8

Table 4.1 (Continued)

Parameters	Value (mm)
P_W	8
G_L	5.5
G_W	20
f_L	7.5
f_W	3
D_L	2.65
D_W	4
E_0	1.5
P_0	3
G_0	4

4.4 RESULTS AND DISCUSSIONS

A graph of the reflection coefficient as a function of frequency for all antennas is shown in Figure 4.3 (Antenna 1–5). Figure 4.3 displays the improvement of the bandwidth-enhanced UWB antenna's reflection coefficient (S_{11}). Here, Antenna 1 exhibits wideband properties with the frequency spectrum of 3.8–12.5 GHz (8.7 GHz), showing an FBW of 106.74%, resulting in a bandwidth ratio of 3.29:1. This initial rectangular Antenna 1 is found to perform badly in Figure 4.3, which falls short of expectations. In order to expand the operating bandwidth of Antenna 2's radiating square patch, which covers the 3.6–18.9 GHz frequency range with an FBW of 136% and a bandwidth ratio of 5.25:1, a rectangular open-ended slit is carved out of the bottom plane right beneath the feed. The radiator's effective length will be extended as a result. In turn, the resonance frequency is decreased, which is a successful method for realizing patch miniaturization and enhancing radiation performance. Antenna 3 is introduced with an FBW of 137.81% and a 10 dB bandwidth encompassing the 3.5–19.01 GHz bands by including an octagonal-shaped slot into the middle of the radiating patch. The operating bandwidth of 10 dB of S_{11} is expanded from 3.5 to 25 GHz by subtracting the radiating edges of the patch in Antenna 4. The antenna under consideration, known as Antenna 5, is shown in the graph, converting the original wideband characteristics to the intended characteristics. The UWB frequency band ranges from 3.1 to 20 GHz, composed of WLAN's conventional IEEE communication bands, WiMAX band, Higher X-Band, 5G band, and various wireless applications. The antenna's resulting –10 dB bandwidth is roughly 16.9 GHz, with an

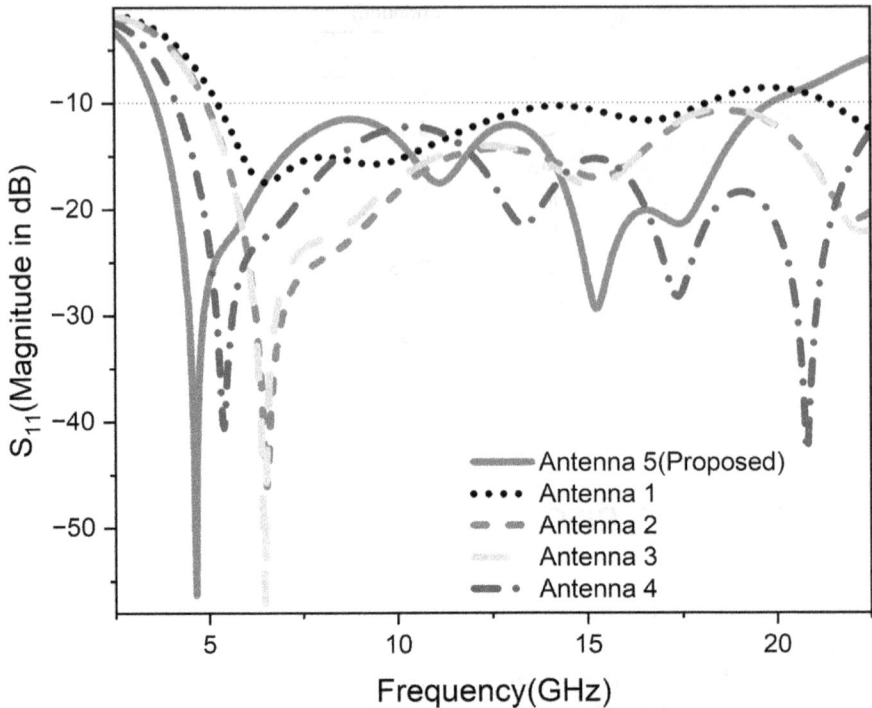

Figure 4.3 Graph of reflection coefficient against frequency for all antennas.

Table 4.2 A Comparison of the Frequency Response Characteristics of all Antennas

Type	Bandwidth (GHz)	Impedance Bandwidth (GHz)	FBW (%)	Max Gain (dBi)
Antenna 1	3.8–12.5	8.7	106.74	5.01
Antenna 2	3.6–18.9	15.3	136	6.2
Antenna 3	3.5–19.01	15.51	137.81	6.35
Antenna 4	3.5–25	21.5	150.87	6.42
Antenna 5	3.1–20	16.9	146.32	6.7

FBW of 146.32 % (3.1–20 GHz). Table 4.2 contains a detailed comparison of various antenna parameters.

Figure 4.4 compares the current distributions for each of the five distinct UWB antenna types operating at 10 GHz. As depicted in Figure 4.4, the antenna's current distributions are significantly wider at the bottom part of the metal patch. Furthermore, the current intensity has improved in the vicinity of the octagonal slot and the uppercut angle section. Accordingly,

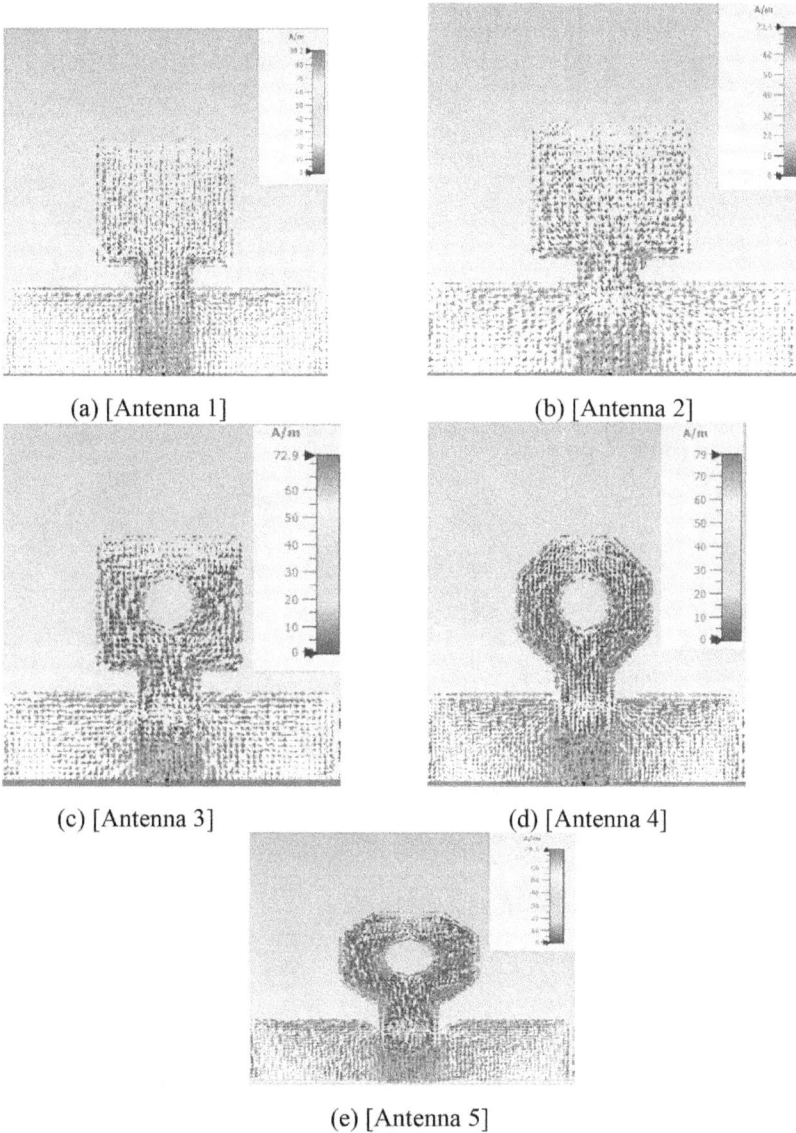

(a) [Antenna 1]   (b) [Antenna 2]

(c) [Antenna 3]   (d) [Antenna 4]

(e) [Antenna 5]

Figure 4.4 The antenna's current distributions at 10 GHz.

current distributions for suggested antenna [Figure 4.4(e)], current density on two beveled ground edges is more significant as compared to Antenna 4.

Figure 4.5 displays the current distributions of the suggested UWB antenna element at 3.5, 5.5, 10.05, 14.05, and 20 GHz.

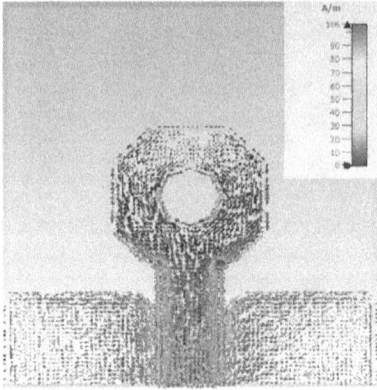

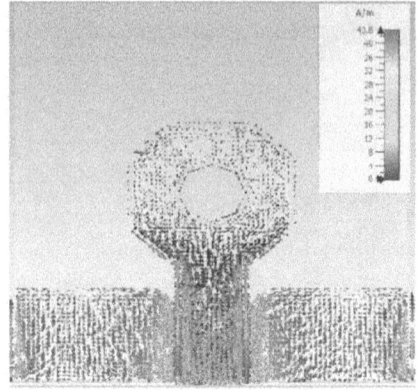

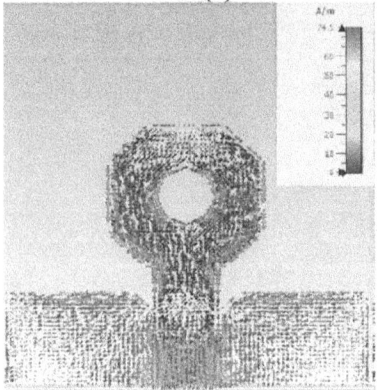

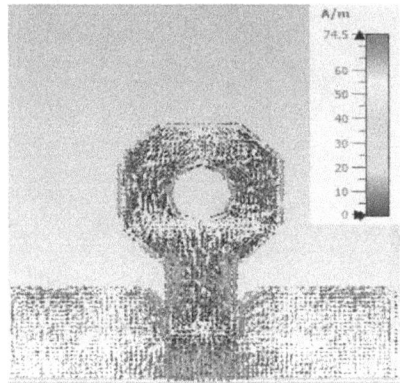

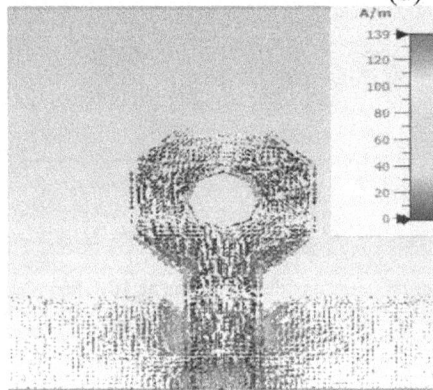

Figure 4.5 The proposed antenna's current distributions at 3.5, 5.5, 10.05, 14.05, and 20 GHz.

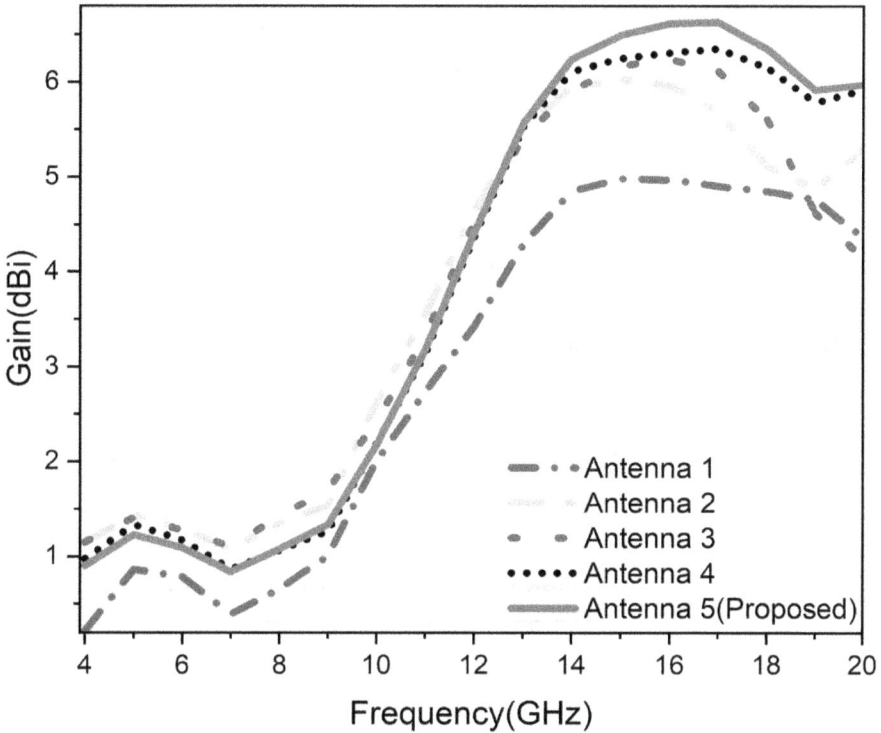

Figure 4.6 Comparison of gain for Antennas 1, 2, 3, 4, and 5.

The current distributions were primarily centered in the feed line, the border of defective ground, and the bottom edge of the curve of the metallic patch. The current distribution at the feed line changes with frequency, even while the distribution's overall area does not. This incident further demonstrates how altering the metallic patch and the DGS can improve the accuracy of the UWB antennas.

Figure 4.6 portrays the peak gain vs. frequency for Antennas 1, 2, 3, 4, and 5, with positive peak gains over the UWB range along the bandwidth, a maximum gain of approximately 6.7 dBi on 17.50 GHz for this proposed antenna.

The efficiency of Antenna 5, shown in Figure 4.7, is higher than 85%, particularly between 11 and 14 GHz.

The measured copolar and cross-polar radiation far-field patterns are shown in Figure 4.8. The picture shows that throughout the full UWB band, an H-plane far-field pattern is essentially omnidirectional. Bidirectional properties resemble monopoles in lower-frequency E-plane patterns.

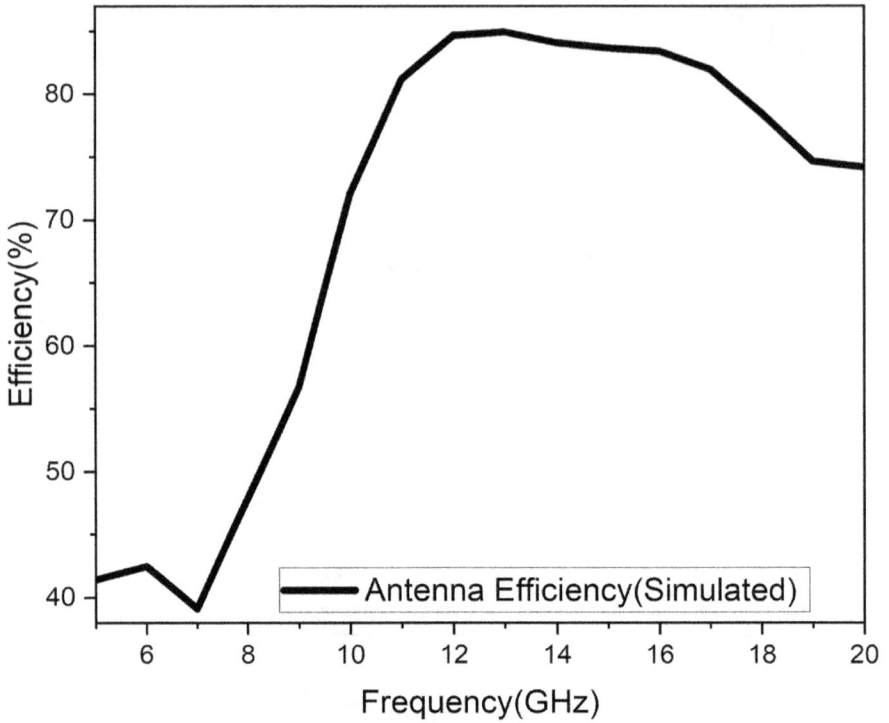

Figure 4.7 Simulated efficiency of Antenna 5.

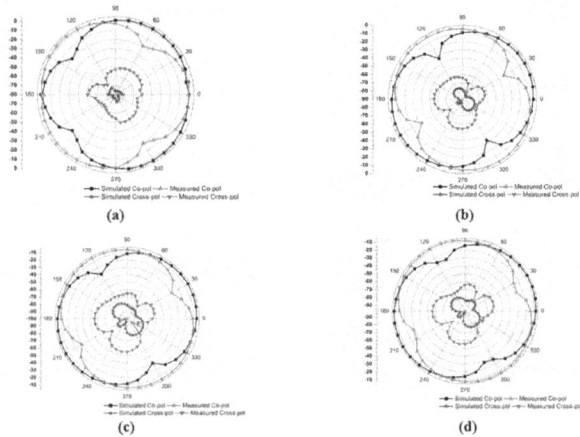

Figure 4.8 The following frequencies were used to measure copolar and cross-polar patterns at (a) 3.20 GHz, (b) 5.50 GHz, (c) 10.05 GHz, and (d) 14.05 GHz.

4.5 PARAMETRIC ANALYSIS

Parametric analysis is carried out to give more design details for antenna engineers. The dimensions of various ground center slot shapes, ground plane lengths, the substrate's dielectric constant, and thickness are various geometrical and electrical factors that significantly affect the antenna's performance.

4.5.1 Impact of Different Ground Plane Lengths

Improvement in S_{11} properties are confirmed for various lengths of optimized grounds. Following the methodical optimization procedure, the optimal ground length is 5.5 mm. This length produced better wideband properties. Figure 4.9 displays the comparison plots of S_{11} for a fully optimized ground structure without slots.

From Figure 4.9, it is clear that the improved ground has a larger impedance bandwidth than the full ground.

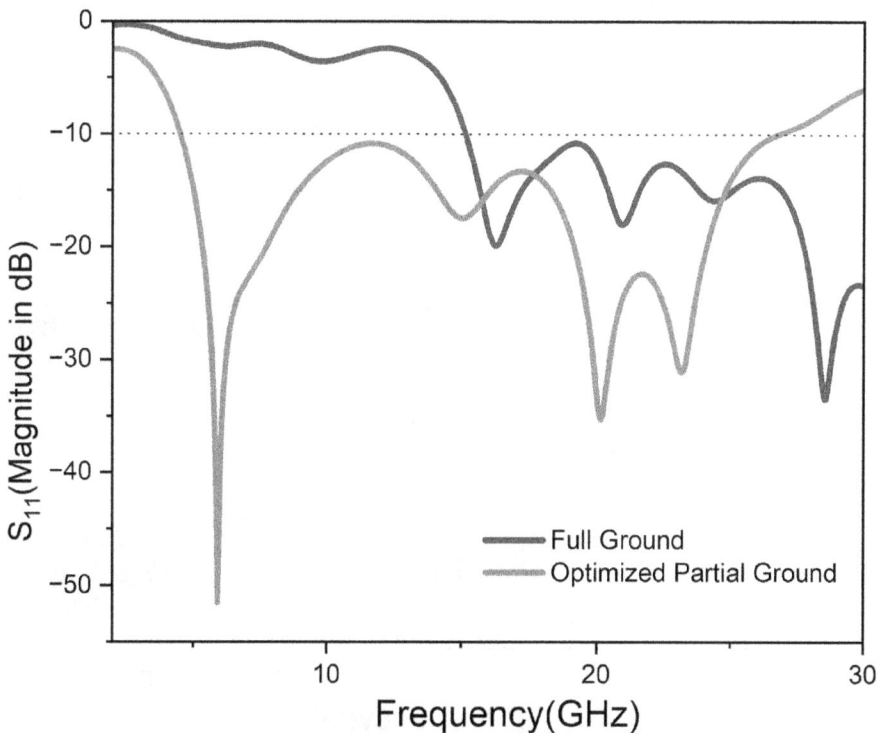

Figure 4.9 The S_{11} comparative plots for optimal and full ground.

4.5.2 Effect of the Substrate's Various Thicknesses

Figure 4.10 displays the reflection coefficient vs. frequency graphs for L = 23 mm and W = 20 mm for three different values of h (0.8 mm, 3.2 mm, and 1.6 mm). The impacts seen with an increase in h from 0.8 mm to 3.2 mm are shown.

The fringing fields from the edges expand as h increases, increasing the extension in length ΔL and, thus, the effective length. This decreases the resonance frequency. In contrast, when h increases, the W/h ratio lowers, causing ε_e to decrease and ultimately raising the resonance frequency. However, the impact of the rise in ΔL outweighs the fall in ε_e. As a result, the resonance frequency is reduced overall.

4.5.3 Effects on the Ideal Ground of Various Ground Center Slot Shapes

The optimized ground is further enhanced by adding rectangular and hexagonal holes to broaden the impedance bandwidth in this configuration. The comparison charts of S_{11} for the optimum ground with various

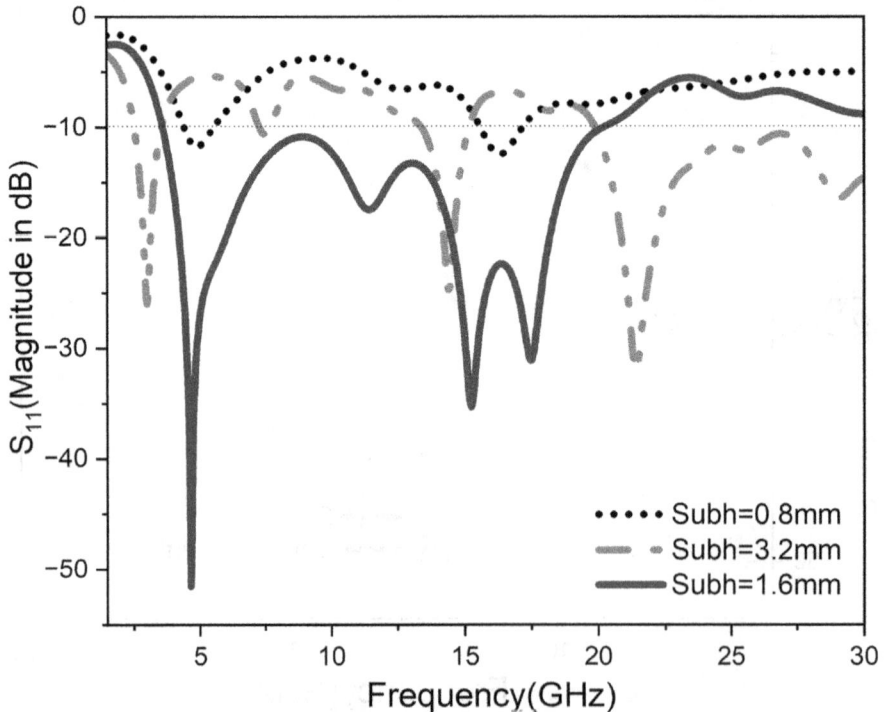

Figure 4.10 The S_{11} comparison graphs for various substrate height values.

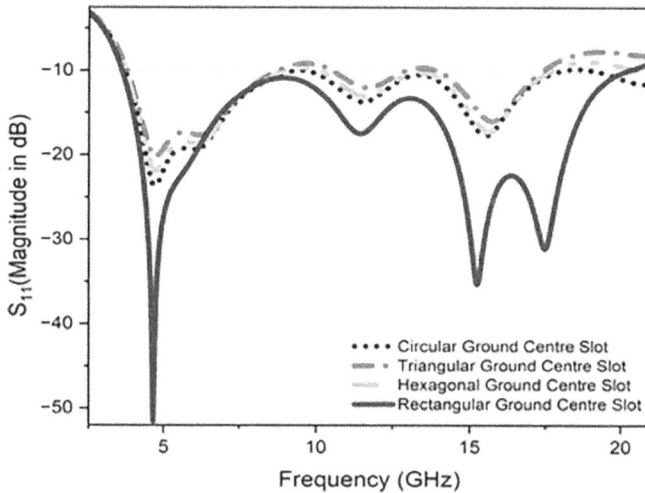

Figure 4.11 Plots of S_{11} comparing the optimal ground with various ground edge slot geometries.

Table 4.3 Optimized Ground Characteristics with Various Edge Slot Shapes

Slots in the Ground	Optimal Slot Dimension (mm)	Operating Range (GHz–GHz)	Impedance Bandwidth (GHz)	S_{11} Minimum (dB)
Circular	2	3.6–9.2	5.6	−23.68
Triangular	2	3.7–8.7	5	−20
Hexagonal	2	3.7–8.9	5.2	−22
Rectangular	2	3.1–20	16.9	−51.5

ground-centered slot geometries are shown in Figure 4.11. The various center slot shapes in the modified ground affecting S_{11} characteristics are also shown in Figure 4.11.

Table 4.3 displays the observations of various centered slot shapes on the suitable ground. Table 4.3 shows that, as mentioned earlier, the described design with the geometries of a central slot in the modified ground considerably increases the operating bandwidth.

4.5.4 Effect of the Different Substrate Type

Teflon and Bakelite, two substrate materials with differing permittivities and properties, are suitable for this particular design and application. However, FR-4 is the primary substrate material in this design since it is sturdy (resistant to tearing), long-lasting, simple to maintain, and reasonably priced. The

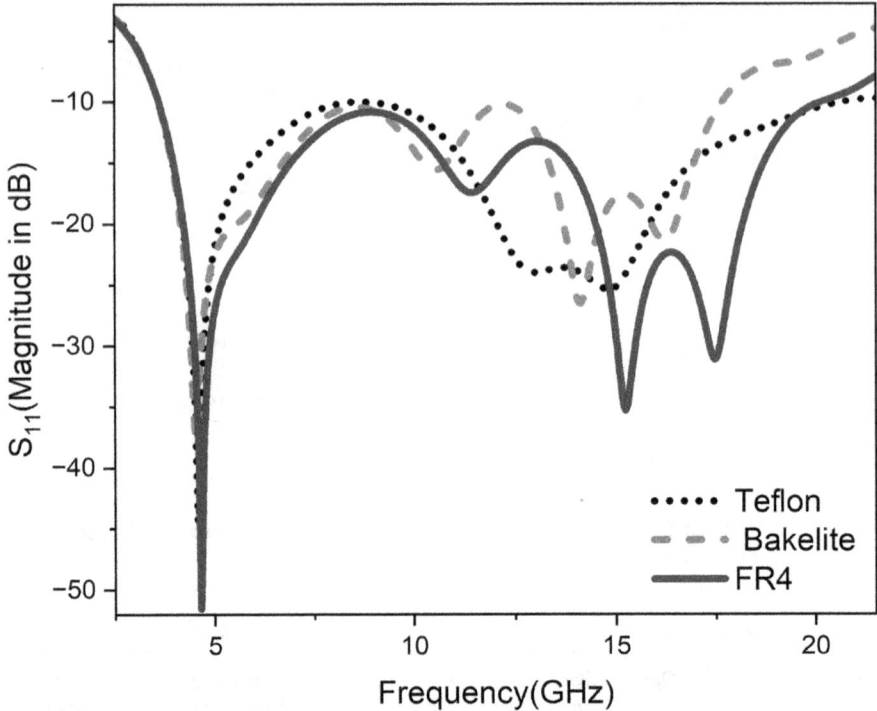

Figure 4.12 The S$_{11}$ comparison graphs for various substrate materials.

S-parameter response for Bakelite, Teflon, and conventional FR-4 epoxy is shown in Figure 4.12 with relative dielectric constants of 2.1, 5, and 4.4.

4.6 MEASUREMENT AND FABRICATION

The suggested UWB monopole antenna (Antenna 5) is constructed and simulated utilizing CST Microwave Studio 2021. This recommended design is made using PCB technology, and the Rohde and Schwarz VNA is utilized to examine the outcomes as measurements and construction of the proposed UWB planar monopole antenna take place. The simulation and measurement findings of the reflection coefficient and the corresponding observed values are provided for the suggested UWB antenna. The verified result and the result of the simulation are frequently in agreement. In manufacturing and verification, there could be some differences between the measured result and the results of the simulation.

Figure 4.13 depicts a view of the suggested constructed UWB antenna. As depicted in Figure 4.14, the measured and the simulated S-parameters

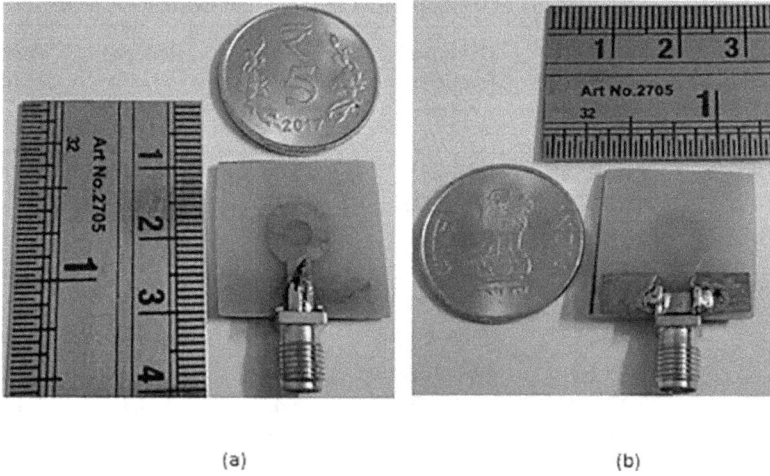

(a)   (b)

Figure 4.13 Fabricated antenna prototype: (a) front view and (b) rear view.

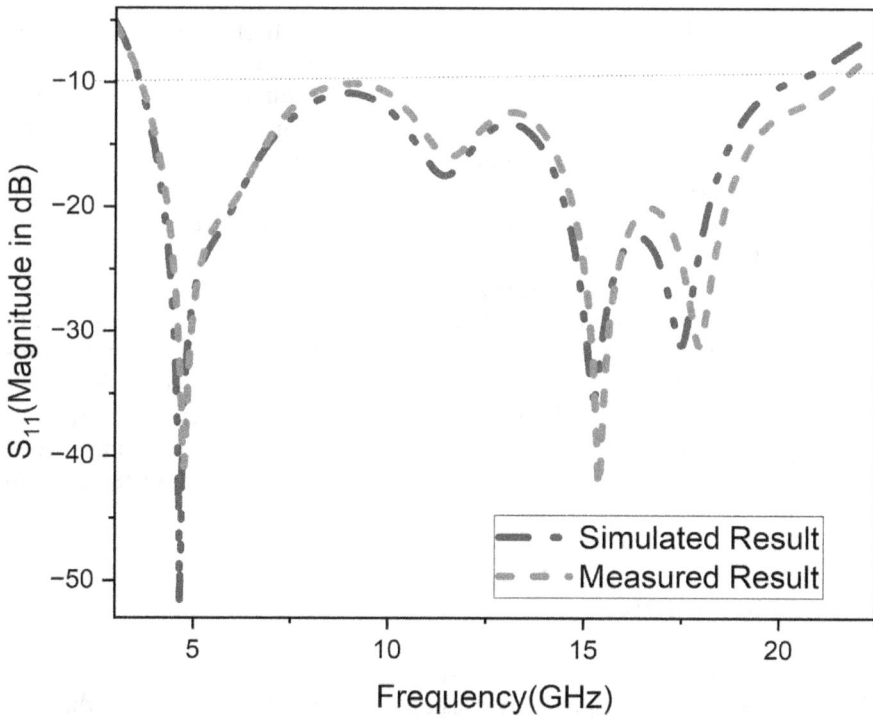

Figure 4.14 Simulated and measured reflection coefficient versus frequency curve for a proposed UWB antenna.

Table 4.4 A Comparative Study of the Suggested Antenna and Existing Antennas

Reference Antennas	Antenna Dimension (mm²)	Operating Frequency Range (GHz)	Impedance Bandwidth (GHz)	FBW (%)	Peak Gain (dBi)
[41]	124 × 110	1.02–24.1	23.08	183.76	7
[43]	40 × 40	2.5–20.25	17.75	156.04	6.1
[44]	20 × 28	2.85–11.85	9	122.45	3.2
[45]	452.39	3.05–12.11	9.06	119.52	8.36
[48]	60 × 60	1.75–7.43	5.68	123.75	16.71
[49]	34 × 25	2.85–20	17.15	150.11	6
[50]	50 × 50	2.6–11.20	8.6	124.64	2.27
[51]	38.87 × 24	3.12–12.54	9.42	120.31	11.30
[62]	70 × 60	2–11	9	138.46	8.52
Proposed Work	23 × 20	3.1–20	16.9	146.32	6.7

of this projected antenna system perfectly match. It should be mentioned that the constructed UWB antenna's measured reflection coefficient ranges from 3.1 to 20 GHz and is below –10 dB. The suggested antenna's measured S-parameters at high frequencies, as shown in Figure 4.14, are not entirely consistent with the predicted findings. However, the measured S_{11} beats the predicted values, being less than –10 decibels throughout the entire bandwidth. The observed and simulated curves have similar trends and display the desired UWB properties.

This suggested UWB miniaturized monopole antenna is then compared to the earlier UWB antennas written about in the literature in Table 4.4. In this case, λ_0 represents a wavelength of the free space operating at the bandwidth of lowest frequency. The table makes it abundantly clear that these advantages are significant despite the specific antennas' better performance regarding radiation patterns, antenna efficiency, gain, etc. The recommended antenna, in comparison, is a compact UWB antenna with an excellent gain and FBR.

4.7 CONCLUSION

This study suggests an innovative, small UWB planar monopole antenna. This reported UWB design comprises an open-ended slit etched at the bottom plane just behind the feed line, octagon-shaped slot at the center of the metallic plane, and a modified square patch with truncated edges. The revised radiator aims to introduce the operating bandwidth, covering from

3.1 to 20 GHz. This suggested antenna measures 20 mm × 23 mm × 1.6 mm. The antenna for the projected UWB is constructed and measured. Simulated and empirically verified performance parameters are used. It has a maximum gain of 6.7 dBi and operates between 11 and 14 GHz. Comparing the obtained findings to the previous literature, it can be determined that this suggested antenna produces good radiation parameters over the UWB band. Therefore, this small antenna has substantial advantages for short-range wireless, handheld, and satellite applications.

Sophisticated wireless communication systems now use superior antenna construction as a standard component. By increasing antenna characteristics and doing more research, antenna gain may be improved.

REFERENCES

[1] H. Nikookar and R. Prasad, Introduction to Ultra Wideband for Wireless Communications, 1st ed. New York, NY: Springer, 2005.

[2] Federal Communications Commission, FCC 02 -214.doc. https://docs.fcc.gov/public/attachments/FCC-02-214A1.pdf.

[3] W. S. Jeon, H. S. Oh, and D. G. Jeong, "Decision of ranging interval for IEEE 802.15.4z UWB ranging devices," IEEE Internet Things J., vol. 8, no. 20, pp. 15628–15638, 2021.

[4] K. S. Gopalan, A. Bansal, and A. R. Kabbinale, "Tracking resurgence of ultrawideband_A standards and certi_cation perspective," in International Conference on Communication Systems & Networks (COMSNETS), January 2022, pp. 4–8.

[5] D. Coppens, A. Shahid, S. Lemey, B. Van Herbruggen, C. Marshall, and E. De Poorter, "An overview of UWB standards and organizations (IEEE 802.15.4, FiRa, Apple): Interoperability aspects and future research directions," IEEE Access, vol. 10, pp. 70219–70241, 2022. doi: 10.1109/ACCESS.2022.3187410.

[6] W. S. Yeoh and W. S. T. Rowe, "An UWB conical monopole antenna for multiservice wireless applications," IEEE Antennas Wireless Propag. Lett., vol. 14, pp. 1085–1088, 2015.

[7] R. V. S. R. Krishna and R. Kumar, "A dual-polarized square-ring slot antenna for UWB, imaging, and radar applications," IEEE Antennas Wireless Propag. Lett., vol. 15, pp. 195–198, 2016.

[8] Y. Wang, T. Huang, D. Ma, P. Shen, J. Hu, and W. Wu, "Ultrawideband (UWB) monopole antenna with dual notched bands by combining electromagnetic-bandgap (EBG) and slot structures," in IEEE MTT-S International Microwave Biomedical Conference (IMBioC), May 2019, pp. 6–8.

[9] S. S. Al-Bawri, H. H. Goh, M. S. Islam, H. Y. Wong, M. F. Jamlos, A. Narbudowicz, M. Jusoh, T. Sabapathy, R. Khan, and M. T. Islam, "Compact ultrawideband monopole antenna loaded with metamaterial," Sensors, vol. 20, no. 3, p. 796, 2020.

[10] N. A. Jan, S. H. Kiani, D. A. Sehrai, M. R. Anjum, A. Iqbal, M. Abdullah, and S. Kim, "Design of a compact monopole antenna for UWB applications," Comput. Mater. Continua, vol. 66, no. 1, pp. 35–44, October 2020.

[11] M. N. Hasan and M. Seo, "A planar 3.4–9 GHz UWB monopole antenna," in IEEE International Symposium on Antennas and Propagation (ISAP), October 2018, pp. 23–26.

[12] B. Roy, S. K. Chowdhury, and A. K. Bhattacharjee, "Symmetrical hexagonal monopole antenna with bandwidth enhancement under UWB operations," Wireless Pers. Commun., vol. 108, no. 2, pp. 853–863, 2019.

[13] S. Ullah, C. Ruan, M. S. Sadiq, T. U. Haq, and W. He, "High efficient and ultra wide band monopole antenna for microwave imaging and communication applications," Sensors, vol. 20, no. 1, p. 115, 2019.

[14] R. Azim, M. T. Islam, and N. Misran, "Compact tapered-shape slot antenna for UWB applications," IEEE Antennas Wireless Propag. Lett., vol. 10, pp. 1190–1193, 2011.

[15] M. M. Alam, R. Azim, N. M. Sobahi, A. I. Khan, and M. T. Islam, "An asymmetric CPW-fed modified bow tie-shaped antenna with parasitic elements for ultra-wideband applications," Int. J. Commun. Syst., vol. 35, no. 9, pp. 1–11, 2022.

[16] S. S. Mirmosaei, S. E. Afjei, E. Mehrshahi, and M. M. Fakharian, "A dual band-notched ultra-wideband monopole antenna with spiral-slots and folded SIR-DGS as notch band structures," Int. J. Microw. Wireless Technol., vol. 8, no. 8, pp. 1197–1206, 2015.

[17] J. Y. Siddiqui, C. Saha, and Y. M. M. Antar, "A novel ultrawideband (UWB) printed antenna with a dual complementary characteristic," IEEE Antennas Wireless Propag. Lett., vol. 14, pp. 974–977, 2015.

[18] H. Liu and Z. Xu, "Design of UWB monopole antenna with dual notched bands using one modified electromagnetic-bandgap structure," Sci. World J., vol. 2013, pp. 1–9, 2013.

[19] M. Mottahir Alam, R. Azim, I. M. Mehedi, and A. I. Khan, "Coplanar waveguide-fed compact planar ultra-wideband antenna with inverted L-shaped and extended U-shaped ground for portable communication devices," Chin. J. Phys., vol. 73, pp. 684–694, 2021.

[20] W. A. Awan, A. Zaidi, N. Hussain, A. Iqbal, and A. Baghdad, "Stub loaded, low profile UWB antenna with independently controllable notch-bands," Microw. Opt. Technol. Lett., vol. 61, no. 11, pp. 2447–2454, 2019.

[21] N. Hussain, M. Jeong, J. Park, S. Rhee, P. Kim, and N. Kim, "A compact size 2.9–23.5 GHz microstrip patch antenna with WLAN band-rejection," Microw. Opt. Technol. Lett., vol. 61, no. 5, pp. 1307–1313, 2019.

[22] B. R. Perli and M. R. Avula, "Design of wideband elliptical ring monopole antenna using characteristic mode analysis," J. Electromagn. Eng. Sci., vol. 21, no. 4, pp. 299–306, 2021.

[23] C.-H. Chang and K.-L. Wong, "Printed \$lambda/8\$ -PIFA for penta-band WWAN operation in the mobile phone," IEEE Trans. Antennas Propag., vol. 57, no. 5, pp. 1373–1381, 2009.

[24] K.-L. Wong and C.-H. Huang, "Compact multiband PIFA with a coupling feed for internal mobile phone antenna," Microw. Opt. Technol. Lett., vol. 50, no. 10, pp. 2487–2491, 2008. https://doi.org/10.1002/mop.23727.

[25] C. Deng, Y. Xie, and P. Li, "CPW-fed planar printed monopole antenna with impedance bandwidth enhanced," IEEE Antennas Wirel. Propag. Lett., vol. 8, pp. 1394–1397, 2009.

[26] J. Y. Siddiqui, C. Saha, and Y. M. M. Antar, "Compact dual-SRR-loaded UWB monopole antenna with dual frequency and wideband notch characteristics," IEEE Antennas Wirel. Propag. Lett., vol. 14, pp. 100–103, 2015.

[27] I. B. Vendik, A. Rusakov, K. Kanjanasit, et al., "Ultrawideband (UWB) planar antenna with single-, dual-, and triple-band notched characteristic based on electric ring resonator," IEEE Antennas Wirel. Propag. Lett., vol. 16, pp. 1597–1600, 2017.

[28] J. Liang, C. C. Chiau, X. Chen, et al., "Printed circular disc monopole antenna for ultra-wideband applications," Electron. Lett., vol. 40, no. 20, pp. 1246–1247, 2004.

[29] J. Liu, S. Zhong, and K. P. Esselle, "A printed elliptical monopole antenna with modified feeding structure for bandwidth enhancement," IEEE Trans. Antennas Propag., vol. 59, no. 2, pp. 667–670, 2011.

[30] A. M. Abbosh and M. E. Bialkowski, "Design of ultrawideband planar monopole antennas of circular and elliptical shape," IEEE Trans. Antennas Propag., vol. 56, no. 1, pp. 17–23, 2008.

[31] M. Gopikrishna, D. D. Krishna, C. K. Anandan, P. Mohanan, and K. Vasudevan, "Design of a compact semi-elliptic monopole slot antenna for UWB systems," IEEE Trans. Antennas Propag., vol. 57, no. 6, pp. 1834–1837, June 2009. doi: 10.1109/TAP.2009.2015850.

[32] M. G. N. Alsath and M. Kanagasabai, "Compact UWB monopole antenna for automotive communications," IEEE Trans. Antennas Propag., vol. 63, no. 9, pp. 4204–4208, 2015.

[33] C.-Y. Hong, C.-W. Ling, I.-Y. Tarn, and S.-J. Chung, "Design of a planar ultra-wideband antenna with a new band-notch structure," IEEE Trans. Antennas Propag., vol. 55, no. 12, pp. 3391–3397, December 2007. doi: 10.1109/TAP.2007.910486.

[34] S. Nikolaou and M. A. B. Abbasi, "Design and development of a compact UWB monopole antenna with easily-controllable return loss," IEEE Trans. Antennas Propag., vol. 65, no. 4, pp. 2063–2067, 2017.

[35] Y. Lu, Y. Huang, H. T. Chattha, et al., "Reducing ground-plane effects on UWB monopole antennas," IEEE Antennas Wirel. Propag. Lett., vol. 10, pp. 147–150, 2011.

[36] M. Ojaroudi, C. Ghobadi, and J. Nourinia, "Small square monopole antenna with inverted T-shaped notch in the ground plane for UWB application," IEEE Antennas Wirel. Propag. Lett., vol. 8, pp. 728–731, 2009.

[37] Z. N. Chen, T. S. P. See, and X. Qing, "Small printed ultrawideband antenna with reduced ground plane effect," IEEE Trans. Antennas Propag., vol. 55, no. 2, pp. 383–388, 2007.

[38] K. Xu, Z. Zhu, H. Li, J. Huangfu, C. Li, and L. Ran, "A printed single-layer UWB monopole antenna with extended ground plane stubs," IEEE Antennas Wirel. Propag. Lett., vol. 12, pp. 237–240, 2013. doi: 10.1109/LAWP.2013.2247555.

[39] D. N. Elsheakh, H. A. Elsadek, E. A. Abdallah, M. F. Iskander, and H. Elhenawy, "Investigated new embedded shapes of electromagnetic bandgap structures and via effect for improved microstrip patch antenna performance," Appl. Phys. A Mater. Sci. Process., vol. 103, pp. 541–545, 2011.

[40] D. Qu, L. Shafai, and A. Foroozesh, "Improving microstrip patch antenna performance using EBG substrates," IEE Proc. Microw. Antennas Propag., vol. 153, no. 6, pp. 558–563, 2006.

[41] J. Liu, S. Zhong, and K. P. Esselle, "A printed elliptical monopole antenna with modified feeding structure for bandwidth enhancement," IEEE Trans. Antennas Propag., vol. 59, no. 2, pp. 667–670, 2011.

[42] A. Bhattacharya, A. De, B. Roy, and A. K. Bhattacharjee, "Investigations on a low-profile, filter backed, printed monopole antenna for UWB communication," Ind J Pure Appl Phys., vol. 58, pp. 106–112, 2020.

[43] A. De, B. Roy, and A. K. Bhattacharjee, "Novel, compact, circular-sectored antenna for ultra-wideband (UWB) communications," Electromagnetics, vol. 40, no. 3, pp. 165–176, 2020.

[44] S. Nikolaou and M. A. B. Abbasi, "Design and development of a compact UWB monopole antenna with easily-controllable return loss," IEEE Trans. Antennas Propag., vol. 65, no. 4, pp. 2063–2067, 2017.

[45] A. De, B. Roy, A. Bhattacharya, G. V. Bharat, and A. K. Bhattacharjee, "Compact UWB monopole antenna with WLAN and X-band satellite filtering characteristics," In 2020 International Conference on Computation, Automation and Knowledge Management (ICCAKM), January 2020, pp. 344–347. IEEE.

[46] H. Kim and C. W. Jung, "Ultra-wideband endfire directional tapered slot antenna using CPW to wide-slot transition," Electronics Letters, vol. 46, no. 17, pp. 1183–1185, 2010.

[47] A. Dastranj and H. Abiri, "Bandwidth enhancement of printed E-shaped slot antennas fed by CPW and microstrip line," IEEE Trans. Antennas Propag., vol. 58, no. 4, pp. 1402–1407, April 2010.

[48] S. Kundu and A. Chatterjee, "Sharp triple-notched ultra wideband antenna with gain augmentation using FSS for ground penetrating radar," Wirel. Pers. Commun., vol. 117, no. 2, pp. 1399–1418, 2021.

[49] M. Gopikrishna, D. D. Krishna, C. K. Anandan, et al., "Design of a compact semi-elliptic monopole slot antenna for UWB systems," IEEE Trans. Antennas Propag., vol. 57, no. 6, pp. 1834–1837.

[50] J. Y. Siddiqui, C. Saha, and Y. M. M. Antar, "Compact SRR loaded UWB circular monopole antenna with frequency notch characteristics," IEEE Trans. Antennas Propag., vol. 62, no. 8, pp. 4015–4020, 2014.

[51] A. De, B. Roy, A. Bhattacharya, and A. K. Bhattacharjee, "Investigations on a circular UWB antenna with Archimedean spiral slot for WLAN/Wi-MAX and satellite X-band filtering feature," Int. J. Microwave Wirel. Technol., vol. 14, no. 6, pp. 781–789, 2022.

[52] D. Gaetano, P. McEvoy, M. Ammann, C. Brannigan, L. Keating, and F. Horgan, "Footwear and wrist communication links using 2.4 GHz and UWB antennas," Electronics, vol. 3, no. 2, pp. 339–350, 2014.

[53] A. Wu and B. Guan, "A compact CPW-fed UWB antenna with dual band-notched characteristics," Int. J. Antennas Propag., vol. 2013, pp. 1–7, 2013.

[54] Y. K. Soni and N. K. Agrawal, "Compact UWB/Bluetooth integrated uniplanar antenna with WLAN notch property," in Proceedings of the ICT and Critical Infrastructure, 48th Annual Convention of Computer Society of India, in Advances in Intelligent Systems and Computing, vol. 248. Cham: Springer, 2014, pp. 459–466.

[55] Y.-B. Yang, F.-S. Zhang, F. Zhang, L. Zhang, and Y.-C. Jiao, "Design of novel wideband monopole antenna with a tunable notched-band for 2.4 GHZ WLAN

and UWB applications," Prog. Electromagn. Res. Lett., vol. 13, pp. 93–102, 2010.

[56] J. Ren, D. Mi, and Y.-Z. Yin, "Compact ultrawideband MIMO antenna with WLAN/UWB bands coverage," Prog. Electromag. Res. C, vol. 50, pp. 121–129, 2014.

[57] F.-T. Zha, S.-X. Gong, G. Liu, H.-Y. Yang, and S.-G. Lin, "Compact slot antenna for 2.4 GHz/UWB with dual band-notched characteristic," Microw. Opt. Technol. Lett., vol. 51, no. 8, pp. 1859–1862, 2009.

[58] Y. Y. Sun, S. W. Cheung, and T. I. Yuk, "An ISM/UWB antenna with offset feeding and slotted ground plane for body-centric communications," J. Elec. Electron. Eng., vol. 1, no. 2, pp. 45–50, 2013.

[59] Ankan Bhattacharya, Bappadittya Roy, Santosh K. Chowdhury, and Anup K. Bhattacharjee, "Computational and experimental analysis of a low-profile, isolation-enhanced, band-notch UWB-MIMO antenna," J. Comput. Electron., vol. 18, pp. 680–688, 2019.

[60] A. De, B. Roy, A. Bhattacharya, and A. K. Bhattacharjee, "Bandwidth-enhanced ultra-wide band wearable textile antenna for various WBAN and Internet of Things (IoT) applications," Radio Sci., vol. 56, p. e2021RS007315, 2021.

[61] G. K. Pandey, H. S. Singh, P. K. Bharti, et al., "UWB monopole antenna with enhanced gain and stable radiation pattern using gate like structures," in International Conference on Microwave and Photonics (ICMAP), Dhanbad, December 2013, pp. 4–7.

Chapter 5

A Printed Array of Nature-Inspired Antennas for IoT and Future 5G Applications

Kalyan Sundar Kola and Anirban Chatterjee

5.1 INTRODUCTION

The Internet of Things (IoT) has grown in popularity as Internet access has been made available to various devices. Ericsson predicts that by the year 2025, more than 5 billion devices will be online simultaneously. Fifth-generation (5G) is growing in popularity as a potential alternative to Long-Term Evolution (LTE) in crowded metropolitan areas due to LTE's low data speeds, constrained bandwidth, and poor quality of service (QoS). On the other hand, 5G provides high throughput, massive capacity, and efficient use of the radio spectrum. The new 5G technology requires outdoor and urban networks and inside systems. Because of these advantages, the IoT and 5G wireless technologies are in great demand, and their graphical representation is shown in Figure 5.1(a) and (b), respectively. In order to link IoT devices in this era of constant connectivity, a new generation of inexpensive and lightweight printed antennas [1, 2] is urgently required for 5G technology. Consequently, a microstrip patch antenna with a broad frequency range is the optimal choice. A small number of articles have looked at the printed antenna and its array.

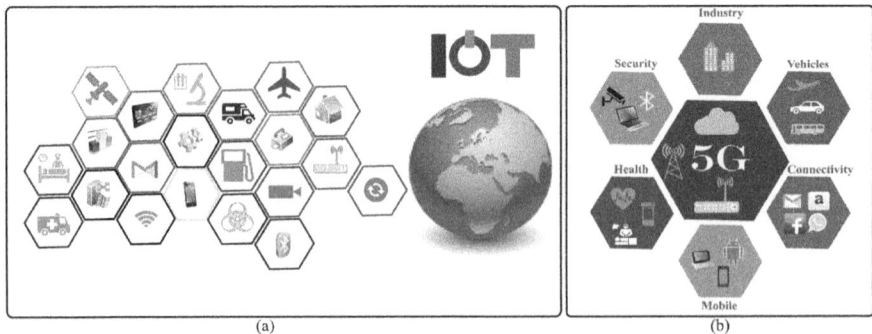

Figure 5.1 Architectural representation of (a) IoT and (b) 5G communication applications.

DOI: 10.1201/9781003459880-5

In the literature [3], Carver *et al.* explored the theoretical and practical implications of printed antennas in the academic literature. As a bonus, Werner *et al.* [4] investigated the development of fractal geometry in antenna construction. Fractal geometry was used to create novel antenna arrays [4]. Several types of multiple-input, multiple-output (MIMO) antennas for wideband and IoT applications have been developed and carried out by Bhattacharya in the literature [5, 6]. Dey *et al.* proposed a wearable ultra-wideband textile antenna for wideband and IoT applications [7]. IoT applications have inspired Al-Sehemi *et al.* [8] to propose and thoroughly examine a broadband waterproof antenna. Ashyap *et al.* [9] created a C-shaped printed antenna made from a laminated fabric material for medical IoT devices. To facilitate the connectivity of IoT devices, Cowsigan *et al.* [10] developed a printed antenna supported by a substrate integrated waveguide (SIW) cavity. Rogers *et al.* [11] provided an ESP8266 antenna module ideal for IoT applications. Future wireless applications may make use of a six-element MIMO antenna, as reported by Sharma *et al.* [12]. Mushtaq *et al.* [13] introduced a T-shaped slotted printed antenna for use in the IoT. For 5G in-building use, Wang *et al.* [14] report the development of a printed meta-material antenna with circular polarization. De *et al.* [15] designed, constructed, and extensively tested a 5G-enabled printed antenna for IoT applications. The dual-band microstrip patch antenna based on an SIW antenna structure was proposed by Singh *et al.* [16] to improve 5G communications. Several pieces of research have recommended using an antenna array to enhance parametric results. A U-slotted patch antenna array was first presented by Wang *et al.* [17]. In [18], Kola *et al.* showed an antenna based on a collection of natural-looking geometries inspired by clover leaves. The author has also designed and studied other microstrip patch antenna arrays [19–23], including those with Christmas tree–, hybrid fractal–, and tulip flower–shaped geometries.

An architectural representation of IoT and the 5G communication systems is carried out in Figure 5.1(a) and (b), respectively. This chapter proposes a fishing hook–shaped printed radiator followed by a two-element array to meet the requirements of such applications. Some of the novel aspects of the planned antennas are outlined here:

- The sole radiator is derived from a mature-based, fishing hook–shaped geometry.
- The sole antenna is capable of offering a wide impedance bandwidth of 3.30 GHz.
- It also offers below −35 dB of cross-polarization suppression along its main beam direction.
- It also offers quite decent absolute gain at both the resonating frequencies.

- To minimize losses, maximize isolation, and enhance bandwidth responsiveness, the feed network for the array was developed using the Wilkinson power divider [24].
- At the optimal radiation point, the array has a wide impedance bandwidth, high gain, and low cross-polarization level.

Apart from the advantages mentioned earlier, the sole radiator and the array have aperture [25, 26] and radiation efficiencies [27–31] of over 63% and 93%, respectively. At appropriate resonating frequencies, both antennas give more than 30 dB/m correction factors [32] and so have modest electromagnetic interference effects. The proposed antenna and the array are both compact, low-profile, easy to fabricate, small-sized, and offer decent parametric outcomes, which meets the minimum requirement for IoT-enabled systems and 5G network–based applications. Therefore, the proposed antennas are the most significant choice for the same.

5.2 THE SOLE RADIATOR

The single monopole antenna for the desired applications has been derived from a natural geometry named 'fishing hook'. Its conventional structure with the names of its different parts is properly depicted in Figure 5.2. Based on this geometry, a look-alike structure has been designed in the SONNET EM simulator. Its front view is presented in Figure 5.3(a). The printed circuit board (PCB) laminated RO-5880 material is used to design the intended geometry. The radiator portion of the structure consists of a couple of semicircle geometries whose radii are fixed like r_1 and r_2. Further, two thinned right-angle triangular-shaped geometries are added to the semicircular structure's top portion. In addition, a thin rectangular $(l_4 \times l_5)$ shaped geometry has been etched from the previously obtained geometry. Further, a square $(l_1 \times l_1)$ geometry is added at the bottom part of the semicircular geometry, which acts like a 'throat' of a fishing hook and from where the microstrip feed-line ('shank') has been connected to excite the proposed geometry. A 50 Ω source port is used to deliver the power to the structure, which is like an 'eye' of a fishing hook. To achieve the desired impedance bandwidth, a monopole structure was created by removing one-third of the antenna's complete ground plane $(L \times W_g)$. The single-layered information of the antenna has been carried out by its vertical view, as depicted in Figure 5.3(b). Using the SONNET EM simulator [33], we run simulations of the whole structure and then double-check them using the Computer Simulation Technology (CST) Microwave Studio. In Table 5.1 we give the specifics of the antenna's dimensions.

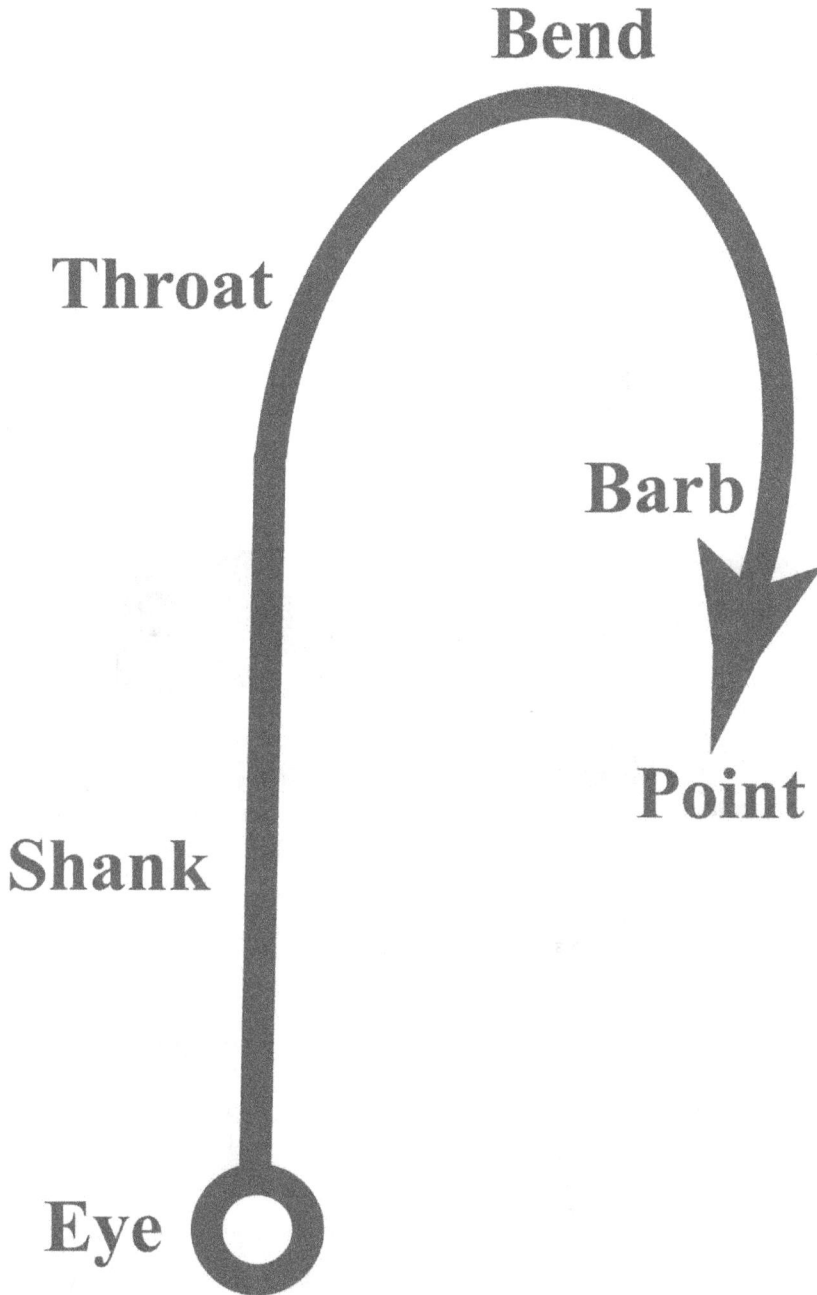

Figure 5.2 A fishing hook and its different parts.

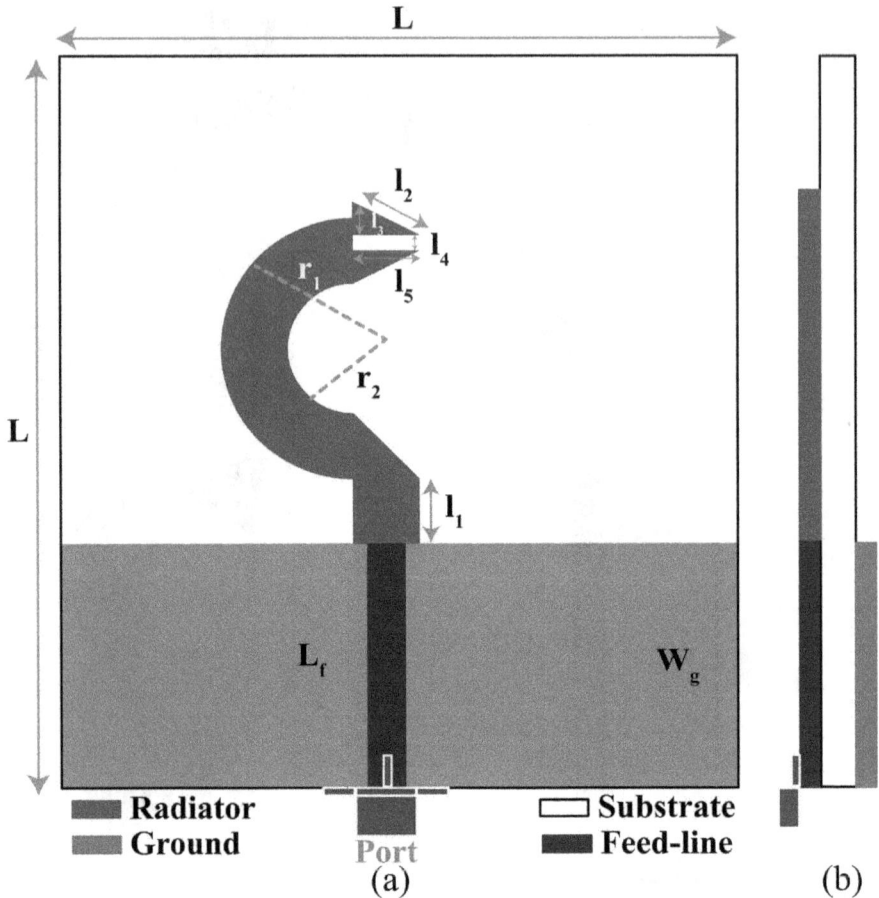

Figure 5.3 Proposed antenna and its (a) horizontal and (b) vertical view.

Table 5.1 Dimension Details of the Intended Antennas (unit:mm)

Coefficient	Value	Coefficient	Value	Coefficient	Value	Coefficient	Value
L	45	r_1	6	r_2	4	l_1	4
l_2	4.4	l_3	2	l_4	1	W_g	15
L_f	15	h_s	0.787	h_t	0.017	L_A	55.50
W_f	65.25	L_G	20.50	R	1	-	-

5.2.1 Contour Length Computation

The effective contour length of the proposed 'fishing hook'–shaped structure can be calculated [33] as follows.

The contour length of the semicircular portion is computed [33] as:

$$P_C = \pi \times (r_1 + r_{21}) \, \text{mm} \tag{5.1}$$

Further, a couple of right triangular–shaped structures have been added with the semicircular portion, and the perimeter of those portions can be computed [33] as follows:

$$P_{TA} = 2 \times \left(l_3 + l_5 + \sqrt{l_3^2 + l_5^2} \right) \text{mm} \tag{5.2}$$

In order to make its 'burb'-like structure, a thin rectangular-shaped geometry has been merged with the recently obtained geometry, and its perimeter is computed [33] as follows:

$$P_R = \left\{ 2 \times (l_4 + l_5) \right\} - l_4 \, \text{mm} \tag{5.3}$$

Lastly, a square structure has been added in the bottom part of the geometry, and its perimeter becomes [33]:

$$P_S = 4 \times l_1 \, \text{mm} \tag{5.4}$$

This allows us to calculate the entire perimeter or contour length [33] of the shape as follows:

$$P_T = P_T + P_{TA} + P_R + P_S \, \text{mm} \tag{5.5}$$

After putting all the parameters' values in eq. (5.1)–eq. (5.5), the computed [33] effective perimeter of the intended geometry becomes 86.28 mm [33].

5.2.2 Analysis of Time-Domain Parameters

In order to get information regarding the signal quality, impedance matching performances, scattering parameters, and group delay characteristics of the proposed sole radiator, the time-domain analysis has been done in the CST Microwave Studio (v2018) platform. For this purpose, a couple of identical antenna elements are placed in two fashions i.e., side-by-side and face-to-face, with a 3-meter space apart, which is the generally preferred distance during measurements. Those antennas are excited with 50 Ω sources individually. Both the arrangements are reflected in Figure 5.4(a) and (b). The applied sinusoidal signal and obtained output signal characteristics

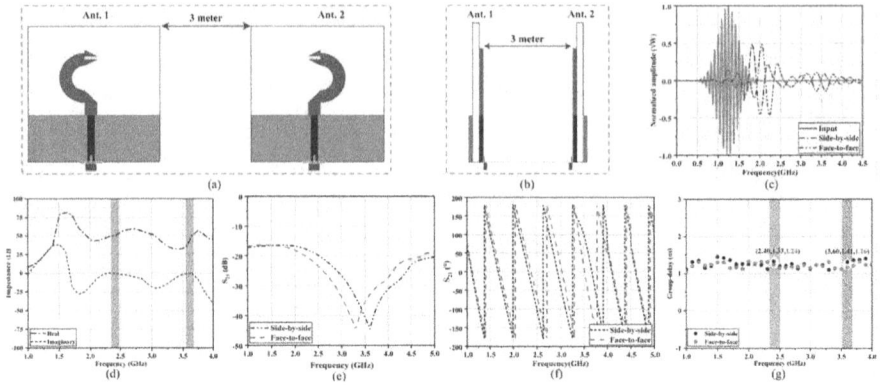

Figure 5.4 Time-domain analysis of the sole radiator: (a) side-by-side, (b) face-to-face arrangement, (c) input-output signal, (d) impedance, (e) S_{21} magnitude, (f) S_{21} phase, and (g) group delay.

are carried out in Figure 5.4(c). The antennas are perfectly matched, which can be clearly observed in Figure 5.4(d). A good amount of isolation has been obtained from the time-domain analysis part, and it can be reflected in Figure 5.4(e). The corresponding phase characteristics for those two simulations have been carried out in Figure 5.4(f). The group delay qualities of the antennas are shown in Figure 5.4(f). It drops below 2 ns, which is desired and widely accepted for wideband applications, especially for IoT and 5G communications. The planned antenna has excellent time-domain properties.

5.3 PROPOSED ARRAY

A linearly polarized array was developed for the intended uses to enhance parametric outcomes. The proposed feed network, followed by the construction of the array, have been carried out in this section.

5.3.1 Construction of the Feed Topology

Figure 5.5 shows a power divider network with three outlets. Each sink port receives the same amount of power from the source via this Wilkinson power divider network. It follows the principle of transmission line theory. The transmission line's characteristic impedance is Z_0, and it is used to transmit power from the source. The power is sent to the quarter-wavelength

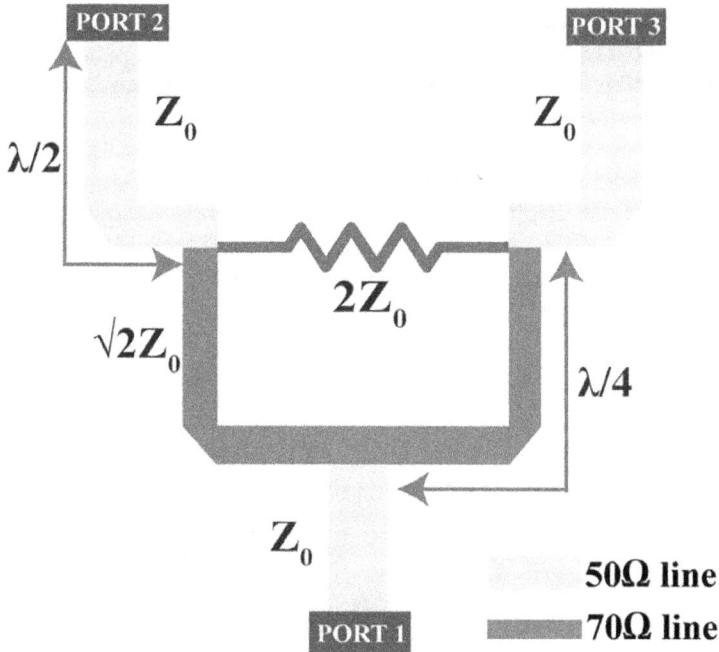

Figure 5.5 Three-way matching networks.

intersection via a transmission line in a similar fashion whose characteristic impedance is $\sqrt{2}Z_0$. Last but not least, the transmission line delivers the power to the sink ports. As illustrated in Figure 5.5, at the junction of the quarter-wave transmission line, a resistor $2Z_0$ is added to achieve proper isolation from the network.

The input impedance (Z_{in}) [1, 28] can be ascertained as follows:

$$Z_{in} = Z_0 \frac{Z_L + jZ_0 tan\beta l}{Z_0 + jZ_L tan\beta l} \tag{5.6}$$

where, Z_L and Z_0 are the as antenna's load and characteristic impedance, respectively, and βl is its length.

5.3.2 Construction of Array Geometry

To achieve better parametric results, the array is built using a few antenna components grouped in a linear form. The proposed linear array is shown in Figure 5.6. It can be clearly observed that the antenna elements are placed face-to-face inside the array geometry. This setup aims to achieve the acceptable radiation pattern in a proper direction, whereas the sole radiator is radiated across at two desired frequencies. To mitigate the antenna's mutual coupling effect and also boost antenna gain, a fixed interelement spacing of 0.5 λ has been devised. Almost 50% of the ground plane has been etched to achieve the wide-impedance bandwidth, which is quite desirable for those applications. The feed topology of the array has been configured by a distinctive Wilkinson power divider [24]. All 50 Ω and 70.7 Ω transmission line lengths have been fixed as the integer multiples of 0.5λ and 0.25λ, respectively. Quarter-wavelength isolation from the feed network is improved by mounting a resistor R in the junction, as shown in Figure 5.6. Table 5.1 provides the quantitative figures for the array's dimension notation.

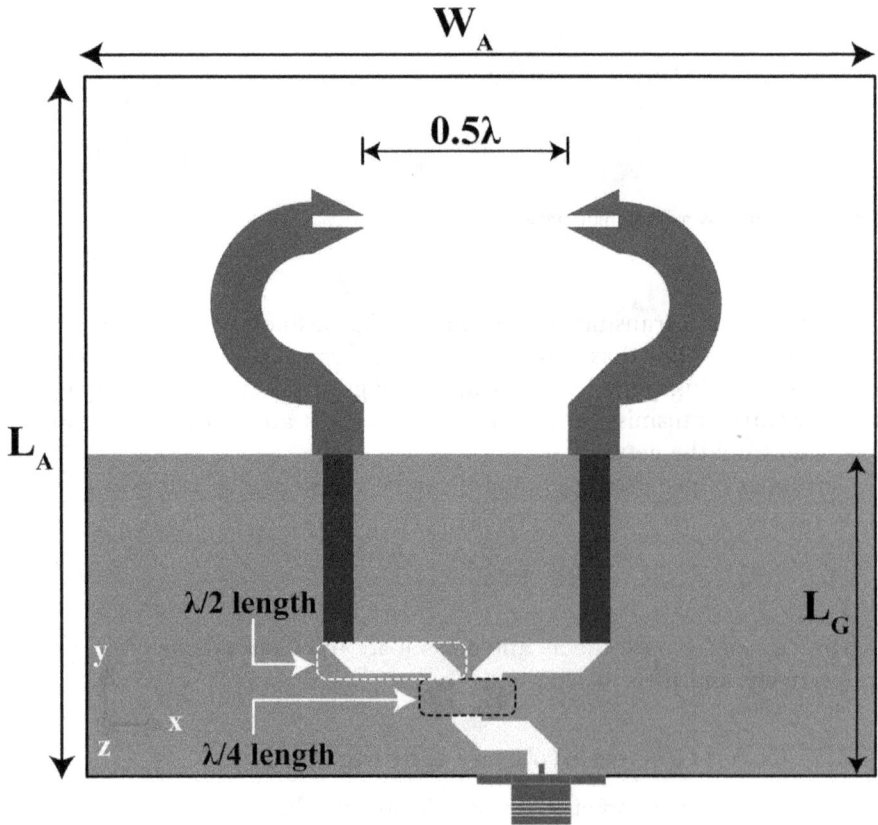

Figure 5.6 Proposed two-element linear array.

5.4 RESULTS AND DISCUSSION

The proposed fishing hook–shaped printed radiator, followed by a linear array, has been designed for IoT and 5G-related applications. Models of both antennas were created in SONNET and then cross-checked in CST Microwave Studio (v2018). A partial ground plane–based planar structured antenna offers wide impedance bandwidth and decent gain for the desired applications under the S-band. This section will cover the discussion about the obtained simulated results of the intended antennas. The sole radiator has two resonances, which are 2.40 GHz and 3.60 GHz, where one is responsible for IoT-related applications and the other is for 5G communications systems. In both the resonances, the antenna offers a return loss of 34.26 and 26.54 dB, respectively, along with a wide return-loss bandwidth of 3.30 GHz, which is depicted in Figure 5.7(a). The absolute gain characteristics of the antenna have been carried out in Figure 5.7(b). It can be observed from Figure 5.7(b) that the antenna offers an immense gain of 7.84 and 8.23 dBi, respectively, at the two desired resonating frequencies. The field patterns (E, H) of the sole radiator for 2.40 and 3.60 GHz resonating frequencies have been carried out in Figure 5.7(c) and (d), respectively. At 2.40 GHz, that antenna offers cross-polarization discrimination as –28.20 dBi at a cut angle of 62 degrees. Similarly, on the

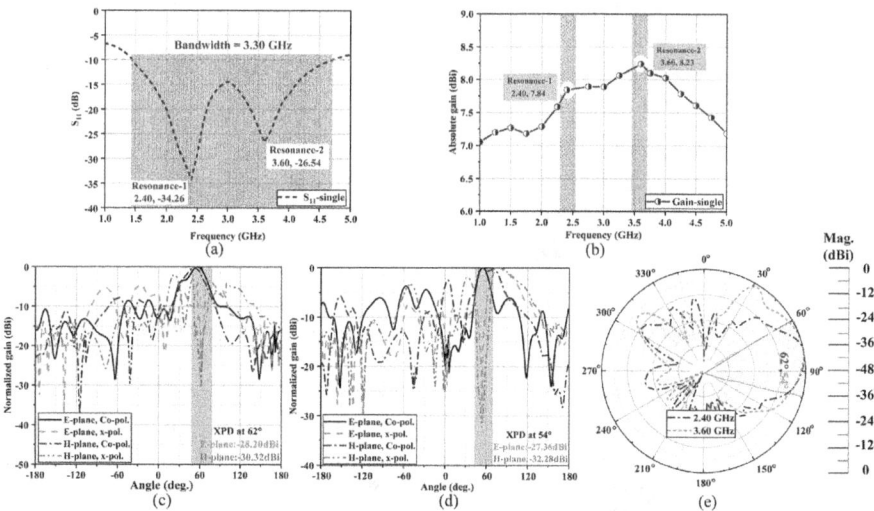

Figure 5.7 Sole radiator's simulated outcomes: (a) S_{11}, (b) gain, (c) E- and H-field patterns for first resonance, (d) E- and H-field patterns for second resonance, and (e) radiation pattern of both the resonances.

Table 5.2 Parametric Performances of the Antenna and the Array

Parameter(s)	Single Antenna		Antenna Array	
	First Resonance	Second Resonance	First Resonance	Second Resonance
f$_r$[GHz]	2.40	3.60	2.41	3.64
S$_{11}$ [dB]	−34.26	−26.54	−31.75	−22.79
BW[GHz]		3.30		2.62
Gain [dBi]	7.84	8.23	11.73	10.68
Directivity [dBi]	9.12	9.68	12.96	12.24
XPD [dBi]	−28.20	−27.36	−25.22	−23.69
Rad. eff. [%]	93	95	97	96
App. eff. [%]	69	72	64	67
CF [dB/m]	30.20	33.16	36.17	32.79

other resonating frequency, it also offers −27.63 dBi cross-polarization at a 54-degree cut angle. Both are desirable and also acceptable for the desired applications in this era. From Figure 5.7(e), it can be clearly observed that the antenna's radiation patterns are symmetrical at the respective cut angles for both frequencies. Table 5.2 has an entry for every numeric value of the antenna parameters.

Apparent efficiency [25, 26] of an antenna is calculated as follows:

$$\eta_{ap} = \frac{D}{D_{max}} \tag{5.7}$$

$$D_{max} = \frac{4 \times \Pi \times A}{\lambda^2} \tag{5.8}$$

where, D and A represent the antenna's directivity and total area, respectively.

The overall surface area of the sole antenna is $45 \times 45\,mm^2$ and the directivity is 5.12 and 9.68 dBi, respectively. As a result, the estimated aperture effectiveness of the antenna becomes 69% for a frequency of 2.40 GHz and 72% for a frequency of 3.60 GHz. The single antenna also gives 93% and 95% radiation efficiency [27–31] for 2.40 and 3.60 GHz, respectively. It can be observed that the sole antenna is highly efficient for IoT and 5G applications.

One of the most essential aspects of an antenna's electromagnetic impact is its electromagnetic compatibility (EMC). Using the surface equivalence technique, the antenna's correction factor (CF) [32] is determined, and the CF is then calculated as follows:

$$CF = \frac{|E^{INC}|}{|V_r|} = 20 \, log\left(\frac{9.73}{\lambda\sqrt{G_a}}\right) \tag{5.9}$$

where, $|E^{INC}| \mapsto$ incident field strength, $|V_r| \rightarrow$ antenna terminal voltage, and $G_a \rightarrow$ antenna's gain. The single antenna gives the correction factor of 30.20 and 33.16 dB/m at two resonating frequencies, and these are widely acceptable for the desired applications.

Improved parametric results at the target frequencies were achieved by developing a Wilkinson power divider [21]–based two-element printed array antenna, the characteristics of which are shown in Figure 5.8. In Figure 5.8(a), we can observe the array's return-loss property graphically. The array has 2.41 and 3.64 GHz resonant frequencies, with a return loss of 31.73 and 22.79 dB, respectively. The array also offers 2.62 GHz impedance bandwidth, which is quite broad and widely acceptable for the desired applications. The absolute gain of the array has been characterized and graphically presented in Figure 5.8(b). It offers an excellent gain of 11.73 and 10.68 dBi at those resonating frequencies. Figure 5.8(c) and (d) shows the field patterns (E, H) of the array at the chosen frequencies. The array's cross-polarization discrimination (XPD) at 0-degree cut angles for 2.41 and 3.64

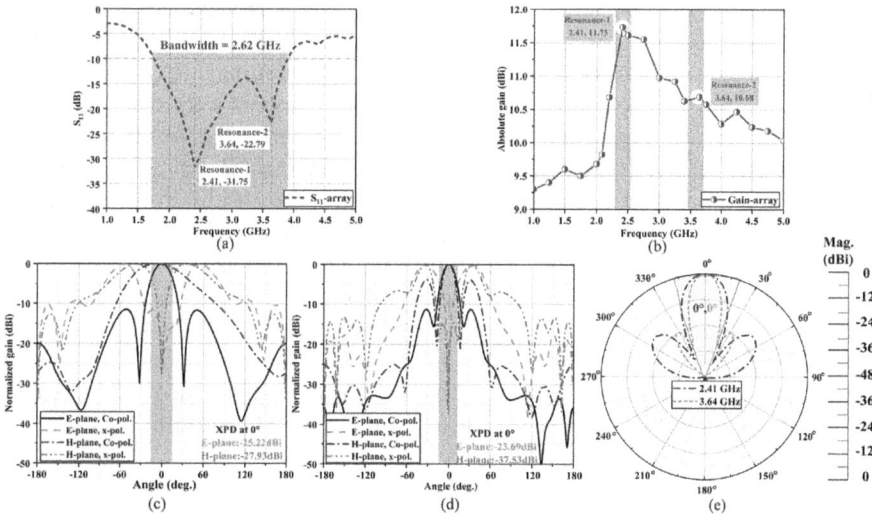

Figure 5.8 Array's simulated outcomes: (a) S_{11}, (b) gain, (c) E- and H-field patterns for first resonance, (d) E- and H-field patterns for second resonance, (e) radiation pattern of both the resonances.

GHz frequencies is –25.22 and –23.69 dBi, respectively. That large amount of XPD is quite desirable and acceptable for IoT and 5G applications. The obtained array radiation pattern is shown in Figure 5.8(e). In addition, it was found that the array's intended symmetric radiation patterns at a 0-degree cut angle for both resonating frequencies were achieved. The array also offers very high radiation efficiency [27–31] of 97% and 96%, respectively. The array's aperture is 55.50×65.25 mm^2, and its directivity is 12.96 and 12.24 dBi. For two different frequencies of operation, the array's aperture efficiency was determined as 64% and 67%, respectively. The array's computed CF [32] values are 36.17 and 32.79 dB/m, which are desired. Table 5.2 has an entry for every numeric value of all the antenna array parameters. Both antennas can support IoT-based home applications and 5G-related high-data-rate communication systems with acceptable performance parameters.

5.5 CONCLUSION

A couple of printed radiator-based microstrip patch antenna arrays have been developed and thoroughly investigated with their parametric outcomes. The sole antenna is derived from a fishing hook–shaped compact structure. It has been designed using regular semicircle, square, rectangular, and square-shaped geometries. A partial ground plane has been incorporated to obtain a wide impedance bandwidth from the antenna. The sole antenna has been analyzed in the time domain to determine its signal quality, isolation, group delay, and other characteristics. To achieve superior parametric performances, the Wilkinson power divider is incorporated to construct the array's feed. The sole radiator offers broad impedance bandwidth, decent gain, minimal cross-polarization level, and better radiation efficiency. Similarly, the array's parametric findings are likewise promising. There is considerable potential for IoT-based home applications and 5G communications with these antennas. An enhanced gain and improved parametric outcomes can be achieved by adding more elements to the array geometry. For 5G communication, beam steering is widely desirable. By incorporating a driver circuit with the proposed array, the main beam direction of it can be controlled, and the overall size of the module becomes very small as the antenna array is very compact and maintains its low-profile characteristics. Similarly, the proposed printed antennas can be used for IoT-based household applications, as they can be easily linked with microwave devices. Hence, both antennas are suitable for IoT and future 5G communications.

REFERENCES

[1] Balanis, C. A., **1997**. *Antenna Theory: Analysis and Design*, New York, Wiley.
[2] Haupt, R. L., **2010**. *Antenna Arrays: A Computational Approach*, New Jersey: Wiley-IEEE Press. https://www.wiley.com/en-us/Antenna+Arrays%3A+A+

Computational+Approach-p-9780470937433#:~:text=Description,-A%20
comprehensive%20tutorial&text=An%20antenna%20array%20is%20
an,signal%20in%20a%20desired%20direction.

[3] Carver, K. R., Mink, J. W., **1981**. "Microstrip antenna technology." *IEEE Transactions on Antennas and Propagation*, AP-29(1), pp. 2–24.

[4] Werner, D. H., Ganguly, S., 2003. "An overview of fractal antenna engineering research." *IEEE Antennas and Propagation Magazine*, 45(1), pp. 38–57.

[5] Bhattacharya, A., Roy, B., Caldeirinha, R., Bhattacharjee, A., 2019. "Low-profile, extremely wideband, dual-band-notched MIMO antenna for UWB applications." *International Journal of Microwave and Wireless Technologies*, 11(7), pp. 719–728.

[6] Bhattacharya, Ankan, Roy, Bappadittya, Chowdhury, Santosh. K., Bhattacharjee, Anup K. **2019**. "Computational and experimental analysis of a low-profile, isolation-enhanced, band-notch UWB-MIMO antenna." *Journal of Computational Electronics*, 18, pp. 680–688.

[7] De, A., Roy, B., Bhattacharya, A., Bhattacharjee, A. K., 2021. "Bandwidth-enhanced ultra-wide band wearable textile antenna for various WBAN and Internet of Things (IoT) applications." *Radio Science*, 56, p. e2021RS007315.

[8] Al-Sehemi, A. G., Al-Ghamdi, A. A., Dishovsky, N. T., Atanasov, N. T., Atanasova, G. L., **2021**. "Design of a flexible waterproof antenna for Internet of Things applications." *Journal of Electromagnetic Waves and Applications*, 35(7), pp. 874–887.

[9] Ashyap, Adel Y. I., Dahlan, S. H., Abidin, Z. Z., Kamarudin, M. R., Majid, H. A., Alduais, Nayef Abdulwahab Mohammed, Dahri, M. Hashim, Alhandi, Somya Abdulkarim, 2021. "C-shaped antenna based artificial magnetic conductor structure for wearable IoT healthcare devices." *Wireless Networks*, 27, pp. 4967–4985.

[10] Cowsigan, S. P., Saraswady, D., 2021. "Substrate integrated waveguide cavity backed antenna for IoT applications." *Journal of Ambient Intelligence and Humanized Computing*, pp. 1–6.

[11] Roges, R., Malik, P. K., 2021. "Planar and printed antennas for Internet of Things-enabled environment: Opportunities and challenges." *International Journal of Communication Systems*, 34(15), pp. 1–32.

[12] Sharma, D., Kanaujia, B. K., Kumar, S., 2021. "Compact multi-standard planar MIMO antenna for IoT/WLAN/Sub-6 GHz/X-band applications." *Wireless Networks*, 27, pp. 2671–2689.

[13] Mushtaq, A., Rajawat, A., Gupta, S. H., 2022. "Design of antenna array based beam repositioning for IoT applications." *Wireless Personal Communications*, 122, pp. 3205–3225.

[14] Wang, Z., Liang, T., Dong, Y., 2021. "Metamaterial-based, compact, wide beam-width circularly polarized antenna for 5G indoor application." *Microwave and Optical Technology Letters*, 63, pp. 2171–2178.

[15] De, A., Roy, B., Bhattacharjee, A. K., 2021. "Miniaturized dual band consumer transceiver antenna for 5G-enabled IoT-based home applications." *International Journal of Communication Systems*, 34, pp. 1–14.

[16] Singh, U., Mishra, R., 2022. "A dual-band high-gain substrate integrated waveguide slot antenna for 5G application." *Progress in Electromagnetics Research C*, 119, pp. 191–200.

[17] Wang, H., Huang, X. B., Fang, D. G., 2008. "A single layer wideband U-slot microstrip patch antenna array." *IEEE Antennas and Wireless Propagation Letters*, 7, pp. 9–12.

[18] Kola, K. S., Chatterjee, A., 2020. "A two-element array with clover-leaf shaped antennas for X-band applications." *7th International Conference on Signal Processing and Integrated Networks (SPIN)*, Noida, pp. 349–354.

[19] Kola, K. S., Chatterjee, A., 2020. "A linear array of Christmas-tree shaped antennas for dbs applications." *International Conference on Computer, Electrical & Communication Engineering (ICCECE)*, Kolkata, pp. 1–6.

[20] Kola, K. S., Chatterjee, A., 2020. "A printed array of high-gain fractal antennas for X-band applications." *International Conference on Communication, Computing and Industry 4.0 (C2I4)*, Bangalore, pp. 1–6.

[21] Kola, K. S., Chatterjee, A., 2021. "A 1 x 2 array of high-gain radiators for direct broadcast satellite (DBS) Services under Ku-band." *8th International Conference on Signal Processing and Integrated Networks (SPIN)*, pp. 297–302.

[22] Kola, K. S., Chatterjee, A., 2021. "An array of tulip-flower shaped printed radiators for direct broadcast satellite (DBS) applications." *Advanced Communication Technologies and Signal Processing (ACTS)*, 2021, pp. 1–6.

[23] Kola, K. S., Chatterjee, A., 2021. "A high-gain and low cross-polarized printed fractal antenna for X-band wireless application." *International Journal of Communication Systems*, 34(10), pp. 1–19.

[24] Wilkinson, E. J., 1960. "An N-way hybrid power divider." *IRE Transactions on Microwave Theory Techniques*, 8(1), pp. 116–118.

[25] Ma, Z., Vandenbosch, A. E., 2012. "Low-cost wideband microstrip arrays with high aperture efficiency." *IEEE Transactions on Antennas and Propagation*, 60(6), pp. 3028–3034.

[26] Vosoogh, A., Kildal, P. S., 2016. "Simple formula for aperture efficiency reduction due to grating lobes in planar phased arrays." *IEEE Transactions on Antennas and Propagation*, 64(6), pp. 2263–2269.

[27] Newman, E. H., Bohley, P., Walter, C. H., 1975. "Two methods for the measurement of antenna efficiency." *IEEE Transactions on Antennas and Propagation*, AP-23(4), pp. 457–461.

[28] Pozar, D. M., Kaufman, B., 1988. "Comparison of three methods for the measurement of printed antenna efficiency." *IEEE Transactions on Antennas and Propagation*, 36(1), 136–139.

[29] Smith, G. S., 1997. "An analysis of the Wheeler method for measuring the radiating efficiency of antennas." *IEEE Transactions on Antennas and Propagation*, 25(4), pp. 552–556.

[30] Chair, R., Luk, K. M., Lee, K. F., 2002. "Radiation efficiency analysis on small antenna by wheeler cap method." *Microwave and Optical Technology Letters*, 33(2), pp. 112–113.

[31] Moharram, M. A., Kishk, A. A., 2016. "MIMO antennas efficiency measurement using wheeler caps." *IEEE Transactions on Antennas and Propagation*, 64(3), pp. 1115–1120.

[32] Paul, C. R., 2006. *Introduction to Electromagnetic Compatibility*, 2nd edition, US: Wiley. https://www.wiley.com/en-us/Introduction+to+Electromagnetic+Compatibility,+2nd+Edition-p-9780471758150.

[33] Kreyszig, E. O., 1983. *Advanced Engineering Mathematics*, New York, Wiley.

Chapter 6

Optimization Algorithms for Reconfigurable Antenna Design

A Review

K. Karthika, K. Anusha, K. Kavitha, and D. Mohana Geetha

6.1 INTRODUCTION

Reconfigurability in an antenna system has drawn a lot of attention recently. E. R. Brown introduced the reconfigurable antenna in 1998. He employed radio-frequency technology-based microelectromechanical systems (RF-MEMS) to offer greater performance and enable new system capabilities. His research provided ultra-low-power dissipation, and very large-scale integration (VLSI) yielded better results than conventional antennas that were available at his time [1]. Reconfigurable antennas provide antenna designers with many degrees of freedom because they can alter their pattern, frequency and polarization. These systems are capable of independently reconfiguring themselves to respond to adjustments or operational needs. In such antennas, reconfiguration is accomplished by attaching and/or detaching different sections of the antenna through switches. Altering the structure alters the surface current distribution. This property redistribution results in the desired change in the antenna's performance and makes it more suitable for different wireless communication platforms. The four key switching mechanisms of reconfigurable antennas are electrical switches (RF-MEMS, varactor diode and PIN diode), reconfigurable materials (graphene plasmonic, thermal switches and liquid crystals), mechanical switches and optical switches. Electrical switches have high isolation, require limited power consumption and their biasing networks need to be properly configured. Optical switches (photoconductive switches) are limited due to integration and power problems, whereas mechanical switches are limited due to structural complexity and size requirements [2]. The tremendous growth in wireless applications demands smart antennas capable of performing multiple operations and that can operate over a wide frequency range. Realizing a multifunctional antenna in a low-profile structure with reduced complexity makes the design more challenging.

In a highly efficient reconfigurable antenna design, complexity is an important aspect that needs to be addressed. Complexity increases expenditure

DOI: 10.1201/9781003459880-6

and reduces the antenna performance. Optimization in antenna design is significant in improving the performance, efficiency and cost-effectiveness of the antenna. By optimizing the radiation characteristics, impedance matching, bandwidth, size and weight of the antenna, it is possible to design a high-performance antenna that meets the requirements of the specific application, leading to a more reliable communication system. The researchers are going for optimization algorithms to get optimal design while reducing complexity without compromising the reliability of the antenna system. Since the early 1990s, researchers have started exploring optimization technique for effective antenna design. Optimization algorithms help in accurate and efficient design as well as improved communication between antennas. Several parameters, which may be either discrete, continuous or both, are typically involved in electromagnetic optimization problems. Constraints are often included in the permissible values. Finding a solution that offers a minimum or maximum global solution while conserving computational resources is the goal of optimization. These approaches are based on evolutionary algorithms (EAs), a stochastic method of the family search inspired by natural biological evolution [3]. Algorithms such as the genetic algorithm (GA) and particle swarm optimization (PSO) are categorized as global optimizers, whereas conventional approaches like quasi-Newton and conjugate gradient techniques are more common in local techniques. The distinction between the two is that the results of local approaches are strongly influenced by the original assumption or starting point and appear to be directly related to the solution domain. Due to the close coupling, local techniques can benefit from the characteristics of the solution space, leading to a reasonably quick convergence to the full local convergence. However, GA places a few restrictions on it and is largely unconstrained by the solution domain and initial conditions.

Global approaches are more robust in terms of solving problems and are significantly good at handling solution spaces with discontinuities, limited parameters and many dimensions with large numbers of potential local maxima [4, 5]. Global procedures often produce a near-global maximum or global maximum rather than a local maximum. It can identify optimal solutions where local techniques are lacking. Global approaches are highly beneficial in tackling new issues when the solution space's nature is mostly unexplored. In this chapter, various optimization algorithms that are crucial to the effective design of reconfigurable antennas are studied and summarized. A short description of each optimization algorithm, the variables influencing the optimization metric for performance improvement and their pros and cons are presented as a part of this work. Section 6.2 discusses evolutionary algorithms. Section 6.3 projects a detailed study of swarm intelligence-based optimization algorithms. The chapter is concluded in Section 6.4.

6.2 EVOLUTIONARY ALGORITHMS

6.2.1 Genetic Algorithm

In an antenna design, different parameters like the position of the feed, the height of the dielectric material, its dielectric constants, the width, the thickness and the length of the patch have an impact on the designed antenna's radiation performance. Conventional optimization approaches are practically difficult to use with so many parameters involved in the antenna design. However, GA optimization is highly efficient and capable of handling a complicated set of parameters. GA was first referred to by J. Holland [6] and was made functional by De Jong [7]. GAs are powerful, stochastic search techniques working on the principles of evolution and natural selection. Compared to random-walk searches, GAs are more effective and offer significantly faster convergence. It is effective in solving problems that are complex, combinatorial and connected. It can be used for non-differentiable and discontinuous functions. Moreover, GA is not limited by the constraints of the search space and is simple to program and implement [4].

In GA, the set of trial solutions are called parameters or populations; a coded form of trial solution is chromosomes, with the parent and child being members of the current and next generation, respectively [8]. The three stages of a typical GA optimization are as follows: initiation, reproduction and generation replacement. Typically, initiation is used to populate a set of randomly generated chromosomes or encoded parameter strings. The current generation is defined as the group of individuals represented by each chromosome. Each individual is designated with a fitness value as a result of executing a fitness function over them. The relation between the GA and the physical problem being optimized is the fitness function [4, 8].

From the current generation, a new generation is created through reproduction. Here, a chosen pair of individuals serving as parents are subjected to crossover and mutation to create a new pair of children [4, 8]. Up until the new generation is fully composed of children, the crossover and mutation operations are repeated. Thus, the new generation replaces the existing generation. A GA can be classified as a generational GA or a steady-state GA. A generational GA is a simple type, where the size of the generation remains constant. In a steady-state GA, the size of a new generation is different from its parent generation and there can be an overlap between the generations [4]. Crossover causes the genes to be rearranged to produce better gene combinations, leading to more suited individuals. Two children are generated because of accepting the parents. The selection comprises a system relating the fitness of a person to the population's average fitness. As a result of selection, parents are selected to take part in the reproduction process based on fitness, which is the measure of an individual's 'goodness' [4]. The mutation is how new genetic material or features are inserted to explore parts of the solution domain.

In the generation-replacement process, the new generation takes the place of the existing generation, and each of the new individuals has their fitness values measured and assigned. The replication process is then repeated if the termination criteria have not been met. The GA works to establish an evolution towards an optimum solution on the chromosome. The GA control parameters are chosen through trial and error [9]. In [10], GA optimization is applied in the antenna design to achieve frequency reconfigurability. Here the entire patch area is segmented into square cells of size 2 mm. Each cell in the patch has either conducting or non-conducting features. GA optimization is applied over it to get the optimal solution by segregating metallic and non-metallic regions contributing to the corresponding resonance. To achieve the desired resonance using GA, first, the constraint functions (6.1) for the targeted bands are formulated in accordance with S_{11} [10].

$$L_f = \begin{cases} |S_{11}(f)|dB, & -10\,dB \le |S_{11}(f)|dB > -5dB \\ -10, & |S_{11}(f)|dB < -10\,dB \\ 0, & |S_{11}(f)|dB \ge -10\,dB \end{cases} \qquad (6.1)$$

Next, the cost function is expressed as in (6.2) [10],

$$Cost = \frac{1}{X_b} \sum_{Xb-1}^{X_b} \left(\frac{\sum_{i=1}^{P} L_{xb}(f_i)}{P} \right) - \frac{1}{X_a} \sum_{Xa-1}^{X_a} \left(\frac{\sum_{j=1}^{Q} L_{xa}(f_j)}{Q} \right) \qquad (6.2)$$

where 'i' denotes the desired resonating frequencies, 'j' denotes the suppressed resonating frequencies, 'X_b' denotes the desired frequency bands and 'X_a' denotes the suppressed frequency bands. 'Q' and 'P' denote the frequency points in the suppressed and targeted bands, respectively. Minimizing the cost function aims to achieve or allow a reflection coefficient below −10 dB in the desired frequency bands [10]. The cost function of the applied GA optimization in [11] determines the size of the slot and the placement of the switches to carry out four modes of operation. The number of operating bands determines the number of terms in the cost function. The cost function suggested in [11] operates well with single objective optimization algorithms and can be useful in designing ultrawideband (UWB) and wideband (WB) antennas with band notching features. GA optimization involved in [12] utilizes a multiobjective optimization algorithm to accomplish frequency reconfiguration with –6 dB bandwidth as a fitness function (6.3).

$$minimize \int_{f_a}^{f_b} \left[S_{11}(f,x) + 6 \right]^+ df, \int_{f_c}^{f_d} \left[S_{11}(f,\tilde{x}) + 6 \right]^+ df \ subject\,to\,x \in \{0,1\}^N \quad (3)$$

where the decision vector x is the same as $\tilde{x}$ except that its switching state is the opposite at the switch position. The frequency range of interest is defined by f_a, f_b, f_c and f_d. S_{11} is expressed in decibel, and 'N' represents decision vector's bit count. Single objective optimization is used to achieve good impedance matching in the two desired bands.

Pros and cons:

- The GA doesn't require a lot of mathematics to solve optimization problems [13].
- The GA provides a great deal of versatility in hybridizing domain-dependent heuristics to apply a particular problem effectively.
- In performing the global search, evolutionary operators make GA effective [13]. Only if the problem has certain convexity properties that effectively ensure that every local optimum is a global one can a global optimum be found.
- Although the GA can locate the solution in the entire domain, it does not readily solve complex constraint issues, particularly for precise constraints, and huge assessments are often time-consuming. The improved GA (enhanced GAs, interval GAs, quantum GAs, etc.,[13, 14]) is suggested to overcome the shortcomings of the GA.

Many improvements and enhancements to the simple GA optimization such as elitism and the use of steady-state algorithms have been developed and used. GA optimization is more often used in electromagnetic applications like the design of antenna arrays, wire antennas, frequency selective surfaces, microwave absorbers, radar target identification and the architecture of wireless networks [4]. However, a GA is subject to a local optimal solution and has a slow rate of convergence in practical applications [14].

6.2.2 Differential Evolution

Differential evolution (DE) was first suggested by R. Storn and K. V. Price in 1995 [15]. DE is a powerful optimization technique used for solving real-valued, non-linear, non-differential and multimodal functions. It is a population-based optimization technique working with a set of solutions changing over generations through selection, generation and replacement methods. Optimizing an antenna using DE usually requires the user to select only three parameters: the size of the population (M), crossover constant (Cr) and scaling constant (F). Like other EAs, DE allows parameters as floating-point variables, which not only permits encoding as real variables but is also mutated conveniently using floating-point arithmetic. Floating-point makes encoding and decoding completely transparent and prevents the user from selecting the resolution constraints. The important features of standard DE are: It follows a generational model, where the current population is replaced when the child population is created completely, and if the child trial vector is inferior to the parent, then the individual is preserved for the next generation, implying the presence of elitism at an individual level [16]. The DE algorithm minimizes or maximizes a given fitness function f (var1, var2, . . . , varN). An N-dimensional vector 'x' gives the possible solutions for an 'N' variable function as (6.4) [17].

$$f(var1, var2, \ldots, varN) = f(x1, x2, \ldots, xN) \tag{6.4}$$

Groups of NP vectors are called populations, whose vectors are called the members of P (5) [17].

$$P = \{x(1), x(2), \ldots, x(NP)\} \tag{6.5}$$

A new generation or population is created from the parent population through recombination and selection. The initial population is generated for every parameter 'i' of x for j = 1 to NP using the function (6.6) [17].

$$X_i(j) = X_{iL} + rand(\beta_i)(X_{iH} - X_{iL}) \tag{6.6}$$

where β denotes the random variable distribution; X_{iH} and X_{iL} denote the upper and lower bounds for a given parameter, respectively. The difference between two vectors selected at random is used in DE to achieve the recombination. DE algorithm requires two constants: 1) the crossover constant CR (usually 0.2–0.8) and 2) the mutation weight $\lambda[0,1]$, determining how much the difference vectors change the original vector. Larger λ values result in increased global value convergence, while smaller values decrease convergence time. However, the smaller λ values decrease the chances of converging to a global solution. After the child population is created, the selection function operates to determine which children and parent populations will be carried onto the next generation. The final step in DE is a stopping function [17].

The DE optimization technique has been utilized extensively in the design and synthesis of RF switches and antenna (feed position optimization). Optimization of the reflector parameters along with the feed array enhances gain and lowers the side lobes in spaceborne synthetic aperture radar (SAR) design. A lot of researchers have worked on improving the DE algorithm for better efficiency and accuracy. Some of the modified versions are fast DE, parallel DE, hybrid DE, quantum DE, DE/and/either-or algorithm, opposition-based and neighborhood-based DE and a lot more. A detailed survey of these algorithms is provided in [18].

Pros:

- Simple.
- Has fewer control parameters without a trade-off in performance.
- Can solve a wide range of problems like multimodal, unimodal, non-separable and separable.
- Space complexity is less and can handle large-scale expensive optimization problems.

Cons:

- Mutation operation lacks selection pressure.
- The ability to search across large distances is limited for clustered populations.
- Still faces premature convergence.

6.2.3 Covariance Matrix Adaptation Evolution Strategy

Covariance matrix adaptation evolution strategy (CMA-ES) is an evolution strategy for solving non-linear functions with many parameters. This was developed by Hansen and Ostermeier. CMA-ES is self-adaptive and does not require parameter tuning. CMA-ES's performance is not impacted by the initial values of the strategy parameters (parameters that parameterize the mutation distribution). The detailed study of the working of general CMA-ES is discussed in [19]. CMA-ES consists of three mechanisms: 1) derandomized adaptation, 2) cumulative step size and 3) covariance matrix adaptation. First, a mean vector $m \in R^n$ is randomly initialized inside its limits $L_k \leq m_k \leq U_k$, for $k = 1,2,...n$, where 'n' represents the number of parameters. Then a covariance matrix $cov = \sigma^2 C$ is defined, where $C \in R^{n \times n}$ is initialized to identity matrix I with σ as the initial step size. Following the initialization, each step in the algorithm samples λ new solutions for the next g + 1 generation from the multivariate normal distribution N (m, cov) as (6.7) [20].

$$X_k^{(g+1)} \sim N\left(X_W^{(g)}, \sigma^{(g)}cov^{(g)}\right) \qquad (6.7)$$

where $X_W^{(g)}$ represents the recombination point, $\sigma^{(g)}$ denotes the global step size adaptation and $cov^{(g)}$ represents the covariance matrix adaptation. The initial value of σ is set generally to 0.5, λ as 4 + [3ln(n)] and μ as [λ / 2]. The loop starts to iteratively run after the distribution has been updated and continues until a stop condition is satisfied. The common stopping standards depend on:

- The maximum iterations possible.
- A constraint on the fitness level to be attained.
- Negative definition of C matrix.
- Tolerance on the smallest standard deviation.

Pros:

- Optimizes a problem space by producing new members around the best mean located in the gaussian hyper-space [20].
- Rotational invariance is useful for a rotationally ill-conditioned search space.
- Robust, effective and quicker than all EAs.

Cons:

- Smaller global step sizes at initial stages lead to worse performance on multimodal functions.
- This may lead to premature convergence as the new members created are scattered around a mean.
- Standard CMA-ES cannot restore useful traits that have been lost.
- Not suitable for problems to find a global minimum present at the edge of the feasible region.

6.3 SWARM INTELLIGENCE-BASED ALGORITHM

6.3.1 Particle Swarm Optimization

The PSO was invented by R.C. Eberhart and James Kennedy. It is a biologically derived algorithm, occupying the gap in nature between evolutionary search and requiring neural processing. Inspired by how flocks of birds or swarms of insects move about in search of food, PSO seeks to determine the most ideal solution in the search space. It includes a very basic concept and can be implemented in a few lines of coding. It is efficient in terms of speed and memory requirements, as it requires simple mathematical operations. PSO is an effective, efficient, robust and accurate algorithm that finds the optimum much easier than traditional methods. It finds the solution to any specific issue in a search space within a period and balances between exploitation and exploration very efficiently. It is fast for non-linear, non-differentiable and multimodal problems. But unlike other EAs, the PSO is based on correlations of biological behavior and not on natural evolution [21]. The PSO algorithm gives each swarm particle equal weight during the entire search process. In each iteration, the position and velocity of each of the particles are modified following the best positions of the particles themselves and the group obtained thus far, thereby not implementing the theory of survival of the fittest. It is used in system design, power allocation of cooperative communication networks, artificial neural networks, machine learning, character and pattern recognition, signal processing, decision-making and identification.

The initial search space has a set of feasible solutions that are randomly placed in the search space. Every single particle of the swarm evaluates its fitness at that point and then starts to move in the search space for a fixed number of iterations in search of better fitness as soon as they are given a velocity. Every individual particle must maintain its location in the search space concerning the fitness function obtained. The movement of the particles is based on their own best location (pbest) as well as on the best location obtained by other particles in the swarm (gbest). The algorithm thus finds the best possible solution to the fitness function in the search space in subsequent iterations. The fitness function is usually defined by a mathematical function to evaluate the system performance and is solely based on the application of the algorithm. A basic PSO uses a swarm of 'n' particles, and the position of every individual particle indicates a possible solution in the search space. The particle updates its conditions based on three factors: its own most optimal solution, its inertia and the swarm's most optimal solution. The velocity and position of the particles change according to (6.8) and (6.9) [21].

$$v_{id}^{k+1} = w v_{id}^{k} + c_1 r_1^{k} \left(pbest_{id}^{k} - x_{id}^{k} \right) + c_2 r_2^{k} \left(gbest_{id}^{k} - x_{id}^{k} \right) \tag{6.8}$$

$$X_{id}^{k+1} = x_{id}^{k} + v_{id}^{k+1} \tag{6.9}$$

where v_{id}^{k} denotes the velocity, X_{id}^{k} denotes position, $pbest_{id}^{k}$ characterizes the personal best position and $gbest_{id}^{k}$ is the global best position (group's best value) of the ith particle at the 'd' dimension in the kth iteration. 'w' is the inertial weight given to the particle's previous position. r_1^{k} and r_2^{k} denote the random function in the range [0,1]. C_1 and C_2 are acceleration constants.

Each particle of the search space is fastened onto the maximum velocity, Vmax, which defines the fineness or resolution by which regions can be sorted out among the current position and the target position [21]. However, particles might miss some good locations if Vmax is too high, or some good places will be out of reach if Vmax is too low. So, the inertial weight ('w') is introduced to eliminate the Vmax dependency. Inertial weight controls the speed of convergence of the algorithm. Higher values of 'w' provide improved global search capabilities, and the reverse results in improved searching of local minima. Acceleration constants C_1 and C_2 relate the speed of particles flying to their own best position and the swarm's fittest location. These acceleration constants stand in for the weighing of acceleration terms towards the gbest and pbest positions. The particle may exceed a fit position if the values are too high, and it may not reach the location if the values are too low. A value of 2 is typically assigned to the constants C_1 and C_2 to shift the times taken toward the pbest of the particle and the gbest of the swarm as equal and by half of the overall time. The particles are attracted to the gbest and pbest positions by adding random numbers. By doing this, the system is prevented from being trapped at local minima or maxima.

Three components make up the velocity update: 1) the ability of the particle to proceed in the same direction as it did during the prior iteration, called momentum; 2) pulling the velocity of the particle towards its pbest, called memory or remembrance or cognitive part; and 3) pulling the velocity of the particle towards the gbest, called cooperation or social knowledge or social part. The fitness evaluation feature of the PSO algorithm, calculated using (6.10), is used to show how effectively a solution satisfies a design parameter [22].

$$F = \sum_{i=0}^{N} W_i f_i \qquad (6.10)$$

In (6.12), the fitness factor is denoted by 'N', the weighing coefficient is expressed by 'W_i' and the ith fitness factor value is represented by 'f_i'. The fitness function to get resonance at a targeted frequency can be obtained using (6.11) [22, 23].

$$Fitness = \min\left(S_{11n}^2\right)_{f_r} \qquad (6.11)$$

where S_{11} represents return loss, 'n' denotes the sample point in the S_{11} Vs frequency and 'f_r' represents the resonating frequency. PSO with velocity mutation can solve multiple complex variables with multiobjective goals in the least possible computational time [24].

Pros:

- Search is done by repeated variation of particle velocity and not by selection or mutation operations.
- Particles move only to optimal locations because of gbest.
- Can be used in scientific and engineering applications.
- Calculations are simple and easy.
- Its parallel execution improves the speed of the process.

Cons:

- An increase in the dimension of the search space results in increased complexity.
- Vulnerable to partial optimism and may lead to inaccuracy.
- Cannot be used for non-coordinate systems due to the lack of dimensionality.
- This can result in a loss of diversity due to high convergence.

6.3.2 Grey Wolf Optimizer

The grey wolf optimizer (GWO) was suggested by Mirjalili in 2014 as a new metaheuristic swarm intelligence-based algorithm. GWO imitates the group hunting behavior and grey wolves' social hierarchy in nature. The grey wolf follows a very strict hierarchy compiled of all members of the pack (group of grey wolves) as illustrated in Figure 6.1.

The alpha wolves (α) hold the top position in the hierarchy and act as the leaders of the pack. They dominate and command other wolves and are accountable for managing the group in hunting, migration and feeding. They need not be the strongest of all, but they are the best at managing the entire pack. The betas (β) are in the second level. Beta wolves act as an advisor to the alpha and help to instruct the lower-level wolves. Betas are suitable candidates for leadership positions when the alpha is sick, injured or dead. The delta wolves (δ) form the third level in the hierarchy, and they act as elders, sentinels, scouts, hunters and caretakers of the pack. They obey the

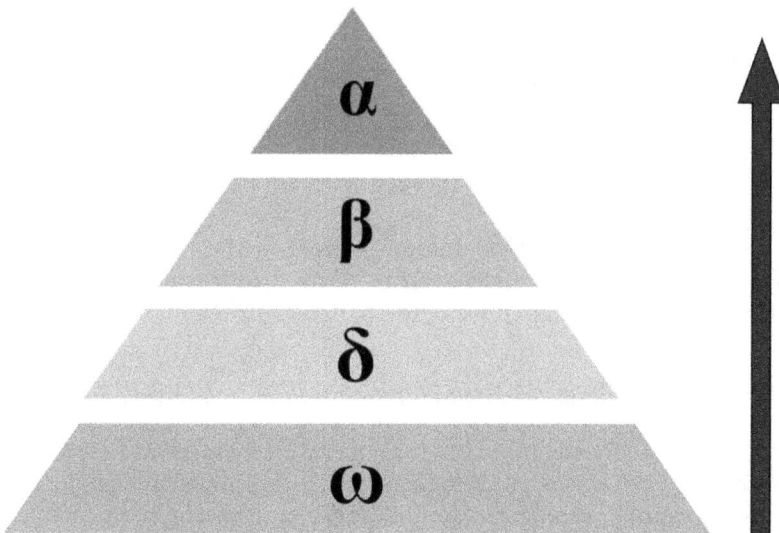

Figure 6.1 Grey wolf's social hierarchy (dominance increases from bottom up) [25].

alphas and betas but dominate the lower-level omegas. The lowest level is occupied by the omegas (ω) which are dominated by other wolves. They act as scapegoats. But they form an essential part of the pack because of their assistance and the pack's desire to maintain their overall organization. The GWO algorithm is given in the following phases.

6.3.2.1 Social Hierarchy

The algorithm starts by defining the objective functions and related parameters. According to the hierarchy of the pack, the mathematical model assumes the best solution found to be alphas, the second-best solution is betas, the third is deltas and the remaining candidate solutions are omegas [26]. Alphas, betas and deltas guide the hunting.

6.3.2.2 Group Hunting

6.3.2.2.1 Encircling Prey

Tracking and encircling the prey is the initial move in a search. Mathematically, the encircling of prey/solution is given using (6.12) and (6.13) [26]:

$$X(t+1) = X_p(t) - A \cdot D \qquad (6.12)$$

$$D = |C \cdot X_p(t) - X(t)| \qquad (6.13)$$

where $X(t+1)$ and $X(t)$ represent the position of a grey wolf at $(t+1)^{th}$ and t^{th} positions and $X_P(t)$ represents the prey's position. C and A are given as the coefficient vectors in (6.14) and (6.15) [26]:

$$C = 2r_2 \qquad (6.14)$$

$$A = (2r_1 \cdot a) - a \qquad (6.15)$$

where r_1 and r_2 represent two random vectors in [0,1], 'a' is an encircling coefficient vector whose values decrease from 2 to 0 over the iterations of the search and are calculated using (6.16) [26].

$$a = 2\left(1 - \frac{t}{T}\right) \qquad (6.16)$$

where 'T' and 't' are the maximum number of iterations and current iterations, respectively. Using (6.12) and (6.13), a candidate solution/grey wolf updates its position around the optimal solution prey. The position of a wolf can be moved by adjusting A and C vectors. The r_1 and r_2 are utilized to simulate various movement velocities and step sizes [26]. This encircling can be also done in n-dimensional space.

6.3.2.2.2 Hunting

The alpha leads the pack together on a hunt. They may select the potential prey and select the break-off of the hunt. The betas and deltas join the alpha in hunting. The omegas may be a caretaker for the newborn or the injured. This group hunting is represented mathematically using (6.17)–(6.19) [26]:

$$D_\alpha = |C_1 \bullet X_{alpha} - X|; D_\beta = |C_2 \bullet X_{beta} - X|; D_\delta = |C_3 \bullet X_{delta} - X| \qquad (6.17)$$

$$X_1 = X_{alpha} - A_1 \bullet D_\alpha; X_2 = X_{beta} - A_2 \bullet D_\beta; X_3 = X_{delta} - A_3 \bullet D_\delta \qquad (6.18)$$

$$X(t+1) = (X_1 + X_2 + X_3) / 3 \qquad (6.19)$$

where X_{alpha}, X_{beta}, X_{delta} and X represent the alpha, beta, delta and omega positions, respectively, in the tth iteration; the updated position of ω in the $(t + 1)$ iteration is denoted by $X(t + 1)$. From Figure 6.2, it is observed that the alpha wolf, beta wolf and delta wolf work to identify the prey's position, and then ω updates its positions accordingly. Likewise, the hunting wolves randomly update their position close to the prey under the guidance of the present top three wolves. Then the hunting wolves start searching for the prey's position and concentrate on attacking [25].

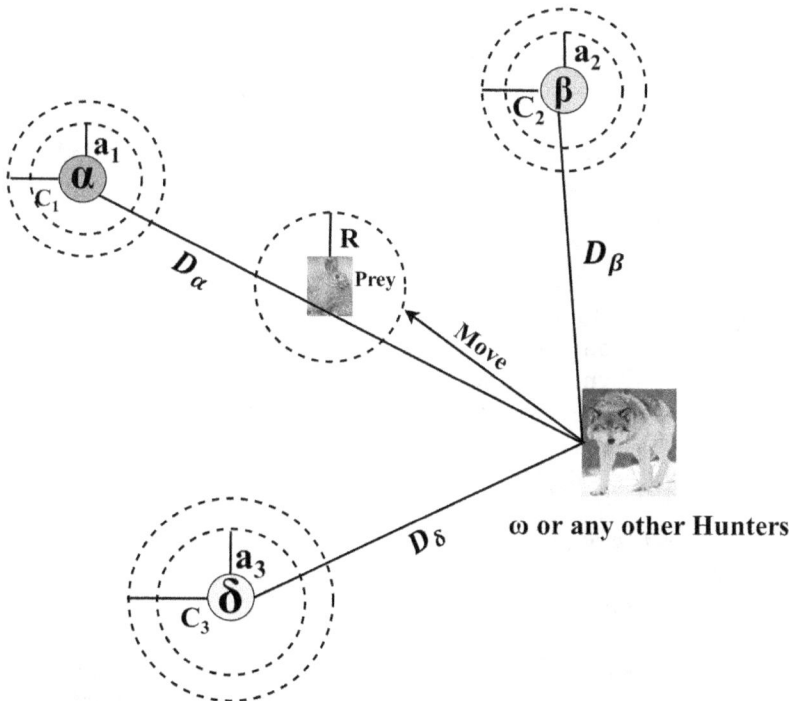

Figure 6.2 Updating of position in GWO [25].

6.3.2.2.3 Attacking Prey (Exploitation)

The wolves strike their target by controlling the 'A' vector using (6.17). It is observed that when $|A|=1$, the wolves surround the prey, and when $|A|<1$, the wolves attack the target by converging toward the prey (exploitation) [25]. Since 'A' depends on 'a', and 'a' decreases linearly between 2 and 0, the A's value also decreases in the range [−2a,2a]. Thus, it is inferred that the wolves may launch closer attacks after more iterations.

6.3.2.2.4 Tracking Prey (Exploration)

When $|A|>1$, the wolves start to diverge and explore other areas in the search space to find better prey (exploration phase). 'C' helps in exploration, and it takes random values in [0,2]. It alters the prey's contribution in deciding the location of omega wolves in the next steps. When $C>1$, the wolves advance quickly toward their prey; when $C<1$, it has a negative effect by slowing down the movement [25]. It can be seen from (6.10) and (6.11) that 'C' has minimal impact on the final stage update of $X(t+1)$ since 'A' reduces linearly from 2 to 0. Both these vectors 'C' and 'A' act as minor disturbances due to their random and adaptive nature, which greatly helps the search to not converge into local optimum.

The exploitation and exploration must be handled properly. The exploitation searches the neighborhood spaces for its solution, while exploration leads to searches extensively to not be trapped in a local minimum. Hence, a balance between both is required or else it may create a premature convergence or a non-optimal solution. The balance is handled by parameter 'A', as values <1 lead to exploitation and values >1 lead to exploration [26]. Over the iterations, the wolves search the position of prey, and they update their values when better values are found. The grey wolf algorithm is ceased when a maximum count of iterations is achieved.

Pros:

- Has fewer tunable parameters (A and C).
- Reduced computation and high performance.
- Eliminates the need for selecting the best operator.
- Easy to implement with less space complexity.

Cons:

- No assurance of premature convergence.
- Slow convergence.
- Low solving precision.

Some of its updated versions, EGWO, IGWO, MOGWO, MAL-IGWO, RW-GWO [27] and a lot more, are proposed frequently and are modified for specific applications such as power flow control and voltage stability enhancement.

Some of the prominent works in reconfigurable antenna design with optimization algorithms are represented in Table 6.1. From the study, we can

Table 6.1 Reconfigurable Antenna Design Implemented Using Optimization Algorithms

Ref. No.	Reconfiguration/ Optimization	Optimized Antenna Parameters	Single Objective/ Multiobjective	Means of Optimization	Control Parameters
[9]	Frequency/GA	Impedance Matching	Single	• Slot shape • Location of switches in the ground	• Population size = 200 • Crossover probability (0.75) • Maximum no. of generations = 50 • Mutation probability (0.35)
[10]	Frequency/GA	Impedance matching	Single	• Free from constraints • Automatic allocation of switches	• Population size = 20 • Crossover probability (100 %) • Maximum no. of generations = 30 • Mutation probability (0.1%)
[11]	Frequency/GA	Bandwidth	Single	• Slot shape • Location of switches in the ground	• Population size = 50 • Crossover probability (1) • Maximum no. of generations = 100 • Mutation probability (0.1)
[12]	Frequency/GA	• Reflection coefficient • No. of switches	Multi-	-	• Population size = 500 • Crossover probability (0.8) • Maximum no. of generations = 200 • Mutation probability (0.01)
[28]	Pattern/DE	Reflection coefficient	Single	• Dimensions of the geometry	• Population size = 10, 20 • Crossover constant = 0.5 • Maximum iteration = 20 • Mutation scale factor F = 0.25

(Continued)

Table 6.1 (Continued)

Ref. No.	Reconfiguration/ Optimization	Optimized Antenna Parameters	Single Objective/ Multiobjective	Means of Optimization	Control Parameters
[24]	Pattern/PSO with velocity mutation	• Low SWR at the input • Tilt in main lobe • High gain • Service area null filling • Outside service area • Low side lobe levels	Multi-	• Dimensions of the Geometry • Array feeding weights	• No. of swarm particles: 20 • Maximum no. of iterations = 2000
[29]	Frequency/PSO	Impedance matching	Single	• Dimensions of the geometry	• No. of swarm particles: 14 • Maximum no. of iterations = 500
[30]	Reconfigurable FSS/hybrid PSO with negative zone	Band pass and band stop filtering action	Multi-	• Geometry Modification	• No. of Swarm particles: 10 • Maximum no. of iterations = 200

observe that GA originates from the natural evolution process. The chromosome represents the solution, and genes represent the parameters of the solution. GA is widely used in wireless applications. However, sometimes it can take a while to discover the optimal solution. DE is an evolutionary algorithm that employs population statistics. It draws inspiration from natural laws on human evolution. DE is a modified version of the GA. The employment of similar operators in different ways distinguishes DE from GA. Compared to other EAs, DE requires only a few parameters to be adjusted and is simple to implement. These features helped it become popular. CMA-ES, a branch of evolution strategy, is useful when there is no direct way to ascertain the parameter set. The PSO algorithm is motivated by birds' flocking behavior. The solution is characterized by the particle. Local best position, global best position, position and velocity are the parameters of the solution.

Optimization is a promising research area in antenna design. There are several possible future improvements that could be made to enhance the performance of antennas further. Some of the potential areas of improvement include artificial intelligence, metamaterials, non-linear optimization and additive manufacturing. These techniques can help in generating optimized antenna designs with unprecedented performance characteristics, making them suitable for a wide range of applications in communication, sensing and other fields. The world of optimization algorithms is so large that it won't fit in this single review. They can be physics-oriented, evolutionary algorithms, or swarm intelligence. The common EAs not addressed in this chapter are evolutionary programming, evolutionary strategy, genetic programming and biogeography-based optimizers. A few of the widely implemented physics-based algorithms are gravitational local search, gravitational search algorithm, charged system search, big bang big crunch, black hole small-world optimization, galaxy-based search optimization and curved space optimization. Similarly, some of the swarm intelligence-based algorithms are artificial fish swarms, wasp swarm algorithms, monkey search, cuckoo search, bee collecting pollen algorithm and firefly algorithm.

6.4 CONCLUSION

Reconfigurable antennas can be used in numerous wireless communication applications. These antennas reduce the system complexity by eliminating the need for additional antennas in a communication system. Optimization algorithms are highly useful in reconfigurable antenna design to minimize the complexity of the structure and feeding mechanism and to enhance its reconfigurability. These meta-heuristic optimization algorithms have become very popular in this modern age and are well known not only in antenna design but also in various fields of study. They are adaptive to various applications and are readily applicable to various types of problems since they only consider the input and output of a problem. Various optimization algorithms are explored in this work. Among these, GA and PSO are quite common in antenna designs. The trust region framework (TRF) algorithm acts as the best candidate for solving multiobjective and multiple-variable problems in antenna design. As per analysis of these optimization methods, it is impossible to identify one optimization algorithm as the best fit among them all. They remain simple algorithms inspired by natural phenomena which can assure the evolution of better algorithms in the future.

REFERENCES

[1] E. R. Brown, "RF-MEMS switches for reconfigurable integrated circuits," *IEEE Trans Microw Theory Tech*, vol. 46, no. 11, Part 2, pp. 1868–1880, 1998. doi: 10.1109/22.734501.

[2] K. Karthika and K. Kavitha, "Reconfigurable antennas for advanced wireless communications: A review," *Wirel. Personal Commun.*, vol. 120, no. 4, pp. 2711–2771, Oct. 01, 2021. doi: 10.1007/s11277-021-08555-4.

[3] G. S. Hornby, J. D. Lohn, and D. S. Linden, "Computer-automated evolution of an X-band antenna for NASA's space technology 5 mission," *Evol Comput*, vol. 19, no. 1, pp. 1–23, 2011. doi: 10.1162/EVCO-a-00005.

[4] J. M. Johnson and Y. Rahmat-Samii, "Genetic algorithms in engineering elec-tromagnetics," *IEEE Antennas Propag Mag*, vol. 39, no. 4, pp. 7–21, 1997. doi: 10.1109/74.632992.

[5] D. Marcano and F. Duran, "Synthesis of antenna arrays using genetic algo-rithms," *IEEE Antennas Propag Mag*, vol. 42, no. 3, pp. 12–20, 2000. doi: 10.1109/74.848944.

[6] J. H. Holland, *Adaptation in Natural and Artificial Systems*. Ann Arbor, MI: University of Michigan Press, 1975.

[7] K. A. De Jong, "An analysis of the behavior of a class of genetic adaptive sys-tems," 1975. https://deepblue.lib.umich.edu/handle/2027.42/4507.

[8] J. M. Johnson and Y. Rahmat-Samii, "Genetic algorithm optimization and its application to antenna design," *IEEE Antennas Propag. Soc.*, vol. 1, pp. 326–329, 1994. doi: 10.1109/aps.1994.407746.

[9] F. Zadehparizi and S. Jam, "Design of frequency reconfigurable patch anten-nas with defected ground structures using genetic algorithm," *Iran. J. Sci. Technol Trans Electr. Eng.*, Vol. 42, no. 4, pp. 485–491, 2018. doi: 10.1007/s40998-018-0065-5.

[10] K. Fertas, F. Fertas, S. Tebache, A. Mansoul, and R. Aksas, "Genetic algorithm based approach for frequency switchable dual-band patch antenna," *Iran. J. Electr. Electr. Eng*, vol. 18, no. 3, 2022. doi: 10.22068/IJEEE.18.3.2454.

[11] F. Zadehparizi and S. Jam, "Frequency reconfigurable antennas design for cog-nitive radio applications with different number of sub-bands based on genetic algorithm," *Wirel Pers Commun*, vol. 98, no. 4, pp. 3431–3441, Feb. 2018. doi: 10.1007/s11277-017-5022-5.

[12] S. Song and R. D. Murch, "An efficient approach for optimizing frequency reconfigurable pixel antennas using genetic algorithms," *IEEE Trans Antennas Propag*, Vol. 62, no. 2, pp. 609–620, 2014. doi: 10.1109/TAP.2013.2293509.

[13] P. Guo, X. Wang, and Y. Han, "The enhanced genetic algorithms for the optimi-zation design," *Proceedings—2010 3rd International Conference on Biomedi-cal Engineering and Informatics, BMEI 2010*, vol. 7, no. Bmei, pp. 2990–2994, 2010. doi: 10.1109/BMEI.2010.5639829.

[14] G. Chen, H. Jiang, and X. Lei, "Reconfigurable antenna design optimization based on improved quantum genetic algorithm," *2014 31th URSI General Assembly and Scientific Symposium, URSi GASS 2014*, no. 7, pp. 3–6, 2014. doi: 10.1109/URSIGASS.2014.6929190.

[15] R. Storn and K. Price, "Differential evolution—a simple and efficient heuristic for global optimization over continuous spaces," *Journal of Global Optimiza-tion*, Vol. 11, no. 4, pp. 341–359, 1997. doi: 10.1023/A:1008202821328.

[16] N. Padhye, P. Mittal, and K. Deb, "Differential evolution: performances and analyses," *2013 IEEE Congress on Evolutionary Computation, CEc 2013*, no. i, pp. 1960–1967, 2013. doi: 10.1109/CEC.2013.6557799.

[17] J. H. Van Sickel, K. Y. Lee, and J. S. Heo, "Differential evolution and its appli-cations to power plant control," in *2007 International Conference on Intel-ligent Systems Applications to Power Systems*, 2007, pp. 1–6. doi: 10.1109/ISAP.2007.4441675.

[18] S. Das and P. N. Suganthan, "Differential evolution: A survey of the state-of-the-art," *IEEE Trans. Evol Comput*, vol. 15, no. 1, pp. 4–31, 2011. doi: 10.1109/TEVC.2010.2059031.

[19] N. Hansen and A. Ostermeier, "Completely derandomized self-adaptation in evolution strategies.," *Evol Comput*, vol. 9, no. 2, pp. 159–195, 2001. doi: 10.1162/106365601750190398.

[20] R. Kundu, R. Mukherjee, S. Debchoudhury, S. Das, P. N. Suganthan, and T. Vasilakos, "Improved CMA-ES with memory based directed individual generation for real parameter optimization," *2013 IEEE Congress on Evolutionary Computation, CEC 2013*, pp. 748–755, 2013. doi: 10.1109/CEC.2013.6557643.

[21] M. Juneja and S. K. Nagar, "Particle swarm optimization algorithm and its parameters: A review," *ICCCCM 2016–2nd IEEE International Conference on Control Computing Communication and Materials*, no. ICCCCM, 2017. doi: 10.1109/ICCCCM.2016.7918233.

[22] Y. K. Choukiker, S. K. Behera, B. K. Pandey, and R. Jyoti, "Optimization of plannar antenna for ISM band using PSO," in *2010 Second International conference on Computing, Communication and Networking Technologies*, 2010, pp. 1–4. doi: 10.1109/ICCCNT.2010.5591601.

[23] F. J. Ares-Pena, J. A. Rodriguez-Gonzalez, E. Villanueva-Lopez, and S. R. Rengarajan, "Genetic algorithms in the design and optimization of antenna array patterns," *IEEE Trans Antennas Propag*, vol. 47, no. 3, pp. 506–510, 1999. doi: 10.1109/8.768786.

[24] I. P. Gravas, Z. D. Zaharis, P. I. Lazaridis, T. V. Yioultsis, N. V. Kantartzis, C. S. Antonopoulos, I. P. Chochliouros, and T. D. Xenos, "Optimal design of aperiodic reconfigurable antenna array suitable for broadcasting applications," *Electronics*, vol. 9, no. 5, 818, May 2020. doi: 10.3390/electronics9050818.

[25] S. Mirjalili, S. M. Mirjalili, and A. Lewis, "Grey wolf optimizer," *Adv. Eng. Soft.*, vol. 69, pp. 46–61, 2014. doi: 10.1016/j.advengsoft.2013.12.007.

[26] X. Li and K. M. Luk, "The grey wolf optimizer and its applications in electromagnetics," *IEEE Trans Antennas Propag*, vol. 68, no. 3, pp. 2186–2197, 2020. doi: 10.1109/TAP.2019.2938703.

[27] M. W. Guo, J. S. Wang, L. F. Zhu, S. S. Guo, and W. Xie, *An Improved Grey Wolf Optimizer Based on Tracking and Seeking Modes to Solve Function Optimization Problems*, vol. 8. IEEE, 2020. doi: 10.1109/ACCESS.2020.2984321.

[28] P. Mahouti, "Design optimization of a pattern reconfigurable microstrip antenna using differential evolution and 3D EM simulation-based neural network model," *Int. J. RF Microw. Comput Aided Eng.*, vol. 29, no. 8, Aug. 2019. doi: 10.1002/mmce.21796.

[29] S. Shoukhath, "Design of MEMS Reconfigurable E-Shaped Patch Antenna Design for Cognitive radio." [Online]. Available: www.ijert.org.

[30] P. Lacouth, A. G. D'Assunção, and A. G. Neto, "Synthesis of new reconfigurable limited size FSS structures using an improved hybrid particle swarm optimization," *J. Microwaves Optoelectron. Electromagn. Appl.*, vol. 18, no. 2, pp. 157–172, 2019. doi: 10.1590/2179-10742019v18i21550.

Chapter 7

Review on Wearable Antennas for IoT, Healthcare, and High-End Applications

Vaibhav Saini, Praful Ranjan, and Ayushi Jain

7.1 INTRODUCTION

Fourth-generation (4G) mobile communication systems now heavily rely on body-centric wireless communication [1]. Body-centric connectivity has a firm place in body area networks (BANs) and personalized network connectivity (PANs) [2]. There are two forms of body connectivity: off-body connectivity and on-body connectivity. Off-body connectivity refers to the radio link between base units or wireless devices located outside the body and devices that are worn on the body. Lastly, in-body connectivity is the exchange of information between on-body nodes and wireless medical devices. In other terms, a wearable antenna is any antenna that is particularly designed to work while being worn. Smartwatches, which frequently have built-in Bluetooth antennas; glasses such as smart glasses, which have Wi-Fi connection and global positioning system antenna arrays [3]; GoPro action cameras, which have Wi-Fi and Bluetooth antennas and are often strapped to users to obtain video clips; and even the fitness tracking sensor, which connects to a smartphone via Bluetooth and is placed in a person's shoe, are some of the most commonly used examples in daily life. Because wearable antennas are increasingly being used in consumer electronics, this chapter is dedicated to explaining the issues connected with wireless antenna design. Wearable, fabric-based antennas are currently the most advanced research subjects in antennas. For all contemporary applications, wearable antennas [4, 5] must typically be lightweight, inexpensive, almost maintenance-free, and require no installation. Many specialized occupational groups use body-centric communication systems, including paramedics, firefighters, and the military. Additionally, wearable antennas can be used for monitoring small children, the elderly, and sportsmen [6].

7.2 WEARABLE ANTENNA TYPES

7.2.1 Traditional Wearable Designs

Traditional antenna layouts such as rectangular dipoles, monopoles, planar inverted F arrays (PIFAs), and previous studies employed microstrip patches

DOI: 10.1201/9781003459880-7

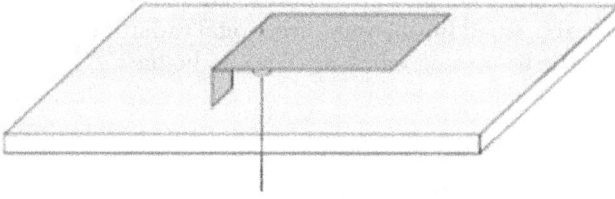

Figure 7.1 Construction of PIFA.

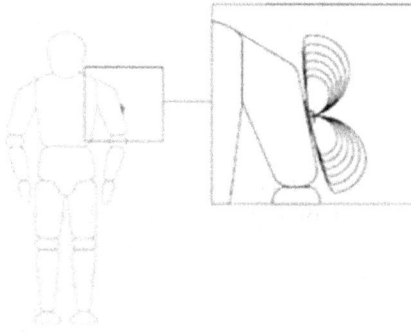

Figure 7.2 Placement possibilities of antennas.

for antenna structural deterioration. Construction of PIFA is shown in Figure 7.1. Printed circuit boards (PCBs) can be used to make planar microstrip antennas [7]. Due to their inexpensive price and ease of production, they became a useful type of antenna. In order to create a wearable antenna that could be attached to a garment's sleeve, Salonen investigated an antenna using a PIFA design. PIFAs have a folded structure that is parallel to the ground plane, like quarter-wave monopole antennas [8]. One potential site for the antenna is on a shirt sleeve. This also highlighted how the bottom layer of the antenna affects the overall direction of the strongest emission. The ground plane protected humans by acting as a barrier, preventing radiation from entering their bodies. To put it another way, the bottom layer served as a radiation reflection. A portable adjustable planar upside-down antenna (flex PIFA) supporting wireless gadgets was created and released based on the identical antenna design principle. The flexible material of the antenna was designed to be installed on a human arm and has a thickness of 0.236 mm, a dielectric constant of 3.29, and a loss tangent of 0.0004. Low profile, 100 MH to 500 MHz operation, omnidirectional azimuth coverage, wideband return loss, and vertical or circular polarization are antenna criteria that influence antenna selection. A wearable antenna should have an omnidirectional radiation pattern so that it can be used with mobile devices and smart apparel. In addition, omnidirectional radiation patterns should be created with little to no side lobes, as they can be harmful to human health.

7.2.2 Textile Design Antenna

It is necessary to check the directional antenna's dependability in different means of application where the user is moving. This dynamism causes the body's orientation to fluctuate constantly. The antenna may continue to receive signals regardless of how the body is oriented because a polarization wave emits energy in all planes, including the horizontal, vertical, and any angular planes in between. Polyimide spacer fabric with a thickness of 6 mm and a permittivity of 1.5 is the substrate utilized in the design [9]. The ground plane and antenna patch were both made of a conductive substance, a woven textile that was nickel-plated. Circular polarization was also used to create a textile antenna for protective apparel in [8] in order to enhance reception in actual use. By positioning the patch's feed point in a circular pattern, the design was able to excite the two orthogonally polarized TM01 and TM10 modes [10].

One of the areas of electromagnetic research that is moving forward the fastest is the electromagnetic band gap (EBG) [11]. The ground plane created with an EBG structure resembles a flawless conductor that is made up of magnets. Applying an EBG layer to the ground plane lowered the antenna's return loss, making it about equal for all resonant frequencies, according to the S_{11} test (at −15 dB). It was totally constructed of textiles. When compared with other traditional planar antenna feeds, the design's aperture-coupled feeding method helped to increase bandwidth.

7.3 CREATING AND DESIGNING OF WEARABLE ANTENNAS

7.3.1 Conductive Substance

Conductive items like textiles are produced by weaving conductive metal or polymer threads into regular cloth. These materials' wearable, sturdy, and flexible qualities made them perfect for incorporation into apparel. One well-known study covered the need for conductive textiles while creating textile antennas. In order to reduce losses, it was preferred that the conductive textile have a low and steady electrical resistance (/square). It was also necessary for the material to be flexible so that the antenna could distort. This chapter illustrates how a different researcher in [12] employed an easy requirement that could be covered in an arm. The substance employed had a 0.125 mm thickness and a 0.05 sq. in. surface resistance of woven conductive fabric type. When building an antenna, choosing the right material is essential for it to be durable and appropriate for the intended purposes.

The substance used in the study in [13], an aramid woven fabric, is heat proof and appropriate for inclusion into firefighter apparel. The conductor in [14] was a nylon that has been metalized and is very conductive. High conductivity was supplied by its three metalized layers (Ni, Cu, and Ag), whereas the surface resistance was just 0.03/square. In addition, the substances offered rusting and pliability resistance, which made it appropriate for use in harsh environments.

7.3.2 Fabrication Method

In contrast to stand-alone antenna testing performed without the presence of a human body, on-body evaluations must be undertaken to evaluate the antenna's performance in various configurations. Depending on how they are used, wearable antennas may have different positions.

The manufactured antenna was evaluated in [15] on the human arm, chest, and in free space. Researchers went so far as to include the human body in the measurement in [13]. According to the findings of previous studies, the antenna, which is located on the backside of the body, provides the most stable spot and will reduce changes in body orientation when compared to certain other anatomical structures such as the arm.

7.4 ANALYSIS REQUIRED FOR WEARABLE ANTENNAS

Characteristic impedance, radiation pattern, gain, and effectiveness are often required for a typical antenna structure. Investigating its bending characteristic is not essential because common planar antennas are flattened. A wearable antenna, on the other hand, demands careful consideration of extra criteria in order to assure the antenna's operation when worn on the body. This section includes additional observations that should be taken in order to evaluate a wearable antenna design.

7.4.1 SAR Modelling

Power quality limits received by the human body must constantly be taken into consideration by engineers and researchers due to open public forums about the fitness consequences of radiation and international regulatory regulations. As a result, wireless devices' specific absorption rate (SAR) has been established. Specific adiation rate limits that are most frequently used are those set by the Institute of Electrical and Electronics Engineers (IEEE) at 1.6 W/kg for every 1 g of tissue and the International Commission on Non-Ionizing Radiation Protection (ICNIRP) at 2 W/kg for every 10 g of tissue [16].

7.4.2 Measurement with Different Bending

Antenna structure measurements should be made in a variety of bending positions. This is done to guarantee that the antenna will work as expected in practical situations, especially when it is attached to circular body parts like an arm. S_{11} tests were performed using various antenna bending scenarios [17]. To study this bending property, the antenna that was wrapped around a plastic cylinder was measured. The frequency of the bending decreased as the bending grew smaller.

7.4.3 On-Body Measurements

In addition to stand-alone antenna tests, which were done without a human body present, on-body evaluations must be performed to evaluate the efficient outcomes of the antenna in various postures of the body [18]. Depending on how they are used, wearable antennas may have different positions. In the manufactured device, measurements were taken on the human arm, chest, and in zero gravity. Researchers even went so far as to measure using the human body.

7.4.4 Significance

The wearable antenna includes a broad range of application fields, such as wearables, the Internet of Things, medical applications, ultrawideband (UWB), telecommunications, defence applications, and electronics. The rapid downsizing of wireless devices has raised the importance of wearable devices in the industry today. A wearable antenna is a component of clothing used for connectivity, including localization and monitoring, portable computers, wireless connectivity, and public safety. Wearable antennas offer the potential for omnipresent monitoring, communication, and energy gathering and storing by utilizing wireless body sensor networks for healthcare and pervasive applications.

7.5 APPLICATION OF WEARABLE ANTENNAS

7.5.1 Healthcare

Active sensors and passive sensors are the two types of sensors used in the wearable category. The classification is ambiguous, and many authors use it the other way around. Although the concept of enabling differs from that of other fields of digital equipment, active sensors need the use of an additional power source to transform an signal (input) into a working signal (output), whereas passive sensors generate self-energy, being evaluated into a beneficial potential outcome [19]. A thermocouple is an example of a passive

Table 7.1 Wearable Biomedical Device Measurements

Evaluation	Parameter	Frequency, Hz
Flow of blood	1–300 mL/s	0–20
BP measurement	0–400 mmHg	0–50
ECG	0.5–5 mV	0.05–150
Electroencephalography	5–300 µV	0.5–150
Electromyography	0.1–5 mV	0–10,000
pH level insights	3–13	0–1
Respiratory rate	2–50 breaths/min	0–10
Temperature rate	32–40°C	0–0.1

sensor that is commonly employed in research settings to monitor body temperature or other phenomena. Patients' vital signs can be picked up by a network of sensors and sent back towards the functional nodes. This section discusses textile sensors used on the human body, their properties for gathering crucial physiological signals, and their incorporation into smart clothing. Sensors must be a basic component of all sensor networks, and advancements in circuit design, sensor fusion, micro-electro-mechanical systems, and nanostructures have a substantial influence on sensor quality. Physiological sensors assess physical signs such as vital signs, regular blood glucose surveillance, bloodstream blood oxygenation, and breathing rate [20]. In addition to core temperature, vital signs, heart rate, and breathing rate, physiological parameters may include measures like posture and movements, blood oxygenation, and electrocardiogram (ECG.). Table 7.1 lists the wearable biomedical device measurements. Body temperature is one of the few medical signs that is either constant or varies extremely slowly. These biosensors are converting dynamic, rather than static, biosignals, which depend on time.

7.5.2 Sports and Fitness

Sports players' performance, comfort, and awareness have all increased as a result of wearing smart materials. Sports have also sparked a lot of research in the smart textile sector, leading to innovations like breathable clothing and moisture-management materials. By absorbing excess heat and releasing it when necessary, phase-change technology, a recent development in smart textiles, has made it possible to regulate body temperature. It is now possible to instantly detect an athlete's metabolic status and biological condition thanks to recent improvements in textile materials. Stitched-in piezoelectric sensors contribute with kinematic analysis, which improves movements and lowers injuries [21].

Furthermore, these sensors provide vital real-time data for tracking performance and are always operational for continuous monitoring. Because they feature an accelerometer, radio frequency identification, wireless module, and motion sensor, smart training socks fall under the category of wearable sensors. Future smart textiles might feature chemical sensors embedded into garments that can extract data from sweat analysis [22].

7.5.3 Internet of Things

The market has predicted that there will be 100 billion connected gadgets through 2030, increasing from the existing rate from around 27 billion. Similar to this, several names were given to the low-power wide area network (LPWAN) concept when it was first launched in the early 1990s, but its connection topologies and networking protocols were the same [23]. Over the past ten years, LPWAN also gained popularity in the Internet of Things (IoT) and was emphasized by SIGFOX technologies. After Long Range entered the picture, cellular operators began to sell IoT devices with connectivity under the names LTE Cat-M and NB-IoT due to its popularity. As shown in Figure 7.3, a few prerequisites determine the quality of LPWAN technologies. According to requirements, several LPWAN technologies are implemented in various application areas based on this capability [24]. Let's go over each of its features in turn. Low-power narrowband IoT technology connects various IoT devices and extends the range of wide-area network technology. Due to these characteristics, the technology can be used for the many applications. In [25], IoT technologies were widely used in many sectors and also in the agriculture sector with more productivity as outcome. The suggested wearable antenna in this research paper runs from 2.40 to 14.88 GHz, with a fractional bandwidth (FBW) of approximately 144.44%. The installation of an open-ended rectangular slit in the antenna array increases the bandwidth [26]. The tiny multiple-input, multiple-output (MIMO) diversity antenna demonstrated in this paper [27] is unquestionably capable for communications, enabling an incredibly broad impedance bandwidth as well as band-notched

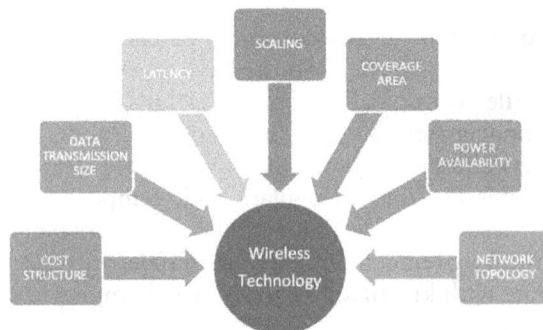

Figure 7.3 Anatomy of wireless technology.

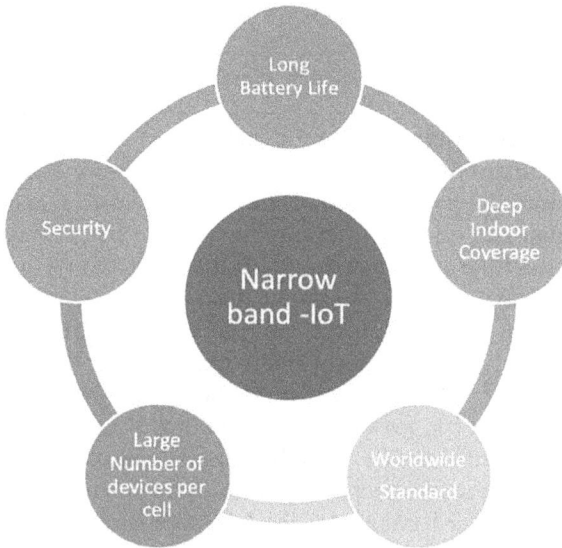

Figure 7.4 Narrowband IoT.

Table 7.2 Applications of Wearable Antennas

Health Monitoring	IoT	Fitness Tracker	Military
• **BP device** • **Glucose check** • **Air filter** • **Hearing aid devices** • **ECG monitoring device**	• Smart watch with health monitoring • Motion detector • Step count technology • Pulse count functionality	• Personal tracker • Calorie counter • Sleep censor • Posture tracking • Active walking	• Smart band • GPS tracker • Gesture-based devices • Finger tracking • Location-sharing device

characteristics for wireless local area networks (WLANs) and WiMAX. An LPWAN technology called NB-IoT was created to meet the demand for deep indoor penetration, extended coverage, and low-power devices. Overall, it can be said that this technology is a blessing for the IoT era because it enables devices to operate with greater network coverage, battery life, and spectrum efficiency. All generations of mobile networks can be compatible with NB-IoT technology, which can also address the issues of coverage and consumption. Additionally, it may be incorporated into all current mobile communication devices, including smartphones, IoT modules, and other chipset devices. For better simulation and mathematical results, industries and labs are using MATLAB and Simulink for automation and better outcomes [28]. Applications of wearable antennas have been reflected in Table 7.2.

7.6 CONCLUSION AND FUTURE SCOPE

According to the research, there are several extra factors that should be taken into account when creating a wearable antenna as opposed to a traditional antenna. A variety of different substances may be utilized to create these modified antennas. To create an antenna design which complies with the wearable antenna gathering analysis, SAR, measurements with various antenna bending, and on-body measurements must be evaluated. The future of these modified wearable antennas is bright and promises to keep up with the expansion of the wireless communication industry. Medical devices that use wireless technology to control internal activities and measure a variety of physiological indicators are rapidly expanding. Some of these devices are implanted, while others are worn on the body. Implanted devices with biosensing and actuation, for example, can control heart function, monitor heart rate, and provide such a valuable smart wearable sensor system for applications in healthcare electrically stimulating neurons and head pressure. Biomedical sensor technology, the IoT, sophisticated wireless network technology, and sports and fitness will all require tight multidisciplinary collaboration. Based on the findings of this chapter, it is feasible to construct antennas with the optimum characteristics and aid antenna engineers in their decision-making by considering significant elements in antenna creation. The creation of elevated fabric substrate antennas capable of overcoming the adverse effects of the human body is the future path for wearable antennas in human interaction applications. When creating antennas for wearable technology, these aspects demand special consideration. While designing wearable antennas, developers should pay close attention to structural deformation, accuracy and precision in antenna production techniques, and size.

REFERENCES

1. Chowdhury, M.Z., Shahjalal, M., Ahmed, S. and Jang, Y.M., 2020. 6G wireless communication systems: Applications, requirements, technologies, challenges, and research directions. IEEE Open Journal of the Communications Society, 1, pp. 957–975.
2. Asam, M., Jamal, T., Adeel, M., Hassan, A., Butt, S.A., Ajaz, A. and Gulzar, M., 2019. Challenges in wireless body area network. International Journal of Advanced Computer Science and Applications, 10(11).
3. Ramdani, S. and Pratama, I., 2019. October. Single Coaxial Feed Microstrip GPS Antenna Aimed at Wearable Device Application. In 2019 IEEE Conference on Antenna Measurements & Applications (CAMA) (pp. 247–250). IEEE.
4. Paracha, K.N., Rahim, S.K.A., Soh, P.J. and Khalily, M., 2019. Wearable antennas: A review of materials, structures, and innovative features for autonomous communication and sensing. IEEE Access, 7, pp. 56694–56712.

5. El Gharbi, M., Fernández-García, R., Ahyoud, S. and Gil, I., 2020. A review of flexible wearable antenna sensors: Design, fabrication methods, and applications. Materials, 13(17), p. 3781.

6. Ghodake, A.P. and Hogade, B.G., 2020. A review of wearable antenna. International Journal of Electronics, 12(1), pp. 13–18.

7. Ali, U., Ullah, S., Shafi, M., Shah, S.A., Shah, I.A. and Flint, J.A., 2019. Design and comparative analysis of conventional and metamaterial-based textile antennas for wearable applications. International Journal of Numerical Modelling: Electronic Networks, Devices and Fields, 32(6), p.e2567.

8. Casula, G.A. and Montisci, G., 2019. A design rule to reduce the human body effect on wearable PIFA antennas. Electronics, 8(2), p. 244.

9. Almohammed, B., Ismail, A. and Sali, A., 2021. Electro-textile wearable antennas in wireless body area networks: Materials, antenna design, manufacturing techniques, and human body consideration—a review. Textile Research Journal, 91(5–6), pp. 646–663.

10. Çelenk, E. and Tokan, N.T., 2022. All-textile on-body antenna for military applications. IEEE Antennas and Wireless Propagation Letters, 21(5), pp. 1065–1069.

11. Ashyap, A.Y., Dahlan, S.H.B., Abidin, Z.Z., Abbasi, M.I., Kamarudin, M.R., Majid, H.A., Dahri, M.H., Jamaluddin, M.H. and Alomainy, A., 2020. An overview of electromagnetic band-gap integrated wearable antennas. IEEE Access, 8, pp. 7641–7658.

12. Tanaka, M. and Jae-Hyeuk, J., 2003. Wearable microstrip antenna. In IEEE Antennas and Propagation Society International Symposium. Digest. Held in conjunction with: USNC/CNC/URSI North American Radio Sci. Meeting (Cat. No. 03CH37450), Columbus, OH (pp. 704–707), vol. 2. doi: 10.1109/APS.2003.1219333.

13. Hertleer, C., Rogier, H., Vallozzi, L. and Declercq, F., 2007. A Textile Antenna Based on High-Performance Fabrics. In The Second European Conference on Antennas and Propagation, EuCAP 2007 (pp. 1–5). Edinburgh. doi: 10.1049/ic.2007.1085.

14. Klemm, M. and Troster, G., 2006. Textile UWB Antenna for On-Body Communications. In 2006 First European Conference on Antennas and Propagation (pp. 1-4). Nice, France. doi: 10.1109/EUCAP.2006.4584865.

15. Santas, J.G., Alomainy, A. and Hao, Y., 2007. Textile Antennas for On-Body Communications: Techniques and Properties. In The Second European Conference on Antennas and Propagation, EuCAP 2007 (pp. 1–4). Edinburgh. doi: 10.1049/ic.2007.1064.

16. Ahmad, A., Faisal, F., Ullah, S. and Choi, D.Y., 2022. Design and SAR analysis of a dual band wearable antenna for WLAN applications. Applied Sciences, 12(18), p. 9218.

17. Zu, H.R., Wu, B., Zhang, Y.H., Zhao, Y.T., Song, R.G. and He, D.P., 2020. Circularly polarized wearable antenna with low profile and low specific absorption rate using highly conductive graphene film. IEEE Antennas and Wireless Propagation Letters, 19(12), pp. 2354–2358.

18. Varma, S., Sharma, S., John, M., Bharadwaj, R., Dhawan, A. and Koul, S.K., 2021. Design and performance analysis of compact wearable textile antennas

for IoT and body-centric communication applications. International Journal of Antennas and Propagation, 2021, pp. 1–12.

19. Mustafa, A.B. and Rajendran, T., 2019. An effective design of wearable antenna with double flexible substrates and defected ground structure for healthcare monitoring system. Journal of Medical Systems, 43, pp. 1–11.

20. Nguyen, T.H., Nguyen, T.L.H. and Vuong, T.P., 2019, March. A printed wearable dual band antenna for remote healthcare monitoring device. In 2019 IEEE-RIVF International Conference on Computing and Communication Technologies (RIVF) (pp. 1–5). IEEE.

21. Rana, M. and Mittal, V., 2020. Wearable sensors for real-time kinematics analysis in sports: A review. IEEE Sensors Journal, 21(2), pp. 1187–1207.

22. Govindan, T., Palaniswamy, S.K., Kanagasabai, M., Kumar, S., Rao, T.R. and Kannappan, L., 2021. RFID-Band Integrated UWB MIMO Antenna for Wearable Applications. In 2021 IEEE International Conference on RFID Technology and Applications (RFID-TA) (pp. 199–202). IEEE.

23. Sabban, A., 2019. Small new wearable antennas for IOT, medical and sport applications. In 2019 13th European Conference on Antennas and Propagation (EuCAP) (pp. 1–5). IEEE.

24. Jacob, S. and Sonadevi, S., 2022. Antenna Design and Analysis for Narrow Band Internet of Things Applications. In Antenna Design for Narrowband IoT: Design, Analysis, and Applications (pp. 162–180). IGI Global.

25. Kumar, A., Ranjan, P. and Saini, V., 2022. Smart irrigation system using IoT. In Agri-Food 4.0. Emerald Publishing Limited.

26. De, A., Roy, B., Bhattacharya, A. and Bhattachaqee, A.K., 2021. Bandwidth-enhanced ultra-wide band wearable textile antenna for various WBAN and Internet of Things (IoT) applications. Radio Science, 56(11), pp. 1–16.

27. Bhattacharya, A., Roy, B., Caldeirinha, R.F. and Bhattacharjee, A.K., 2019. Low-profile, extremely wideband, dual-band-notched MIMO antenna for UWB applications. International Journal of Microwave and Wireless Technologies, 11(7), pp. 719–728.

28. Saini, V., Shah, P. and Sekhar, R., 2022, December. MATLAB and Simulink for Building Automation. In 2022 IEEE Bombay Section Signature Conference (IBSSC) (pp. 1–6). IEEE.

Chapter 8

A Review of Design Challenges of Metamaterial-Inspired Body-Worn Antennas

Moumita Bose, Aparna Kundu, Abhishek Sarkhel, and Ujjal Chakraborty

8.1 INTRODUCTION

With the introduction of 5G networks, global data demand has increased and will continue to rise in the near future [1]. As such, conformal, handy, and portable wireless devices are in high demand; among them, unlicensed frequency bands like Industrial Scientific and Medical, WiMAX, Wi-Fi, WLAN, etc., are of lucrative application for local wireless communication, sensing, the Internet of Things, human consensus, and 5G mobile integrated devices. The future of upcoming technology is handy, robust, and trending in use, which has called for flexible antenna [2] and wearable devices to be developed. Figure 8.1 [3] depicts the adoption of wearable devices and the architecture of Internet of Things [4], which also includes wearable communication in 5G. In our daily life, clothing is a very essential part, and integrating electronic or electrical devices along with textiles results in smart clothing, which can find application in the medical sector, defence sector, and many more [5]. The demand for wearable antennas is increasing progressively, as they are light in weight, easy to carry, and have an in-house fabrication facility. Its flexible nature makes it perform better in different bending situations when placed on the human body. Metamaterial used for designing wearable antennas provides miniaturisation facilities as well as a decrease in back-radiation to improve the specific absorption rate. When exposed to electromagnetic waves, the human body suffers greatly [6–8]. In our day-to-day lives, this metamaterial-based flexible wearable antennas can be used in different health monitoring applications [9] and safety-based systems. Use of metamaterial can improve antenna performance for narrowband [10–12], dual band [13–16], and wideband [17–19] antennas by minimising back-radiation and the specific absorption rate. Article [20] provides an uneven ultra-wideband antenna on denim material for wireless body area network application, while [21] provides material features of autonomous communication and sensing of wearable antennas.

The aim of this chapter is to cover some recent works on metamaterial-inspired wearable antennas and their practical design challenges. Some

DOI: 10.1201/9781003459880-8

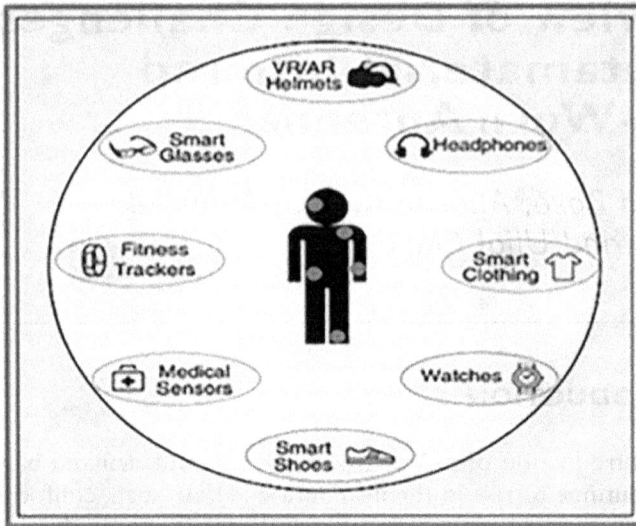

Figure 8.1 Wearable device adoptions: from a medical sensor to entertainment helmet [3].

application-oriented structures are considered to encounter different design challenges by invoking appropriate excitation and boundary conditions for obtaining the negative permittivity, permeability, and refractive index at desired frequencies. The selection of a flexible substrate is also important. Wearable antennas require high bandwidth with small dimensions. The wearable metamaterial antennas must comply with Institute of Electrical and Electronics Engineers (IEEE) C95.1–1999 and IEEE C95.1–2005 standards for an average specific absorption rate value that should be 1.6 W/kg for 1 g of tissue and 2 W/kg for 10 g of tissue. Metamaterial-based wearable antennas are designed such that their efficiencies, radiation patterns, and impedance [22] should not degrade when they come into contact with the human body, even in motion. The bending analysis has to be taken into account for a tolerable degree of freedom. This chapter collectively assembles metamaterial-inspired antennas, which will pave the way for wireless portable communication, sensing physical miniaturization, and specific absorption rate for the future 5G and Internet of Things wearable devices.

8.2 METAMATERIAL-INSPIRED DUAL WIDEBAND WEARABLE ANTENNA FOR WIRELESS APPLICATIONS

Figure 8.2 depicts a dual-band wearable antenna inspired by metamaterial on a jeans substrate for on-body and off-body communication in the GSM 1800 MHz, WLAN, and WiMAX bands, with detailed dimensions given in

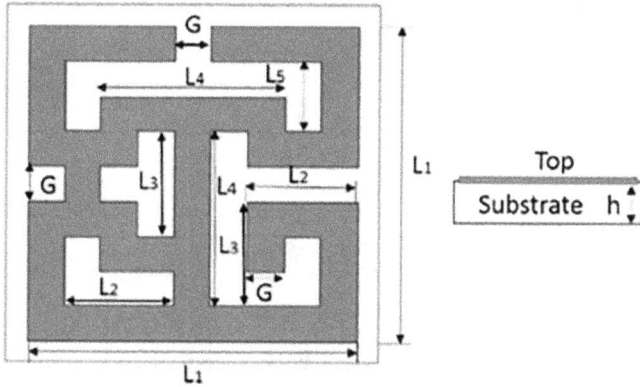

Figure 8.2a Geometry of metamaterial unit cell [23].

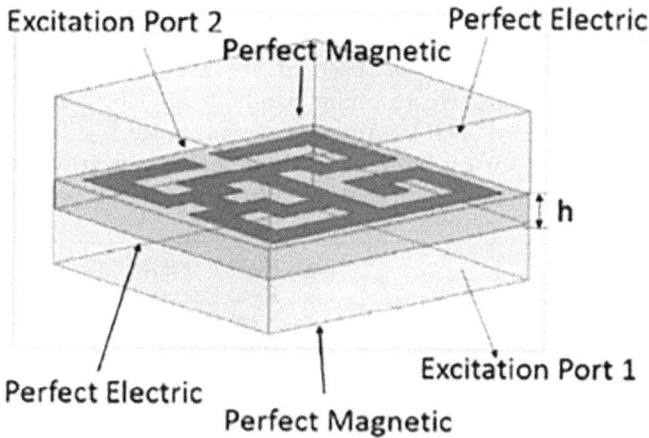

Figure 8.2b Analysis setup of unit cell in HFSS [23].

article [23]. The antenna's performance deteriorated when it was placed on the surface of the human body.

This structure was simulated using Ansys HFSS with proper boundary conditions and excitation to obtain the desired magnitude and phase of the scattering parameter. This scattering parameter is used for the extraction of material parameters like permittivity, permeability, and refractive index using MATLAB. For designing double-negative (DNG) metamaterial, Richards and Ziolkowski's DNG media theory is used for small losses where material parameters follow the following relations as given in Equations (8.1–8.3) [23]:

$$E_r \approx \left(\frac{2}{jk_0 d}\right)\left(\frac{1-S_{21}-S_{11}}{1+S_{21}+S_{11}}\right) \tag{8.1}$$

$$\mu_r \approx \left(\frac{2}{jk_0 d}\right)\left(\frac{1-S_{21}+S_{11}}{1+S_{11}-S_{11}}\right) \tag{8.2}$$

$$n \approx -|E_r \mu_r|^{0.5}\left[1+j\frac{1}{2}\left(\frac{E'}{j|E_r|E_0}+\frac{\mu'}{j|\mu_r|\mu_0}\right)\right] \tag{8.3}$$

The characteristics of unit cells are studied by varying the height and dimension of the antenna substrate. A change in permeability and permittivity is observed as height changes. The consideration was to obtain negative permeability and permittivity in the desired band. The simulated result for material parameters versus frequency for 1.5 mm substrate height is shown in Figure 8.3a. The effect of changing the substrate dimension on S_{11} and the upper and lower bandwidth was observed, as shown in Figure 8.3b. It signifies that an antenna has two resonant bands, from 1.6 GHz to 2.56 GHz and from 4.24 GHz to 7 GHz. Antenna performance is affected by bending. When it is used for on-body or off-body communication, it should be conformal to the human body. The human body's material property changes

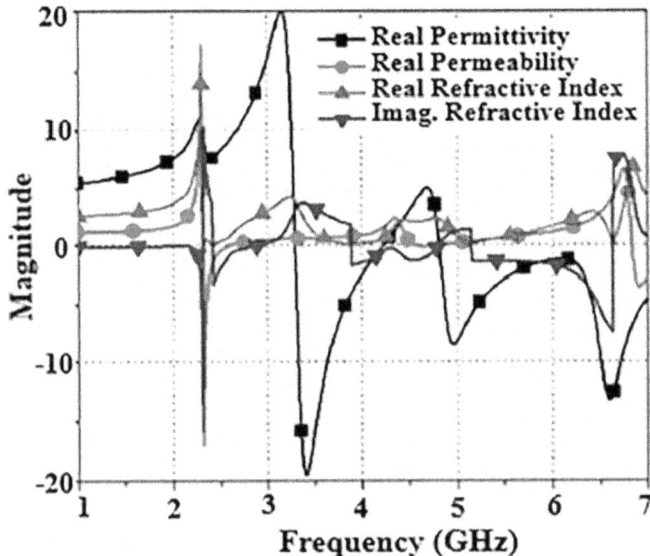

Figure 8.3a Variation of metamaterial parameters with frequency at substrate height 1.5 mm [23].

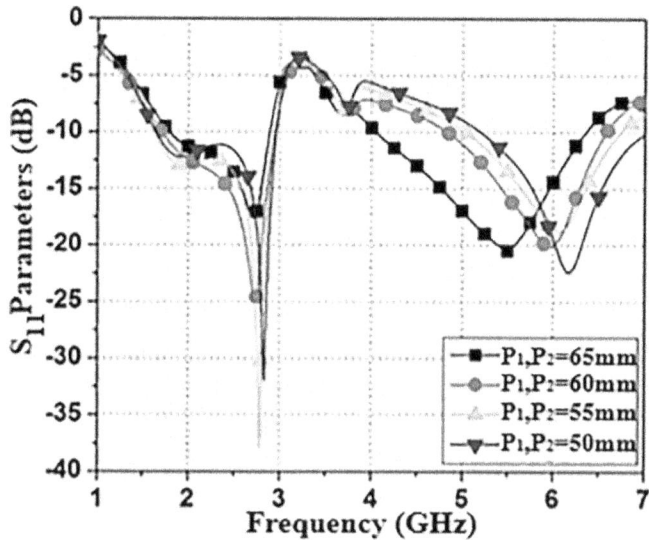

Figure 8.3b Simulated S_{11} parameter for different substrate dimensions [23].

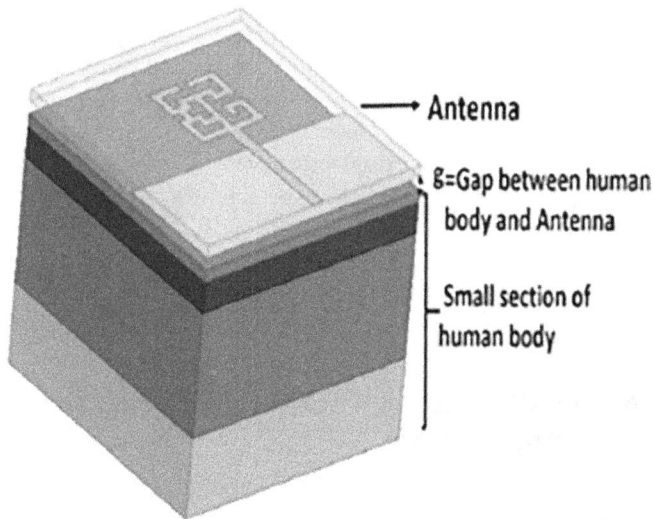

Figure 8.3c Antenna mounted on cubic human tissue model [23].

with frequency, so analysing the effect of a multiband antenna on it is a challenging job. To study the on-body and off-body effects of this metamaterial antenna, a computational simulation was done on an equivalent section of a human body model, as shown in Figure 8.3c. Its material properties at 2.4 GHz and effect on S_{11} are studied with the antenna mounted on a

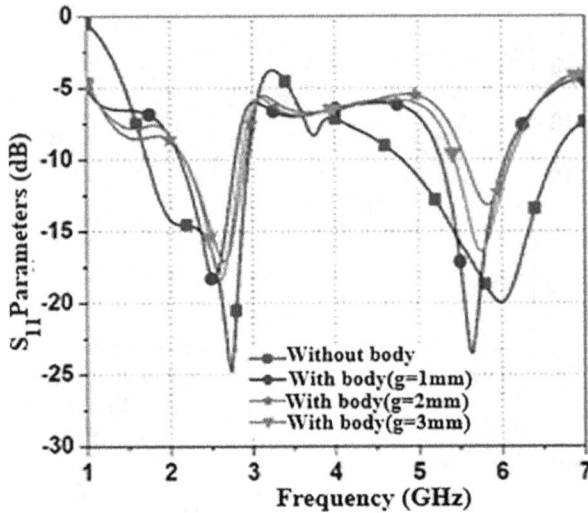

Figure 8.3d Simulated S$_{11}$ parameter with on-body and off-body effect [23].

cubic human model with a gap of 'g' between the human body and antenna to encounter the practical challenges of a conformal antenna. The mutual coupling between the antenna and the human body affects antenna performance, as illustrated in Figure 8.3d.

Article [24] describes another dual-band wearable antenna on a jeans substrate. In this paper, the antenna with a split-ring resonator operates at 2.32 GHz and 3.52 GHz, as visualised in Figure 8.4b. The interior and exterior radius of a circular ring resonator are 5.14 mm and 5.43 mm, respectively. The drawback of this kind of structure for dual resonant bands is its overall increased size.

8.2.1 A Novel Wearable Metamaterial Fractal Antenna for Wireless Applications

Figure 8.5c depicts the steps of evolution of a novel multiband wearable fractal metamaterial antenna along with top and ground geometry (see Figure 8.5a and b) designed for GPS, WiMAX, and Wi-Fi band applications [25]. The paper reveals the effect of a spiral metamaterial in the ground plane upon a fractal patch makes permittivity, permeability, and refractive index negative, as shown in Figure 8.5d. It consequently reduces the specific absorption rate and makes antenna radiation less absorbed by human tissue, which is a prime requirement for wearable applications. Another advantage of using spiral metamaterial with a fractal antenna is that it

Figure 8.4a Dual band fabricated antenna with SRR [24].

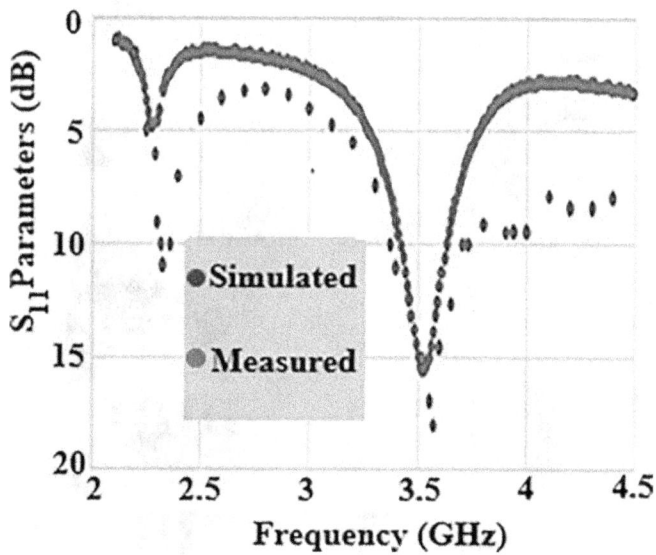

Figure 8.4b Simulated and measured S_{11} with SRR [24].

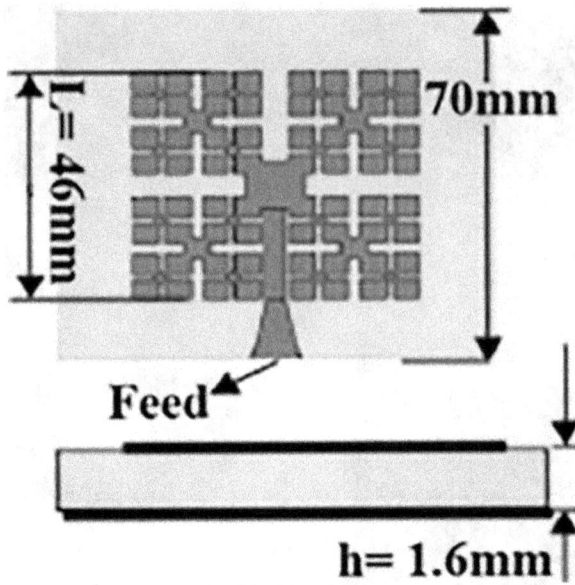

Figure 8.5a The fractal antenna with third iteration [25].

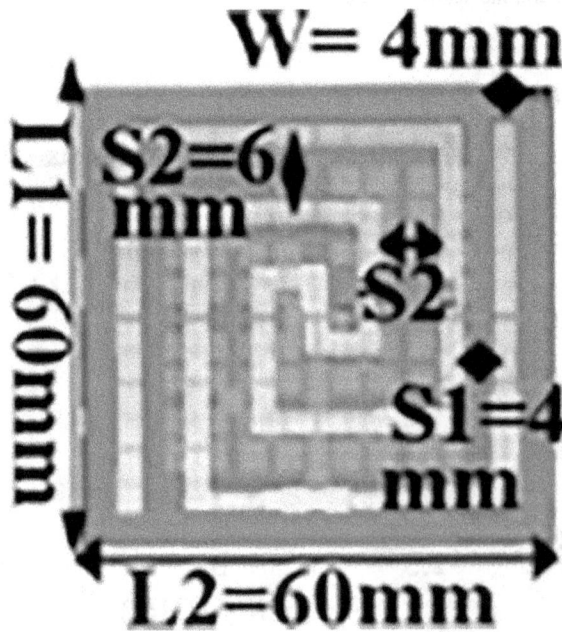

Figure 8.5b The fractal antenna's ground geometry [25].

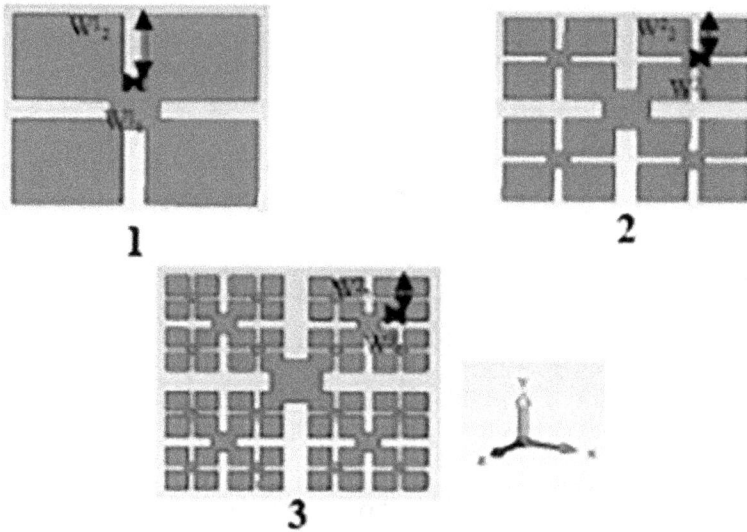

Figure 8.5c Steps of evolution [25].

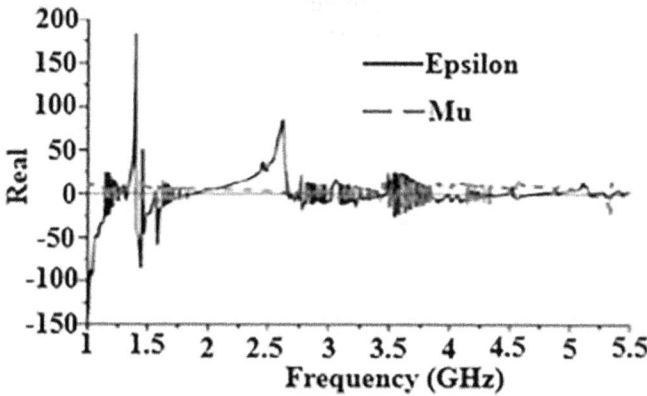

Figure 8.5d MATLAB simulated metamaterial parameters [25].

improves bandwidth performance, as shown in Figure 8.5e. In this design, the third iteration is used to increase the path length of the current, causing a shift in resonance frequency towards lower bands and improving bandwidth. The iteration length of the fractal antenna is calculated using the equation denoted in [25]. It has a maximum specific absorption rate value of 0.925 W/Kg, which is well suited for wearable purposes and satisfies international safety standards at resonant frequency. This type of antenna is ideal for use in life jackets during the search for a human body in the event of an accident.

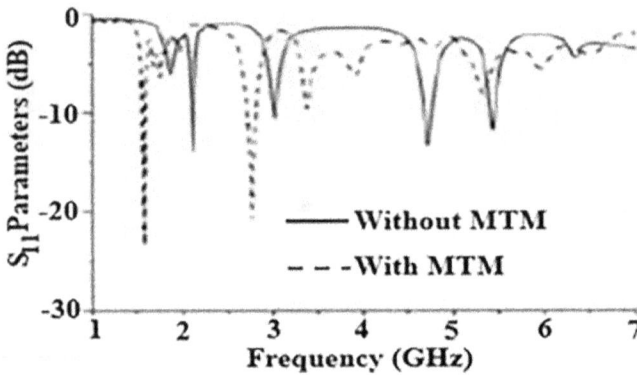

Figure 8.5e Variation of S_{11} with frequency for third iteration of the wearable antenna (with and without) metamaterials [25].

8.2.2 Radiation Pattern Reconfigurable Wearable Antenna Based on Metamaterial Structure

In [26], a radiation pattern reconfigurable metamaterial-based antenna in the ISM band (2.4 GHz) is designed as in Figure 8.6a. A dispersion diagram using Equations (8.4 and 8.5) [26] is used to analyse the resonance of an inductor-loaded patch antenna between the zeroth and positive first-order modes.

$$\mathrm{n\,p} = \frac{1 - S_{11}S_{22} + S_{12}S_{21}}{2S_{21}} \tag{8.4}$$

$$\mathrm{mn\,p} = \pi m \tag{8.5}$$

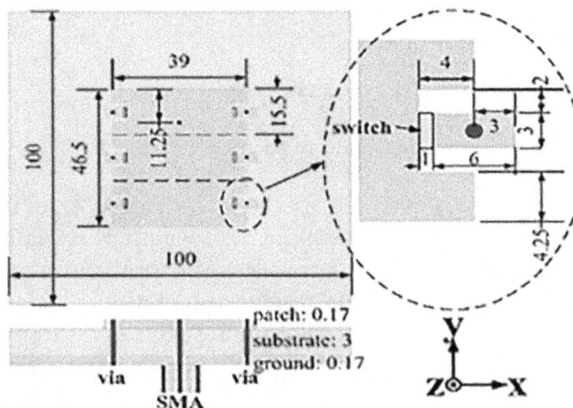

Figure 8.6a Reconfigurable antenna [26].

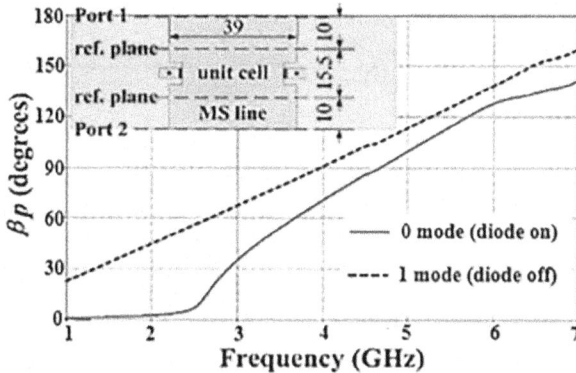

Figure 8.6b Dispersion diagram of unit cell [26].

where β, p, m, and n are phase constants, unit cell size, antenna having the number of unit cells, and resonance modes, respectively. The dispersion diagram recovered with calculated S-parameter values considering a unit cell is presented in Figure 8.6b.

The transmission line vias were connected to the edges of the patch through switches. When the switches are turned off, a disconnected transmission line causes the antenna to function as a simple patch and radiate in broadside mode. The patch has a half-wavelength of 2.45 GHz, an n of one, and an m of three. Thus, each unit cell has a 60-degree delay. When switches are on, shunt inductors connected as loaded vias generate a phase delay of 0 degrees at 2.45 GHz and the antenna operates in zero resonance mode, having a uniform distributed electric field throughout the patch. Now the antenna radiates in an omnidirectional pattern. Using switches, the patch is converted to metamaterial, and it resonates with the same 2.45 GHz frequency, but its radiation pattern changes from a broadside to an omnidirectional pattern, and metamaterial accounts for diversity in the radiation pattern.

The main difficulty with this type of antenna being conformal to the human body is that it must withstand bending without changing its resonance frequency. This is achieved by not reducing the effective magnetic current length of the patch while bending. This type of antenna is well suited for off-body (omnidirectional) and on-body (broadside) use by switching radiation pattern reconfiguration.

8.2.3 Compact All-Textile, Dual-Band Antenna Loaded with Metamaterial-Inspired Structure

This wearable antenna [27] shown in Figure 8.7a proposes an alternate and feasible solution for dual band operation using metamaterial for wireless local area network (WLAN) applications. Ways were adopted for dual band

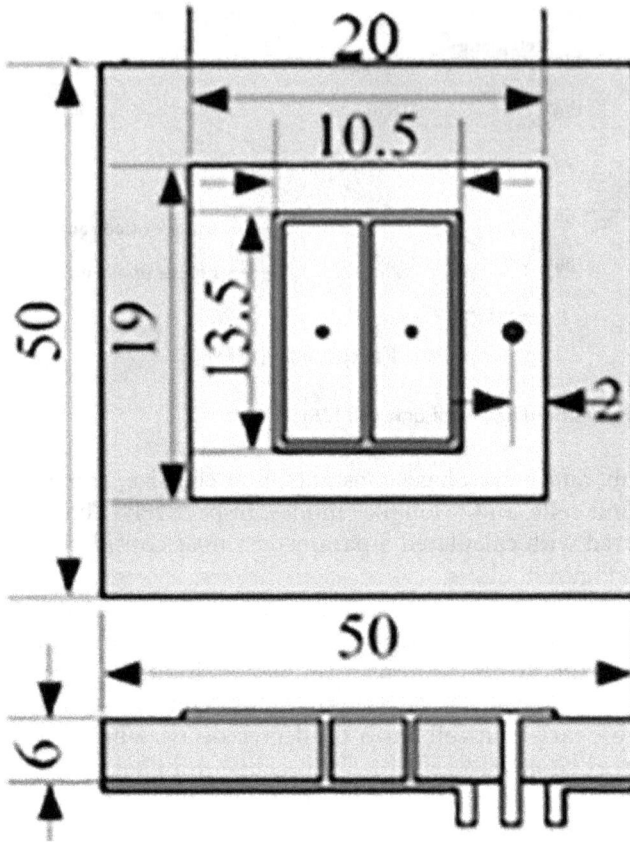

Figure 8.7a The antenna [28].

operation, like using parasitic patches and staked patches. Patches with slot topology have shortcomings like a significant increase in size, complexity in fabrication, etc. With the use of flexible substrates and slots, the current distribution is affected, resulting in radiation distortion. Side radiation increases coupling between human body and antenna. In this design, the right-handed structure (outer patch) is loaded with a composite right- or left-handed transmission line (inner mushroom-like structure) acting as metamaterial structure with a patch. With the introduction of a composite right- or left-handed transmission line loaded metamaterial, not only is miniaturisation obtained but also it opens possibilities of multifrequency operation i.e. dual band operation in conjunction with the right-hand structure at lower frequencies, as well as the structure's compactness and simplicity, make it ideal for integration in a handheld wearable or portable device. The design challenges of operating at low and high frequencies with a symmetrical radiation pattern

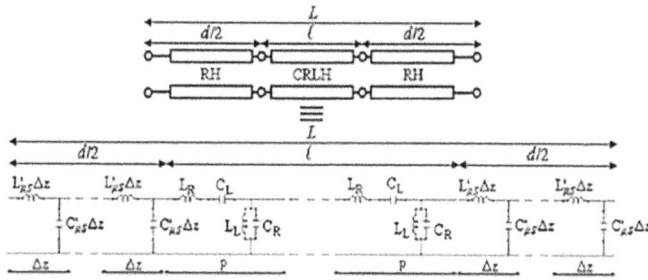

Figure 8.7b 1D CRLH-TL and its equivalent circuit [28].

can be understood by deducing equivalent circuits, as shown in Figure 8.7b, from a transmission line model for similar kinds of structures [28].

In this transmission line model $L'_{\mu S}$, $C'_{\mu S}$ are inductance and capacitance per unit length and are of a microstrip structure with length d/2 on both sides. Along with LH (left handed), the effect is contributed by coupling between adjacent cells (C_L) and via to the ground inductance (L_L), whereas the RH effect is due to the current flux (L_R) and parallel plate capacitor (C_R) between the metal patch mushroom and ground plane constituting the composite right- or left-handed area as a whole of unit cell length (P). The rectangular patch increases the coupling (C_L), which helps to lowers the frequency of operation at resonance for metamaterial. M is the number of unit cells and ℓ is total length of composite right- or left-handed transmission line such that $\ell = Mp$ and total length $L = d + \ell$. With the help of this model dispersion relation from eigen-frequencies, working frequencies can be determined, or once the antenna is ready, a scattering parameter may be used for it. One of the necessary conditions for a homogeneous structure of composite right- or left-handed is the size of the unit cell ($p<<\lambda_g/4$) i.e. the mushroom structure should be much smaller than the guided wavelength. As from the equivalent circuit of a unit cell, at low frequencies L_R and C_R tend to minimise leaving C_L and L_L which is left-handed and allows left-handed propagation, whereas at higher frequencies C_L and L_L tend to minimise, and it behaves like a right-handed structure and right-handed propagation prevails. The introduction of composite right- or left-handed (mushrooms) in the right-hand (patch) structure makes the dispersion curve flatter as compared to composite right- or left-handed alone; consequently, the axial ratio between the working frequencies reduces remarkably and a similar dipolar mode can be achieved at lower frequencies, which is an area of interest. The eigenfrequencies at resonance under impedance matching condition must satisfy the following Equation (8.6) [28].

$$\beta_n L = \beta_n^{RH} d + \beta_n^{CRLH} p M = n\pi \qquad (8.6)$$

where β_n represents the propagation and n represents resonant indices with −1 as the first fundamental mode supporting LH propagation and +1 as the first fundamental mode with RH propagation.

In this chapter two symmetrical radiation patterns were observed at 2.45 GHz for n = −1 and 5.15 GHz for n = +1. In this type of structure, lower bands are sensitive to the size of the inner mushroom and higher bands are influenced by patch sizes. For assured WLAN operation, a tolerance is to be imposed on the outer patch and the inner patch, i.e. here ±0.5 mm ±0.5 mm (outer) and ±0.2 mm ±0.2 mm (inner). This reveals that as the bending radius decreases, its resonance is stable for both the lower and upper band and a further decrease in radius shifts up both the resonances to a higher frequency, but this shift is insignificantly low in percent and accounts for the robustness of this antenna under deformity. The specific absorption rate simulated on various positions was found to be in limit for standards and generally, the specific absorption rate value is higher at 5.2 GHz as compared to 2.4 GHz because of the intrinsic properties of human model tissue and conductivity of tissue increases at higher frequencies. This property makes this antenna especially useful for off-body wireless communication.

8.2.4 A Compact Triple-Band Metamaterial-Inspired Antenna for Wearable Applications

In this paper, the prospect of the use of organic materials for RF applications was explored. In [29] a Hilbert metamaterial-based printed antenna on organic indium phosphide (INP) substrate was designed for energy harvesting depicted in Figure 8.8a. It is an ultra-wideband antenna operating in a band from 2.4 to 10 GHz. In the energy harvesting process, energy is captured from external sources, stored, and used in sensors or wearable electronic devices. In this article, the exponential tapered coplanar waveguide feed used Willis-Sinha formulae, whose characteristics impedance Z_o as given by [29] is:

$$Z_0(y) = Z_i exp^{\left\{\left[\frac{y}{L} - \frac{0.2405}{2\Pi} \sin\left(\frac{2\Pi y}{L}\right)\right] \ln\left(\frac{Z_p}{Z_i}\right)\right\}} \tag{8.7}$$

From 50 Ω SMA port Z_i is the input impedance, L is the length, along the feed line y is length at any point, and patch impedance is Z_p. At any point on the transmission line width 'w' is given by [29]:

$$w(y) = w_0 exp^{Ay_1} \tag{8.8}$$

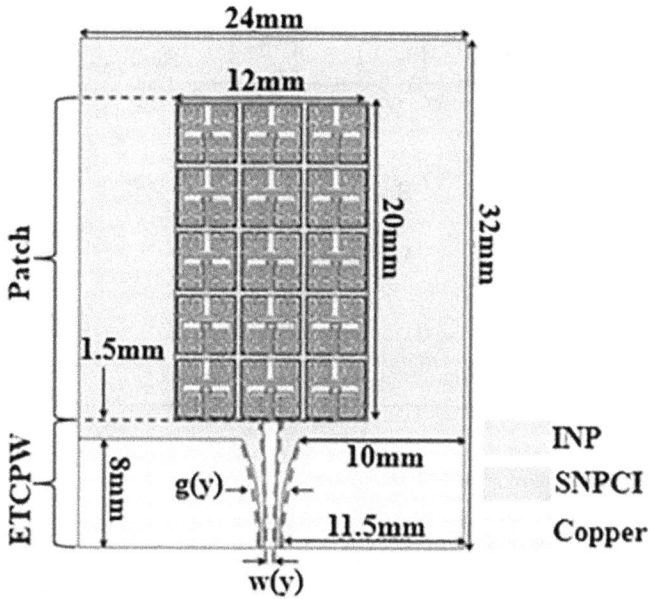

Figure 8.8a Antenna geometry [29].

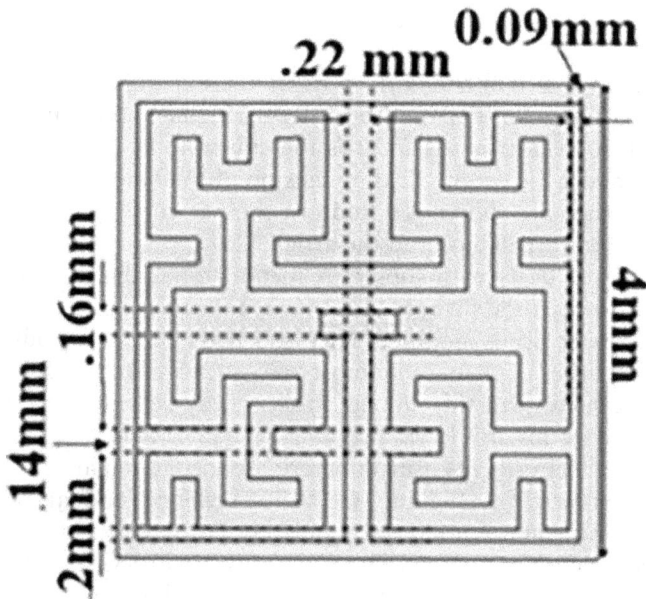

Figure 8.8b Metamaterial unit cell [29].

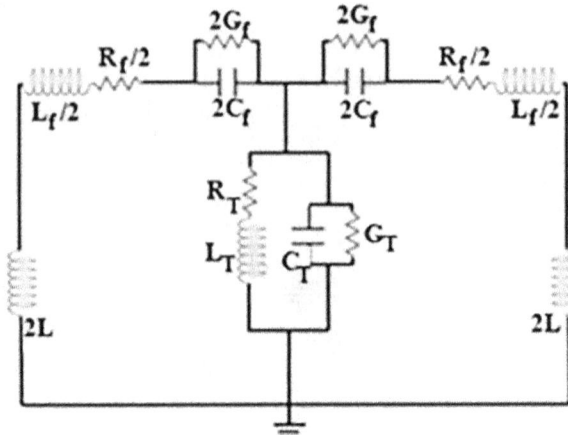

Figure 8.8c Equivalent circuit of Hilbert structure [29].

The gap g is given by:

$$g(y) = (w) + g_0 exp^{By_2} \tag{8.9}$$

$$g_0 = \frac{t}{2\Pi\mu_r\varepsilon_r}\left[1 + \ln\left(\frac{8\Pi w_0}{t}\right)\right] \tag{8.10}$$

The microstrip line width has $A = 0.12$ as the exponential factor and the exponential factor for gap width is $B = 0.18$, the initial microstrip line thickness at 50 Ω port is w_0 and t is the substrate height at E_r and μ_r. In Figure 8.8b and c the geometrical dimension of the Hilbert unit cell and its equivalent circuit are shown. To find out the parameters of the equivalent circuit, equations are depicted in the article [29].

Another paper deals with substrate, metamaterial characterisation, conductor, and surface losses whose effective E_r and μ_r retrieved from the simulated S parameter using modified 'Nicolson Ross-Weir' methods shows no negative, but either of them are almost zero at 2.4 GHz and 5.8 GHz [30]. Different parameters for this Hilbert structure can be found using the equations provided in this article.

This article [31] presents a metamaterial-loaded, circular polarised (CP), and circular polarised web-guide (CPW) feed, wearable, dual band, planar antenna designed on a liquid crystal polymer (LCP) substrate, as shown in Figure 8.9a, to work in WiMAX (3.5 GHz), WLAN (5.8 GHz), and vehicular communication bands (5.9 GHz). The conformal, flexible behaviour of this dual-band antenna finds good correlation with simulation and measurement values of return loss and axial ratio.

From [31] it is found that the antenna has low deformability, high thermal stability, and fabrication is simple. The antenna's current response is simulated and measured for different bending angles: 30, 45, 60, 90, and 120 degrees and for LH and RH circular polarisation. Here by altering the measure of the feed line, design optimisation is done. Introduction of a split ring resonator (SRR) on the feed line results in bandwidth enhancement. Simulation and measurement are done for different substrate heights as in Figure 8.9b. Figure 8.9c shows a dual-band characteristic and reflection coefficient for different bending angles.

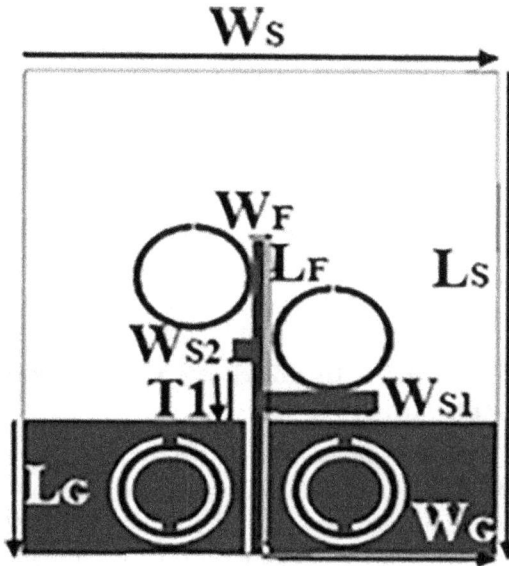

Figure 8.9a Antenna, [31].

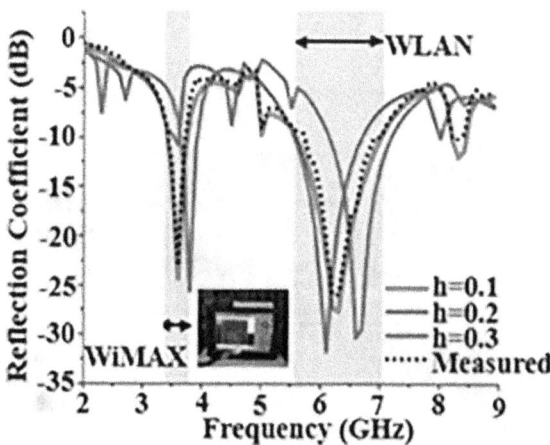

Figure 8.9b Measured and simulated S_{11} for different thickness of the substrate [31].

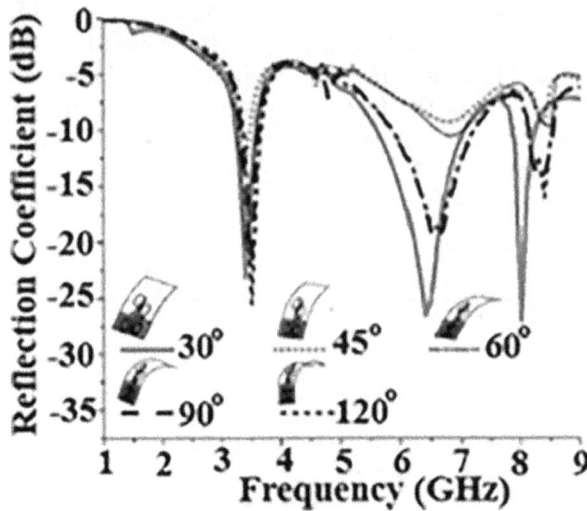

Figure 8.9c Measured reflection coefficients for different bending angles [31].

8.2.5 Metamaterial-Embedded Wearable Rectangular Microstrip-Patch Antenna

This paper reveals a metamaterial implanted wearable antenna as displayed in Figure 8.10 designed on a polyester substrate to work in the IEEE 802.11a WLAN band having a resonant frequency of 5.10 GHz [32]. Upon the rectangular patch, slots are introduced to reduce the bending effect and

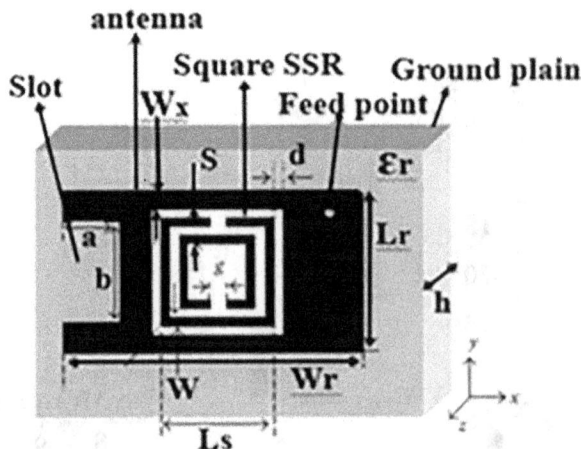

Figure 8.10 SRR-embedded wearable antenna [32].

resonance frequency, but it results in impedance mismatch. To overcome this, an SRR was built inside the slot. The measurements for designing the antenna are given in reference [32]. Use of metamaterial in antennas generates subwavelength resonance, which results in considerable reduction in size [33, 34]. The SRR is magnetically coupled with a slotted rectangular patch to form a liquid crystal (LC) resonator. A small shift in the position of the SRR causes a considerable change in resonance frequency. The bending effect of the antenna is experimented upon different radii of curvature and found that, with more bending, the curvature resonance frequency shifts towards a higher frequency as the resonant length of antenna is reduced.

8.2.6 Metasurface-Enabled Hepta-Band Compact Antenna for Wearable Applications

This study proposes a hepta-band antenna with inductive ground plane, reactive impedance surface (metasurface-RIS), slotted patch, and 50-ohm SMA (subminiature version A) connector depicted in Figure 8.11a [35]. Designing

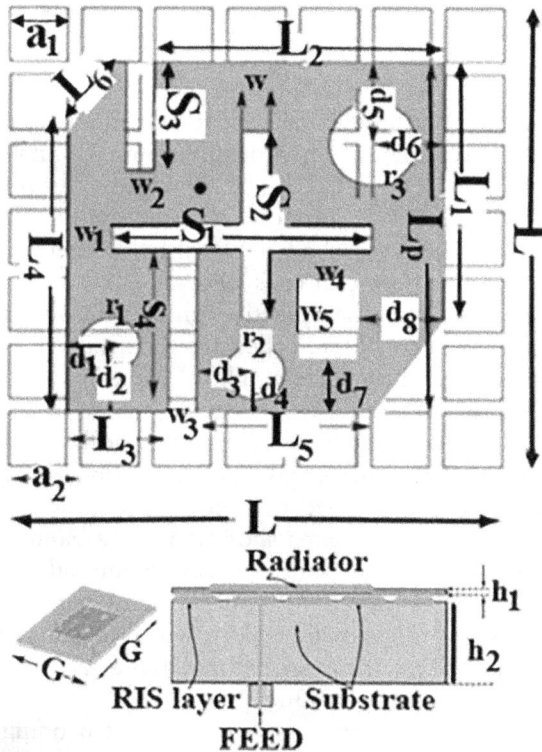

Figure 8.11a Hepta-band antenna top and side view [35].

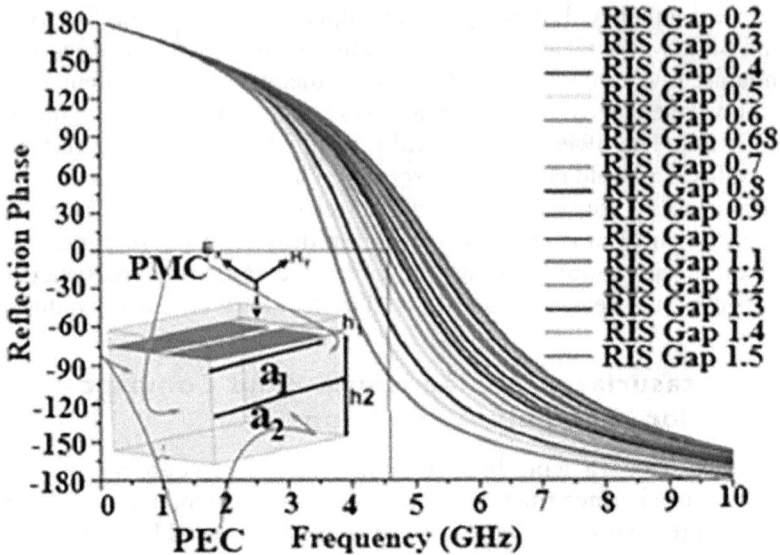

Figure 8.11b Reflection phase of the RIS unit cell for gap variation (anticipated gap of 0.68 mm) [35].

a multiband antenna is a perplexing task to accomplish on the vicinity of the human body. This antenna is low profile, compact, and works in GSM 1.8 GHz, WLAN 2.45 GHz, LTE 2.5 GHz, Navigation 2.96 GHz, Wi-MAX 3.5 GHz, sub-6 GHz 5G band 3.4–4.2 GHz, and satellite 4.5 GHz applications. Impedance mismatches are taken care of by the metasurface layer. The characteristics of metasurface unit cells were studied with proper boundary conditions. In Figure 8.11b, the reflection phase versus frequencies plot is shown. The zero reflection phase, called the artificial magnetic conductor (AMC), has frequencies in the left or right of it that make the metasurface inductive or capacitive in nature.

Initially without a metasurface and with a very low substrate height, the antenna has a resonance around 4.4 GHz; with an increase in substrate height up to 3 mm, the antenna showed a triple-band characteristic, and when the metasurface is used, the antenna showed a hepta-band characteristic as shown in Figure 8.11c. The metasurface reduces mutual coupling between the patch and ground; hence a better impedance matching is achieved and the metasurface can store magnetic energy, which increases the inductance of patch and cancels the electric energy stored in the near field, which results in miniaturisation of the antenna [36].

The antenna performance studied under different bending conditions on human phantoms and found that the S_{11} lies below 10 dB. The maximum specific absorption rate value for all bands is .0296 W/kg.

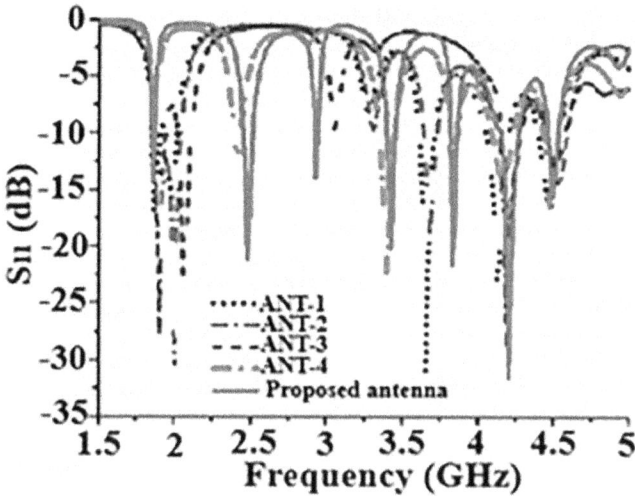

Figure 8.11c Consequence of inductive ground (RIS) and PEC ground on S_{11} for different substrate thicknesses [35].

8.2.7 An Ultra-Wideband, Low-SAR, Flexible, Metasurface-Enabled Antenna for WBAN Applications

In this paper [37], a metasurface-enabled, ultra-wideband miniaturised, flexible, lightweight antenna designed and simulated with a low specific absorption rate and high gain for wireless body area application is depicted in Figure 8.12. Wireless body area network (WBAN) antennas find applications in the health sector, tracking and navigation, wearable computation rescue and emergency services, and many more [37–39]. Here, a 7 × 7 square shape metamaterial unit cell array designed to work as a metasurface is shown in Figure 8.12b to reduce back-radiation and specific absorption rate values with increased gain and front-to-back ratio (FBR). The specific absorption rate effect was simulated on a cubic human body tissue model using a CST simulator, but no bending analysis was given. By using a metasurface, the specific absorption rate value is reduced by 97% and peak gain is increased by 98% with a demerit in lower radiation efficiency.

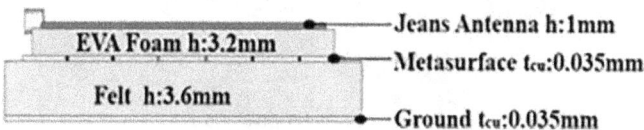

Figure 8.12a Cross-sectional view of antenna placed on the metamaterial [37].

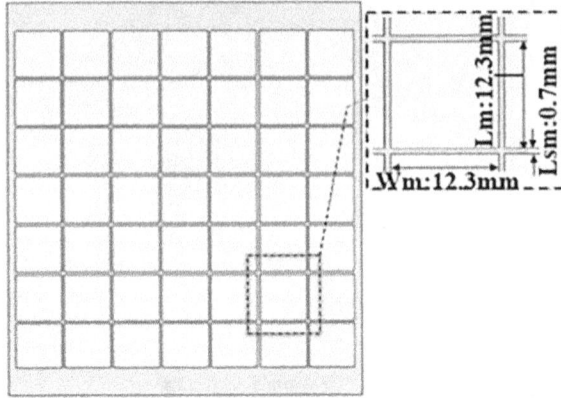

Figure 8.12b Dimension of metasurface structures [37].

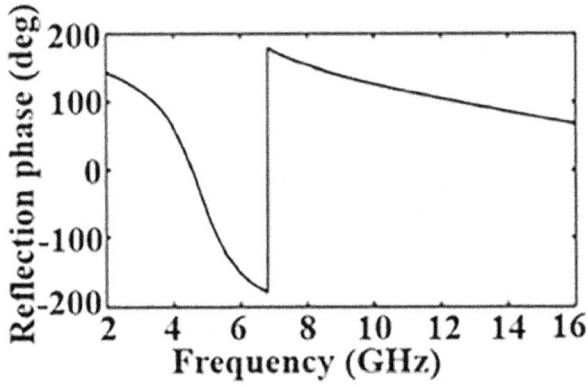

Figure 8.12c Reflection phase of metamaterial [37].

Figure 8.12d Fabricated antenna on jeans [37].

Figure 8.12e Fabricated metasurface [37].

8.3 SUMMARY

In the coming era wearable antennas will play a pivotal role in potable wireless communication and their compatibility with future technology like 5G and Internet of Things. Space to fit, commercial low cost, multiband operation, radiation pattern reconfigurability, simplicity, robustness, wide bandwidth, high gain, low specific absorption rate, and conformal nature will be key points to adapt with vast operating environment requirements where a fast data rate, security, low latency, and multi-input/-output capability to work with software-based equipment are fundamental requirements. This review intended to cover those aspects, in a preview of design approaches and difficulties. This started with a unique antenna which acts as a radiator and resonator with a dual band of operation with significant miniaturization of $0.1\lambda \times 0.1\lambda$ and is suitable for multiple environments for on-body and off-body communication. It shows the way to extract metamaterial parameters for low losses with the help of scattering parameters and MATLAB. This metamaterial-based wearable antenna shows the effect of the metamaterial when used on the body with different parametric variations and bending analysis. After that the effect of spiral metamaterial on a fractal patch benefitted by improvement in bandwidth and specific absorption rate reduction. The radiation pattern can be controlled by switching vias for incorporating diversity in radiation. It shows how the extracted S parameter was used for the calculation of a dispersion diagram. The crucial aspect while designing the same structure is zero-phase delay per unit cell, which can be observed from the dispersion diagram. A slight change in dimension can increase the delay, thereby simultaneously affecting the radiation pattern reconfigurability, so care must be adapted for almost zero-phase delay per unit cell from characterisation of the dispersion diagram for metamaterial. While performing bending analysis to avoid a shift in resonance frequency, the correct direction has to be considered so that the effect of the magnetic current length does not reduce. Next a compact, all-fabric antenna which offered simplicity, robustness, miniaturisation, and multifrequency operation approach with dissymmetric radiation pattern along with the loading of CRLH-TL in the right-handed structure was discussed. It helped to flatten the dispersion curve for wireless applications with singularity in field distribution at 2.4 GHz and 5.8 GHz frequencies. The main design apprehension in this is the length and width of the inner mushroom acting as a metamaterial which can affect the resonance frequency for a lower mode so its dimension are varied with sensitivity analysis for the proposed antenna frequency versus ratio of width to length of the mushroom. A slight change in width or length of the CRLH-TL can increase the coupling capacitance C_L and result in a shift of resonance to a lower frequency. To counter its effect, tolerance is imposed for guaranteed WLAN operation. Then a Hilbert-shaped unit radiator was arranged in array fashion to obtain an ultra-wideband on organic material for green recycling and energy harvesting. Here the printed metamaterial antenna was excited by co-planar waveguide (CPW) feed. The radiation pattern was found to be omnidirectional, which is desired for energy harvesting applications. Another Hilbert-based antenna had an end-fire radiation pattern, which is suitable for changing operating systems and environment at a frequency of interest. This section uses a special fishnet structure on the patch loaded with Hilbert-shape metamaterial with fractal ground to obtain a steerable radiation pattern from 3.3 to 3.9 GHz and due to use of fractal ground and patch leads to anisotropic characteristics at 5.8 GHz resulting in the endfire pattern. To characterise anisotropic behaviour, dispersion diagrams may be calculated and successive mode stop bands can be estimated to utilise its properties from an application point of view. This section also reveals the fabrication difficulties like conductor losses, surface losses, and way to minimise it; also the dimension of the curve should be chosen to avoid cross lines and match with the resolution of print traces for reasonable performance. This type of wearable antenna finds application for short-range communication between handheld wireless device receivers and wearable sensors or nearby station and energy harvesting. A circularly polarised flexible antenna designed on an LCP substrate shows conformal flexible bending, which makes the antenna suitable for wearing. The ring resonators were used in the design to get circular polarisation and enhance bandwidth. A change in feed line does not affect the lower frequency band, but the upper band. In bending analysis, the antenna shows a triple-band response for 30-degree and 45-degree angles. Moreover, the design process is simple. The metamaterial embedded rectangular microstrip patch antenna uses slot removal from the patch and loading SRR to excite lower resonant frequency to work in WLAN band. At first, the rectangular antenna's length and width were designed to work at 8.48 GHz and then the slot is introduced, resulting in reduced resonant length with poor impedance matching. Then for impedance matching, it was loaded with an SRR designed for negative permeability i.e. mu-negative media (MNG) in the range of 8.35–8.7 GHz. This loading provides good impedance matching and uniform current distribution due to which large electric fields are induced across the split gaps and mutual capacitance between split rings. In addition, the SRR and edge of the rectangular patch induce mutual inductance, resulting in modification of the resonant frequency to 5.10 GHz and thereby miniaturisation is achieved. The gap between the SRR and inner edge of the rectangular patch is an important parameter, as slight variations can affect the impedance matching and resonant frequency of WLAN operations. This type of antenna, with an increase in bending, results in a shift of resonance to a higher frequency due to a reduction in resonant length. This antenna uses a simple technique for miniaturisation on the same patch. A metasurface-enabled hepta-band antenna used a metasurface structure to get better impedance matching, seven resonance frequencies, and better FBR and gain. Initially a perfect electric conductor (PEC) ground was used with open and short loaded slots, with two different substrate heights, but the antenna was not showing a good resonance effect. When the array of 7×7 periodic square patches (metasurface) is used, the resonant modes of RIS and slots interact with each other and generate the hepta-bands. From the frequency versus reflection phase diagram we can see the metasurface has zero reflection phase at 4.604 GHz frequency and works as a polymer matrix composites (PMC). All the resonance frequencies are lying to the left of the zero reflection phase of the metasurface structure, resulting in an inductive nature. A decrease in patch dimension of the metasurface causes a shifting in resonance frequency towards higher bands. The antenna was tested under different bending conditions and maintained the value of S_{11} below –10 dB, and its specific absorption rate values are well suited for wearable applications. On the other hand, an ultra-wideband, low specific absorption rate, flexible, metasurface-enabled antenna has added advantages like high data rate, less interference, and low power spectral density which are striking traits for battery-operated wearable technology. This antenna adopted a metamaterial reflector for an ultra-wideband antenna to reduce the specific absorption rate and increase gain directivity. The design consideration for such a wide band was a challenging task, overcome by taking into account for the impedance match of the optimum metamaterial with an antenna by observing the S_{11} parameter. The unit cell of metamaterial on felt substrates can be characterised by a reflection phase graph. To avoid any electrical contact, ethylene vinyl acetate copolymer (EVA) foam was used as a separator. The radiation efficiency is compromised due to impedance matching, dielectric, and surface wave losses. This type of antenna is suitable for high-speed short-distance wireless communication technologies.

A comparison is given in Table 8.1 of different metamaterial-based wearable antennas showing their specifications, operating frequency, and applications. Metamaterial-based reconfigurable, steerable, wearable antennas with novel structures can be designed further for different on-body and off-body applications in the future. There is a future scope of work to integrate metamaterial in a substrate-integrated waveguide for smart antennas for implantable medical devices, etc. Metamaterial-based wearable antennas have future prospects in IoT and 5G applications. The metamaterial-based wearable antennas have optimistic forthcoming applications in the coming era of the smart world.

Table 8.1 Comparison of Different Metamaterial Antenna Designs

Ref. No.	Antenna Parameters (dimension, band width, gain)	Substrate Used	No. of Operating Frequency Bands	Operating Frequency (GHz)	MTM Type	SAR	Loss Tangent	Applications
[23]	60 × 60 × 2 mm³, 960 MHz, 2760 MHz, 1.8 dBi, 5.2 dBi maximum gain	Jeans	Dual band	2.4, 5.2, 5.8	DNG	—	.02	WLAN, WiMAX 2.3, GSM 1800
[25]	70 × 70 × 1.6 mm³, 80 MHz, 200 MHz, 130 MHz, 180 MHz, max gain 3.56 dBi	FR4	Multibands	1.57, 2.7, 3.4, 5.3	DNG	1.1 W/Kg for 1 gm tissue	.02	GPS, WiMAX, Wi-Fi
[26]	100 × 100 × 3.34 mm³, 119 MHz, maximum 3.9 dB	Felt	Single band	2.4	SNG	.01 W/Kg	.044	ISM

Table 8.1 (Continued)

Ref. No.	Antenna Parameters (dimension, band width, gain)	Substrate Used	No. of Operating Frequency Bands	Operating Frequency (GHz)	MTM Type	SAR	Loss Tangent	Applications
[27]	50 × 50 × 6.34 mm³. 130 MHz, 698 MHz, maximum gain is 6.2 dB	Felt	Dual band	2.45, 5.15	—	.69 W/Kg	.044	WLAN
[31]	38 × 32 × 0.1 mm³, 380 MHz, 1200 MHz	LCP	Dual band	3.5, 5.9	—	—	.0025	WiMAX, WLAN
[32]	30 × 15 × 1 mm³, 97 MHz BW, 4.92 dBi gain	Polyester	Single band	5.10	MNG	—	.01	WLAN
[35]	61.3 × 61.3 × 3 mm³, 80 MHz, 130 MHz, 80 MHz, 150 MHz, 110 MHz, 20 MHz, 140 MHz, peak gain 6.9 dBi	Flexible polyimide	Hepta-band	1.875, 2.45, 2.96, 3.5, 3.875, 4.25, 4.53	Inductive RIS metasurface	Maximum 0.0296 W/kg	.002	GSM, WLAN, LTE, Wi-MAX, sub-6 GHz 5G Band, navigation, satellite
[37]	58 × 80 × 1 mm³, 8.9 GHz BW, peak gain 9.1 dB	Jeans	Tri-band	4, 7, 10	_	0.086, 0.198, and 0.103 W/ kg	0.026, 0.005	WBAN

REFERENCES

[1] M. Agiwal, A. Roy, N. Saxena, "Next generation 5G wireless networks: A comprehensive survey," IEEE Communications Surveys & Tutorials, p. 1, 2016.

[2] P. Salonen, M. Keskilammi, J. Rantanen, L. Sydanheimo, "A novel bluetooth antenna on flexible substrate for smart clothing," Systems, Man, and Cybernetics, 2001 IEEE International Conference on, Tucson, AZ, vol. 2, pp. 789–794, 2001.

[3] H. Sun, Z. Zhang, R. Q. Hu, Y. Qian, "Wearable communications in 5G: Challenges and enabling technologies," IEEE Vehicular Technology Magazine, 13(3), pp. 100–109, Sept. 2018, doi: 10.1109/MVT.2018.2810317.

[4] L. Farhan, S. T. Shukur, A. E. Alissa, M. Alrweg, U. Raza, R. Kharel, "A survey on the challenges and opportunities of the Internet of Things (IoT)," 2017

Eleventh International Conference on Sensing Technology (ICST), doi: 10.1109/ICSensT.2017.8304465.

[5] S. Jiang, O. Stange, F. Bätcke, S. Sultanova, L. Sabantina, "Applications of smart clothing—a brief overview," Communications in Development and Assembling of Textile Products, 2, pp. 123–140, 2021, doi: 10.25367/cdatp.2021.2.p123-140.

[6] G.-P. Gao, B. Hu, S.-F. Wang, C. Yang, "Wearable circular ring slot antenna with EBG structure for wireless body area network," IEEE Antennas and Wireless Propagation Letters, 17(3), pp. 434–437, 2018.

[7] Z. H. Jiang, D. E. Brocker, P. E. Sieber, D. H. Werner, "A compact, low-profile metasurface-enabled antenna for wearable medical body-area network devices," IEEE Transactions on Antennas and Propagation, 62(8), pp. 4021–4030, 2014.

[8] K. Kamardin, M. K. A. Rahim, P. S. Hall, N. A. Samsuri, T. A. Latef, M. H. Ullah, "Textile artificial magnetic conductor jacket for transmission enhancement between antennas under bending and wetness measurements," Journal of Applied Physics A, 122(4), pp. 423, 2016.

[9] IEEE 802.11 Wireless LAN Working Group [Online]. Available: www.ieee802.org/11/.

[10] Z. H. Jiang, D. E. Brocker, P. E. Sieber, D. H. Werner, "A compact, low-profile metasurface-enabled antenna for wearable medicalbody-area network devices," IEEE Transactions on Antennas and Propagation, 62(8), pp. 4021–4030, 2014.

[11] U. Ali, S. Ullah, J. Khan, M. Shafi, B. Kamal, A. Basir, J. A. Flint, R. D. Seager, "Design and SAR analysis of wearable antenna on various parts of human body, using conventional and artificial ground planes," Journal of Electrical Engineering & Technology, 12(1), pp. 317–328, 2017.

[12] Z. H. Jiang, D. H. Werner, "Robust low-profile metasurface-enabled wearable antennas for off-body communications," 2014 8th European Conference on Antennas and Propagation (EuCAP). IEEE, 2014, pp. 21–24.

[13] S. M. Saeed, C. A. Balanis, C. R. Birtcher, A. C. Durgun, H. N. Shaman, "Wearable flexible reconfigurable antenna integrated with artificial magnetic conductor," IEEE Antennas and Wireless Propagation Letters, 16, pp. 2396–2399, 2017.

[14] G. Gao, B. Hu, S. Wang, C. Yang, "Wearable planar inverted-F antenna with stable characteristic and low specific absorption rate," Microwave and Optical Technology Letters, 60(4), pp. 876–882, 2018.

[15] F. N. Giman, P. J. Soh, M. F. Jamlos, H. Lago, A. A. Al-Hadi, M. Abdulmalek, N. Abdulaziz, "Conformal dual-band textile antenna with metasurface for WBAN application," Applied Physics A, 123(1), 32, 2017.

[16] S. Yan, P. J. Soh, G. A. Vandenbosch, "Wearable dual-band magneto-electric dipole antenna for WBAN/WLAN applications," IEEE Transactions on Antennas and Propagation, 63(9), pp. 4165–4169, 2015.

[17] H. Malekpoor, S. Jam, "Improved radiation performance of low profile printed slot antenna using wideband planar AMC surface," IEEE Transactions on Antennas and Propagation, 64(11), pp. 4626–4638, 2016.

[18] X. Liu, Y. Di, H. Liu, Z. Wu, M. M. Tentzeris, "A planar windmill like broadband antenna equipped with artificial magnetic conductor for off-body communications," IEEE Transactions on Antennas and Propagation, 15, pp. 64–67, 2016.

[19] F. Wang, T. Arslan, "A wearable ultra-wideband monopole antenna with flexible artificial magnetic conductor," 2016 Loughborough Antennas & Propagation Conference (LAPC). IEEE, 2016, pp. 1–5.

[20] A. De, B. Roy, A. Bhattacharya, A. K. Bhattacharjee, "Bandwidth-enhanced ultra-wide band wearable textile antenna for various WBAN and Internet of Things (IoT) applications," Radio Science, 56, p. e2021RS007315, 2021, https://doi.org/10.1029/2021RS007315.

[21] K. N. Paracha, S. K. Abdul Rahim, P. J. Soh, "Wearable antennas: A review of materials, structures and innovative features for autonomous communication and sensing," IEEE Access, 7, pp. 56694–56712, 2019.

[22] S. Bhattacharjee, M. Midya, S. R. Bhadra Chaudhuri, M. Mitra, "Impedance Matching Technique for Wearable Antennas Using Metamaterial Ground," 2018 IEEE MTT-S International Microwave and RF Conference (IMaRC), 1–4.

[23] S. Roy, U. Chakraborty, "Metamaterial based dual wide band wearable Antenna for wireless applications," Wireless Personal Communications: An International Journal, 2019, https://doi.org/10.1007/s11277-019-06206-3.

[24] F. Raval, S. Purohit, Y. P. Kosta, "Dual-band wearable antenna using split ring resonator," Waves in Random and Complex Media, 26(2), pp. 235–242, 2016, doi: 10.1080/17455030.2015.1137374.

[25] M. I. Ahmed, Mona F. Ahmed, A. A. Shaalan, "A novel wearable metamaterial fractal antenna for wireless applications," IEEE Middle East Conference on Antennas and Propagation (MECAP), 2016, doi: 10.1109/MECAP.2016.7790096.

[26] S. Yan, Guy A. E. Vandenbosch, "Radiation pattern reconfigurable wearable antenna based on metamaterial structure," IEEE Antennas and Wireless Propagation Letters, 15, 2016, doi: 10.1109/LAWP.2016.2528299.

[27] Sen Yan, Ping Jack Soh and Guy A. E. Vandenbosch, "Compact All-Textile Dual-band Antenna Loaded with Metamaterial Inspired Structure," IEEE Antennas and Wireless Propagation Letters, 14. doi: 10.1109/LAWP.2014.2370254.

[28] F. J. Herraiz-Martinez, V. Gonzalez-Posadas, L. E. Garcia-Munoz, D. Segovia-Vargas, "Multifrequency and dual-mode patch antennas partially filled with left-handed structures," IEEE Transactions on Antennas and Propagation, 56, no. 8, pp. 2527–2539, 2008. doi: 10.1109/TAP.2008.927518.

[29] T. A. Elwi, "Printed microwave metamaterial-antenna circuitries on nickel oxide polymerized palm fiber substrates," Scientific Reports, 9, p. 2174, 2019. doi: 10.1038/s41598-019-39736-8.

[30] T. A. Elwi, Z. Asaad Abdul Hassain, O. Almukhtar Tawfeeq, "Hilbert metamaterial printed antenna based on organic substrates for energy harvesting," IET Microwaves, Antennas & Propagation Research Article, doi: 10.1049/iet-map.2018.5948.

[31] M. V. Rao, B. T. Phani Madhav, T. Anilkumar, B. Prudhvinadh, "Circularly polarized flexible antenna on liquid crystal polymer substrate material with metamaterial loading," First published 10 October 2019, https://doi.org/10.1002/mop.32088.

[32] J. G. Joshi, S. S. Pattnaik, S. Devi, "Metamaterial embedded wearable rectangular microstrip patch antenna," International Journal of Antennas and Propagation, 2012, Article 974315, 9 pages, doi: 10.1155/2012/974315.

[33] A. Al'u, F. Bilotti, N. Engheta, L. Vegni, "Subwavelength, compact, resonant patch antennas loaded with metamaterials," IEEE Transactions on Antennas and Propagation, 55(1), pp. 13–25, 2007.

[34] P. Y. Chen, A. Alu, "Dual-mode miniaturized elliptical patch antenna with μ-negative metamaterials," IEEE Antennas & Wireless Propagation Letters, 9, pp. 351–354, 2010.

[35] M. Alam, M. Siddique, B. K. Kanaujia, M. T. Beg, S. Kumar, K. Rambabu, "Meta-surface enabled hepta-band compact antenna for wearable applications," IET Microwaves, Antennas & Propagation, 13(13), pp. 2372–2379, doi: 10.1049/iet-map.2018.6212.

[36] Y. Dong, H. Toyao, T. Itoh, "Compact circularly-polarized patch antenna loaded with metamaterial structures," IEEE Transactions on Antennas and Propagation, 59(11), pp. 4329–4333, 2011.

[37] H. Yalduz, B. Koç, L. Kuzu, M. Turkmen, "An ultrawide band low SAR flexible metasurface enabled antenna for WBAN applications," Applied Physics A, 125, p. 609, 2019, https://doi.org/10.1007/s00339-019-2902-4

[38] M. A. B. Abbasi, S. S. Nikolaou, M. A. Antoniades, M. N. Stevanović, P. Vryonides, "Compact EBG-backed planar monopole for BAN wearable applications," IEEE Transactions on Antennas and Propagation, 65(2), 453–463, 2017.

[39] S. M. Saeed, C. A. Balanis, C. R. Birtcher, A. C. Durgun, H. N. Shaman, "Wearable flexible reconfigurable antenna integrated with artificial magnetic conductor," IEEE Antennas and Wireless Propagation Letters, 16, 2396–2399, 2017.

Chapter 9

An Efficient Wearable Antenna Deploying Different Geometry for Wireless Capsule Endoscopy

Vanitha Rani Rentapalli and Bappadittya Roy

9.1 INTRODUCTION

The wearable antennas have tremendous activation power with the human body, as shown in [1, 2]. With this antenna, a high dielectric constant and losses have been determined to effect the efficiency of the antenna. Wireless capsule endoscopy (WCE) has fascinated increasing consideration for diagnosing causes of gastrointestinal diseases for patients [3]. The WCE was designed mainly to identify the disorders that were caused due to cancer and polyps [4]. In this WCE method a capsule is equipped with a camera and that connection is swallowed into the patient's intestine. From here, the antenna would collect the image from the human body and convey it to external system; there it would store the required data related to the human body [5, 6]. Placing an antenna inside the human body causes huge challenges, because the antenna factors like size, bandwidth and security need to be taken into consideration for capturing human tissues of a body [7]. An endoscopy system with high transmit image data and resolution could be capable to get the wider bandwidths. So always the system needs to process the human data with a high transmission data rate [8, 9]. The bandwidth of a conformal antenna is large for printing the outer wall of the capsule for exhibiting favorable data acquisition conditions to the wearable antenna, but the expanding of the bandwidth is not considered all the time due to security reasons; hence the body area network needs special attention to operate at a high frequency of operation. An ultra-wide bandwidth (UWB) endoscope antenna system encompassing chirp spread spectrum modulation (CSSM) determined the required data rate, which could be more than 10 Mbps, along with the measured low-power consumption. Moreover, the design of a UWB antenna for incorporating with a capsule is fundamentally a complex task because the size of the capsule is electrically small when compared to the proposed antenna [10]. But for an elongated wideband operation, the measurable quarter-pole antennas are bounded with the capsule; this would be the perfect parameter to retain their cylindrical volume and to inhabit the space inside the capsule [11]. This elongation rises the size of the capsule, even though it makes it hard to

DOI: 10.1201/9781003459880-9

swallow. The capsule endoscopy system should need a transmitter with compact size and need to consume ultralow power for offering an inclusive wide bandwidth. So, it would be enhanced for spectral signal variation through the human body [12]. The Wireless Application Protocol is used for future body area network communications to investigate telemedicine for capsule endoscopy systems [13, 14]. The high-quality images which are transmitted at a high quality rate are being restricted by the capsule's signal to noise ratio (SNR) of the system, which are spaced between the capsule and the receiver [15]. The outer wall of the loop antennas has been proposed for endoscopy capsule systems, which incorporates capsule geometry in coordination with the antenna and could give a better performance for the vital development of the capsule endoscopy in the best proposed way as compared with the proposed antenna [16]. The capsule endoscopy antenna is fabricated spectrally with the inner portions of the capsule, which deliberately evades direct contact with human tissues. The selected bandwidth for the endoscopy system is not that much wider for transmission of image data with high resolution. WCE is a painless technique to eliminate uncomfortable procedures in the endoscopy process [17]. This process does not inhibit the normal actions of the patient. This method interacts with the distant areas of human intestine internally which could not be possible with the traditional endoscopy. A novel differentially fed compact dual-band implantable antenna for biotelemetry applications has been proposed in [18]. The techniques that are used for capsule endoscopy are radio frequency (RF) signals, magnetic fields and the other image processing methods [19]. The research on scientific technical visualization brought more advantageous developments towards capsule endoscopy resolved methods [20, 21]. The research group has been formed to invent body area networks in order to inculcate communication standards and [22] to enhance the power devices operating near the human body. The capsule endoscopy technology is implementing highly powerful implanting techniques for high results. In this chapter we were going to describe a systematical dual-loop antenna, having UWB performance for wearable body area network applications [23–25]. The system that has been designed for ultra-band applications could operate in industrial, scientific and medical applications [26, 27]. The main function of designing a UWB antenna is to resolve the frequency shift located in high-speed data transmission. This can be attained by adding the patch and feeding for spinning the track of an antenna as enlightened in [28, 29]. The advanced antenna for the specific function would be manufactured and measured to attain required band of the system [30]. An elongated surface-mountable dual-loop antenna with determined UWB variations of capsule endoscopy have been discussed; the design and its significant performance of the implantable capsule antenna are mentioned in Section 9.1. Section 9.2 explores the performance of the antenna and the impact of the capsule for defining its importance and shell thickness on its evitable variation. Wearable antenna design principles along with spectrum formations are mentioned in

Section 9.3. Measurement setup and the results of a capsule endoscopy design are shown in Section 9.4. Design challenges and salient features are mentioned in Section 9.5, and finally the conclusion is provided in Section 6.6.

9.2 CAPSULE ANTENNA DESIGN

The proposed system was considered a capsule antenna with some specific function that allows the size 10×15 mm^2 (inside diameter and length) to enhance the inner parameters of the capsule as shown in Figure 9.1(a) [31].

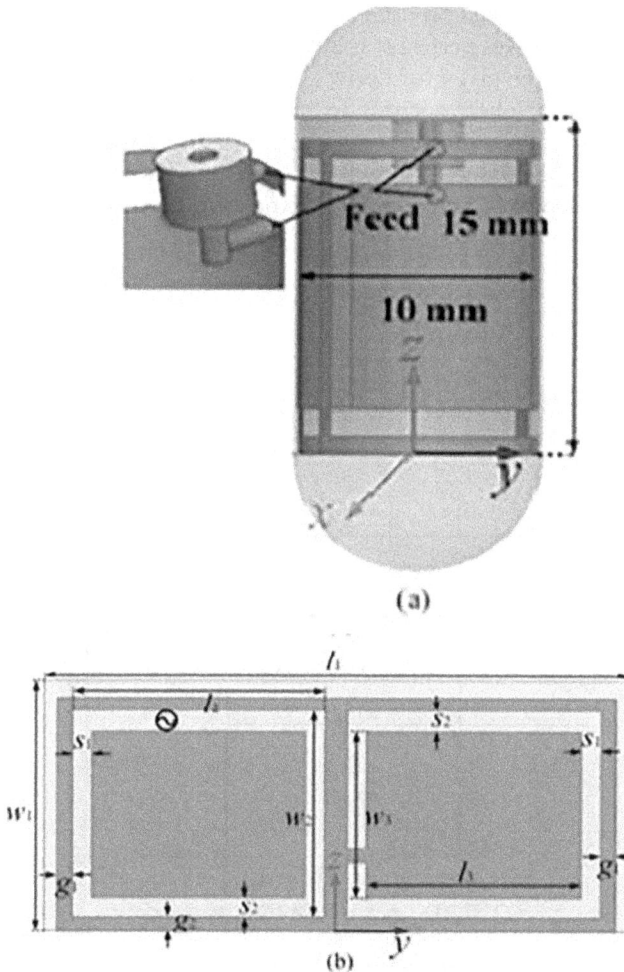

Figure 9.1 Proposed design of ultra-wideband antenna [1]. (a) Envelope of capsule antenna. (b) The wrapping conditions of an antenna.

The central part of the capsule antenna is aligned to the rectangular patch at the center of the feeding patch, which can be visualized distinctively in the top left portion of an antenna [32, 33]. Antennas for biomedical applications have been proposed in [34, 35]. The capsule antenna was first designed on Rogers 5880, t_r-2.2, tan δ = 0.008 and thickness 0.254, as given in Figure 9.1(b). The dielectric properties that were obtained from the design were very close to the air by maintaining the inner diameter 10 mm [36]. The dual-core capsule antenna connected to the assigned patch near the feeding loop also had same size and could function as a loop. The loop that is joined to a rectangular patch can attain different purposes as shown in [37, 38]. The feeding patch and outer patch are used for to balance the impedance matching in a proposed antenna system. A loading coil (LC) is introduced for an additional resonance in the circuit, which gives perfect stability to the system. The parameters of an antenna that have been given for perfect optimization are shown in Table 9.1. The simulation atmosphere surrounded by an antenna is a muscle box with dimensions 60 mm × 60 mm × 70 mm, which would always concur with the variations in the body tissue as shown in Figure 9.2. The muscle box is always oriented with the capsule antenna for specific loop variation [39, 40].

Table 9.1 Detailed Dimensions in Millimeters [1]

Parameter	Value	Parameter	Value
L_1	31.4	W_1	15
L_2	13.6	W_2	12.4
L_3	11.5	W_3	10
S_1	1.05	S_2	1.2
G_1	0.8	G_2	0.8

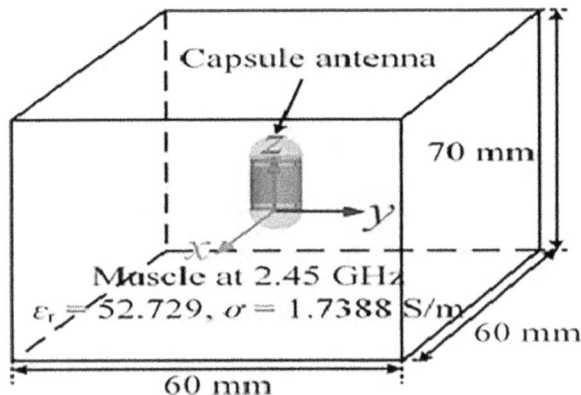

Figure 9.2 Simulation characteristic arrangement of an antenna using HFSS [1].

9.2.1 Working Principle

The working principle and its presentation of the proposed antenna would are shown in Figure 9.3(a). Structure 1 and structure 2 of the antenna are formed like a perfect pattern to embed the same as the proposed antenna [41]. The two structural representations of an antenna are being analyzed to represent the principle of the proposed antenna design [42, 43]. The capsule antenna delivers an important variation for preserving of the orientation along the Z axis; with this formation the capsule is aligned equally by the XOZ plane as mentioned in [44]. The parameters of both the structures remain unchanged as shown in [1]. Adding a patch along with varying the feed revolution enhances the impedance matching of structure 1 and 2 as shown in Figure 9.3(b) that gives good stability to the system [45]. Moreover, by inserting an LC loading inside the loop, additional resonance would occur

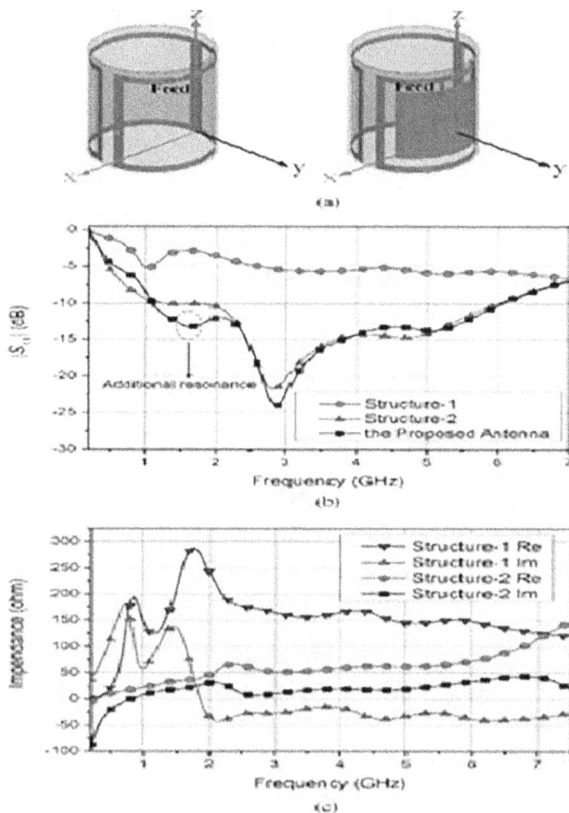

Figure 9.3 Differences among the structures 1 and 2 and assessment between them compared to the proposed antenna. (a) Two functional erections. (b) Measured result | S_{11} | for the two structures compared to the proposed antenna. (c) Input impedance of two structural parameters [3].

that improves the performance of the system. The input impedance is the known parameter for both the structures, especially while considering the operation of an antenna [46, 47]. The structural performance would be calculated with HFSS and compared in Figure 9.3(c). It was shown from the figure that the impedance of structure 1 varies from 126 ohms to 288 ohms and ranges from 1.11 to 6.03 GHz, whereas the variation of structure 2, from 25 to 72 ohms [48]. It may be mentioned here that 50 ohms is the impedance value in usual cases. The performance of the two structures along with impedance-varying parameters decides the function of the capsule antenna endoscopy for UWB applications as shown in [49, 50]

9.2.2 Effects of the Slot on the Antenna

The operating condition of an antenna would be evaluated with the slots of a capsule in Y and Z directions. The resonant frequencies from the LC circuit are marked at f_1, f_2 and f_3. If there is any variation in slot s_1, the resonant frequencies remain unchanged; then the impedance matching of f_1 and f_2 improves gradually [51, 52]. The proposed antenna would vary with the adjusting of s_1 without changing the f_1, f_2 and f_3 for proper impedance matching that would vary in the circuit system [53, 54]. Furthermore, the consequence of s_2 as a variable constraint shows the changes in impedance matching; these changes would be shown in the radiation pattern. The slots of an antenna depict the function of a wearable antenna design; the major radiation pattern and its radiating properties are mainly exhibited in the structure of an antenna. The main function of an antenna design is to radiate electromagnetic energy in all the directions equally by giving proper preference to the other parameters of an antenna, including E and H planes. The capsule antenna which is injected into the human body can detect the abnormalities in the human tissues of a body.

If S_2 decreases at a certain extent, the impedance matching of f_1 would suffer with rapid deterioration. This reduces the resonant frequency impedance matching of f_2 at the certain rate, even the function of f_3 remains the same [55, 56]. The working frequency of a capsule antenna kept at 2.05 GHz for perfect consideration of operation at $S_1 = 1.05$ mm, $S_2 = 1.2$ mm is considered [57, 58]. This condition exhibits the pertaining conditions of an antenna for better considerable variations detailed in [59, 60].

9.3 WEARABLE ANTENNA DESIGN

The wearable antenna for dual-mode capsule antenna operation, as shown in Figure 9.4, in a body area network at 2.45 GHz for medical and scientific strategies has been included for presentation in a specific manner. Most of

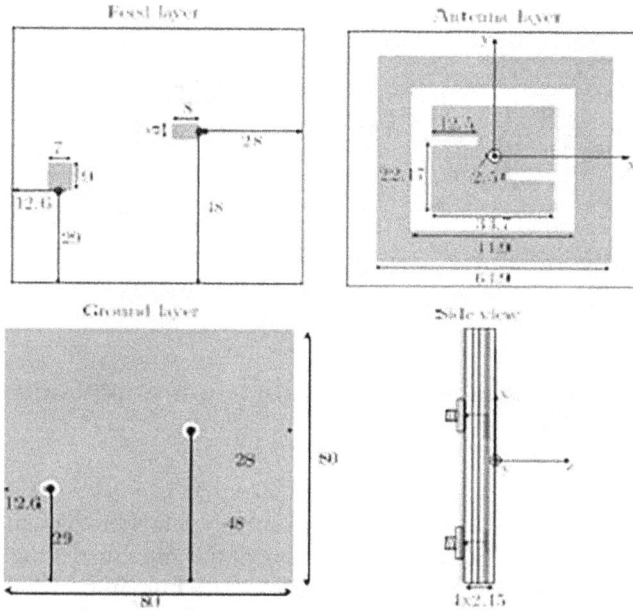

Figure 9.4 Wearable antenna geometry for three metallic layers [4].

the advantageous health care technologies have been developed for the specific standards of body area networks (BANs) [61]. As per recent advances, BANs were implemented for health and emerging technologies based on the requirements of specific application implementations. Wearable antennas are the required components for BANs [62]. But comparatively the radiation characteristics of the human body changes were different to evaluate the creation of radiation properties shown in Figure 9.5. The on-body manner uses a monopole-like omnidirectional pattern, whereas off-body representation uses the broadside radiation pattern [63, 64]. But for the configured sustained implementations, the dual-mode implementations use a wearable microstrip antenna at 2.45 GHz. The on-body and off-body mode of a wearable antenna could easily gain the performance of the radiating elements of the capsule endoscopy system; the modes were replaced with a ring and a patch as radiating elements [65, 66].

The wearable antenna design properties would achieve the perfect accurate radiation efficiency to illustrate the validation strategy for on-body and off-body BAN scientific communication systems [67]. The antenna features electrocoupling for perfect magnetic stability between the human body and an antenna. Design of an ultra-wideband system for in-body wireless communications has been proposed in [68]. Wearable antenna are the essential components in wide BANS (WBANs) for capsule endoscopy due to their enhancing

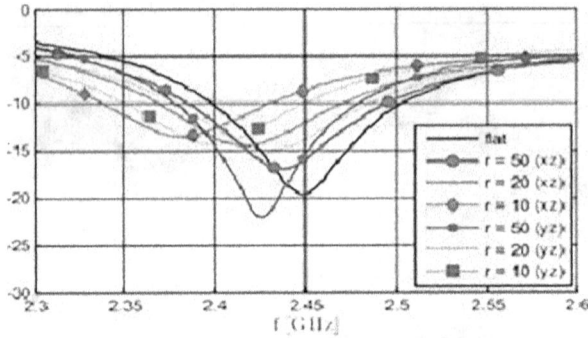

Figure 9.5 S parameter of the measured dual-band antenna for different bending radius (r in cm) [2].

features depicted as flexibility, ease of incorporation into wear and patient relief. The capsule endoscopy antenna mainly depends on reduction of specific absorption rate. It prevents the coupling of the antenna through dielectric layers through muscle and the internal tissues of a human body, which could change the performance of a system. The wearable antenna design gives information about the radiation pattern of a reference design from on- or off-body implementations. The structure of three metallic layers for feeding capacitors provide antenna input matching with proper radiation characteristics as shown in Figure 9.6. The designing of an antenna with proper radiation efficiency connected to the ring width could assure the good efficiencies of the ring and patch of a wearable capsule antenna as mentioned in [69, 70].

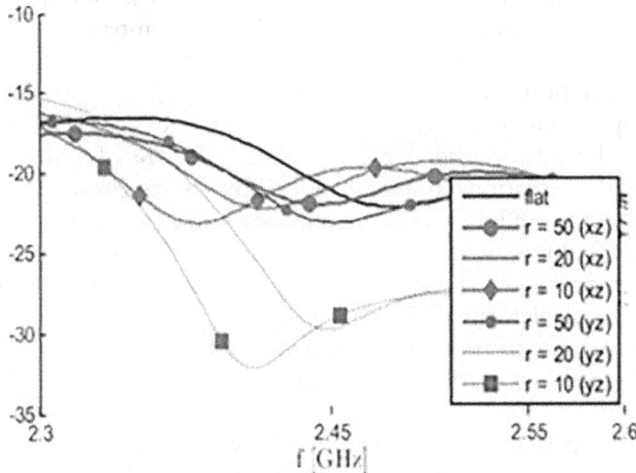

Figure 9.6 The radiation pattern of a wearable antenna at the frequencies at (a) 2.45 GHz and (b) 5.85 GHz [9].

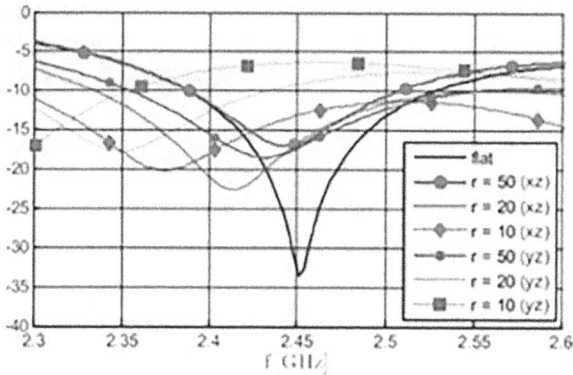

Figure 9.6 (Continued)

9.4 SIMULATION RESULTS

The capsule antenna that needs to be fabricated with its measurement setup is shown in Figure 9.7. The evaluation between the re-creation and the calculated results are shown in Figure 9.8. The frequency of the procedure and the measured bandwidth would maintain a particular region, even though the contours of the plots are inconsistent [71, 72]. Two attributes need to be considered for the operation of an antenna with a given range. One is coaxial cable with a rated extent of 10 cm is used during the evaluation. Secondly, a coaxial weld power cable creates a larger variation on the input impedance [73]. The simulated far-field wearable antenna could attain the bandwidth 2.45 GHz at gain of −12.6 dB, which is the desired bandwidth for the perfect operation of a capsule endoscopy as shown in Figure 9.9. As per medical observations of a patient, the massive specific absorption rate (SAR) of a

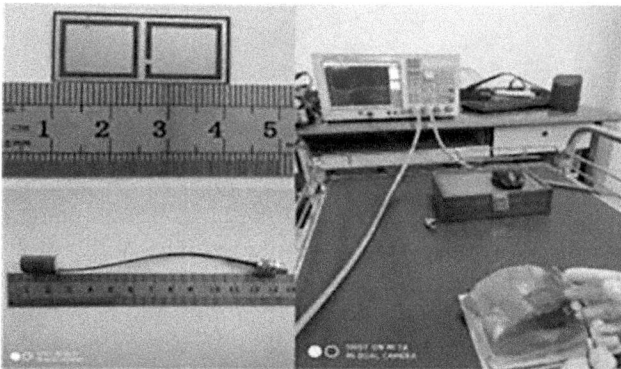

Figure 9.7 Fabricated ultrawideband antenna and the measurement setup [2].

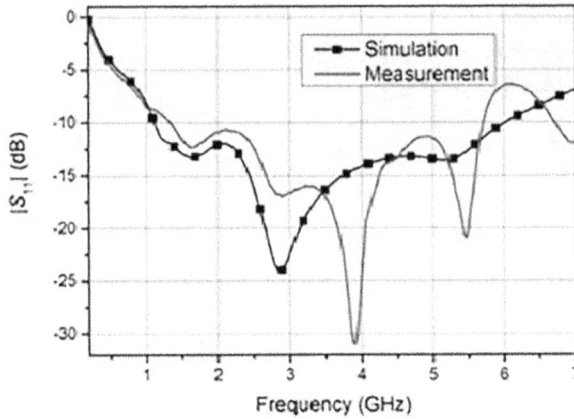

Figure 9.8 The S parameter characteristics |S₁₁| capsule antenna at 2.45 GHz [1].

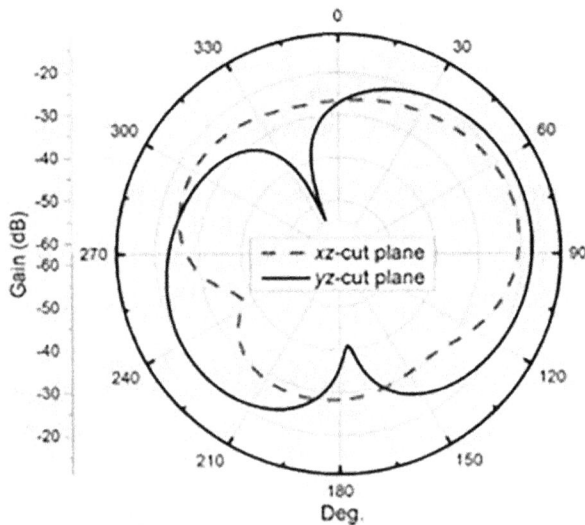

Figure 9.9 The gain plot for the measured proposed of the proposed antenna [17].

capsule antenna should not exceed 1.6 W/kg [74]. If the antenna is operating at 1 W and the SAR muscle tissue should be 215.7 W/Kg at 2.45 GHz, it anonymously decreases the input power rating up to 7.42 mW. The comparison between our work with all the specifications and previously designed antennas variations are shown in Table 9.2. As per simulation results, we could predict that the proposed wearable antenna could achieve satisfactory bandwidth-varying response as shown in Figure 9.10 and Figure 9.11, and

Table 9.2 Comparisons among the Antenna Type, Resonance Frequency, Impedance Band Width and Gain of a Capsule Antenna [1]

Ref.	Antenna Type	Frequency (GHz)	Band (MHz)	Gain dBi
[1]	Outer loop wall	0.5	260	
[2]	Outer loop wall	0.403	541	−31.5
[3]	Outer loop wall	0.5	2200	
[4]	Outer loop wall	0.433	795	−35
[5]	Inner loop wall	2.45	4310	−27.2
[6]	Inner wall patch	0.9150	81	−21
[7]	Inner wall dipole	0.402	158	−37
[8]	Inner wall	0.4	31	−31
[9]	Inner wall dipole	2.45	150	−18.5
[10]	Inner wall	0.402	>15	−22

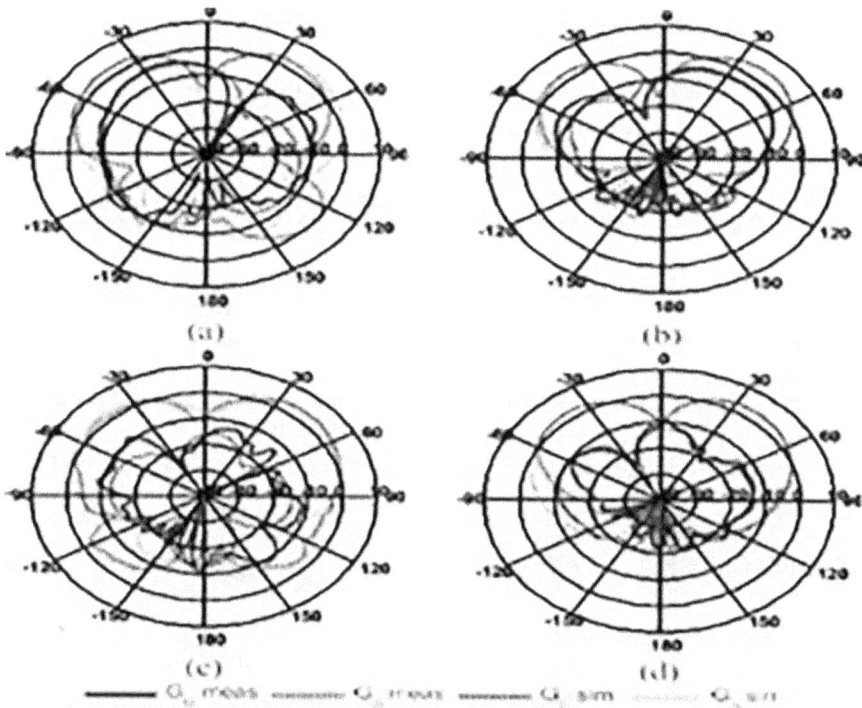

Figure 9.10 The radiation pattern of a human body in elongated mode at 2.45 GHz (a) and (b) yz plane and (c) and (d) xz plane [11].

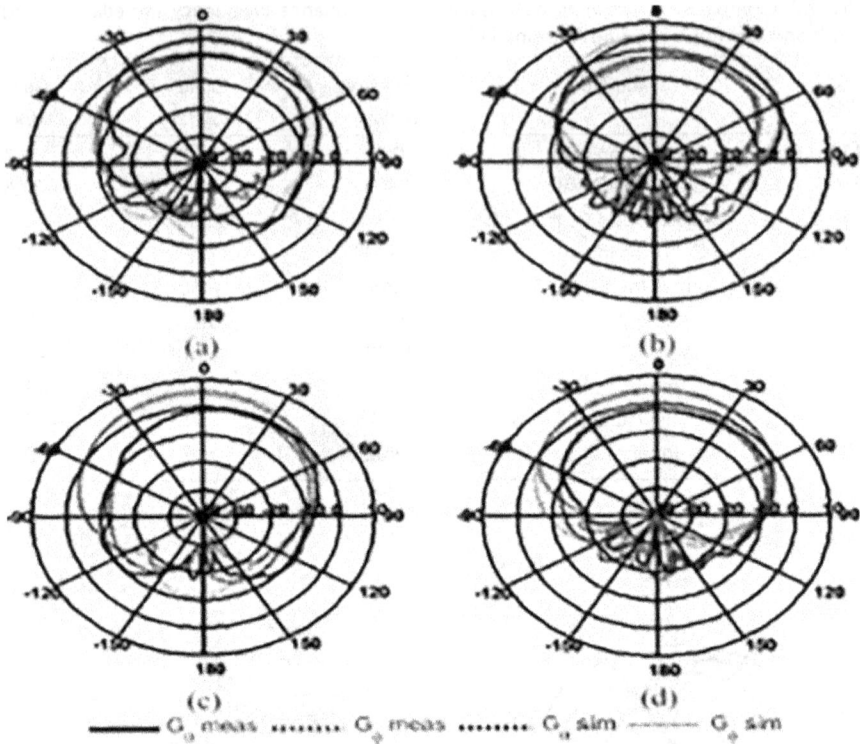

Figure 9.11 The radiation pattern at diminished mode at 2.45 GHz (a) and (b) yz plane and (c) and (d) xz plane [15].

the S parameters shown from these variations demonstrate the exhibited obvious advantage would arise in impedance matching over the other antennas. The simulation results have described the operating performance of a wireless wearable capsule endoscopy system for medical observation of a patient with body disorders in a specified manner. With capsule endoscopy, the human body tissues would be detected and corrected as shown in [75].

9.5 DESIGN CHALLENGES

As we design any kind of system, we need to consider design challenges for particular applications. The design methodologies are changing from time to time; with this, the scientific technological research is also improving [76]. To design any kind of system, including health care, sports, child monitoring and consumer electronics, with wearable antenna, first we need to consider the system designing parameters. If we have surplus knowledge in scientific and technical standards, we can invent anything new with our progressive research knowledge. Integration of wireless functions in electronics advances the fabrication process and attractive features of flexible antenna categories [77–79]. Nowadays researchers are facing problems in implementing new scientific technology, because there might be a problem with a lack of knowledge in using sensors and network adaptations which could not coincide with the technology design principles [80]. A wearable antenna designer needs to gain lots of knowledge on all aspects before implementing system methodologies. As per the latest technology the wearable electronics need to gain the significant interest from academia and industrial applications [81–83]. The design and realization of wearable antennas should integrate with the significant research area of interest. The antenna design should aware of flexibility, robustness and tolerance to extreme reproductions. If a researcher implements any kind of technology for developing of antennas with flexible technologies, then only the design parameters would reach up to the mark [84–88].

9.6 CONCLUSION

In this chapter, we have presented a UWB wearable capsule antenna for a wireless endoscopy system, which was investigated as per the antenna size with an adequate bandwidth. The dual-loop symmetrical wearable antenna with the feeding patch at the middle of the feeding variation was presented. The impedance matching of the system should match the direction of the feed rotated by 90 degrees, which is the necessary design parameter for the perfect operation of an antenna. A dual-loop antenna is presented on the outer side of the capsule, configured mainly to diminish the magnitude of the capsule by inner antenna design. The proposed antenna uses an omnidirectional radiation pattern for obtaining balanced endoscopy communications. To add a surplus resonance to the capsule design, an LC loading is provided for an antenna with UWB performance. A dual-loop wearable antenna has been created for BAN application to get a good arrangement between the simulation and experimental results. It was clearly demonstrated that particular attainment of a wearable antenna concept is suitable for a BAN that further resembles to human body communication systems. This chapter outlines the design of a dual-mode capsule antenna with wideband applications.

The capsule endoscopy antenna helps us to detect problems in the human body and could easily sense the inner human tissues; with that connected circuit, the performance of the design would be easily identified and verified with the HFSS. The working mechanism and measurements of the dual antenna were investigated in a special implanted device. The results indicate that the proposed antenna has a satisfactory response in case of enormous bandwidth and exhibits various developing proportions in gain over other antennas.

REFERENCES

[1] Tu-Tuan Le and Tae-Yeoul Yun, "Miniaturization of a dual-band wearable antenna for WBAN applications," IEEE Antennas and Wave Propagation Letters, vol. 19, no. 8, August 2020.

[2] Jiangli Shang and Ying Yu, "An ultra-wideband capsule antenna for biomedical applications," IEEE Antennas and Wireless Propagation Letters, vol. 18, no. 12, December 2019.

[3] M. S. Miah, A. N. Khan, C. Icheln, K. Haneda, and K. Takizawa, "Antenna system design for improved wireless capsule endoscope links at 433MHz," IEEE Transactions on Antennas and Propagation, vol. 67, no. 4, pp. 2687–2699, April 2019.

[4] Martina Barbi, Concepcion Garcia-Pardo, and Andrea Nev Erez, "UWB RSS based localisation for capsule endoscopy using a multilayer phantom and in VIVO measurements," IEEE Transactions on Antennas and Propagation, vol. 67, no. 8, August 2019.

[5] J. Wang, M. Leach, E. G. Lim, Z. Wang, R. Pei, and Y. Huang, "An implantable and conformal antenna for wireless capsule endoscopy," IEEE Antennas Wireless Propagation Letters, vol. 17, no. 7, pp. 1153–1157, July 2018.

[6] Ke Zhang and Changrong Liu, "A conformal differentially fed antenna for ingestible capsule system," IEEE Transactions on Antennas and Propagation, vol. 66, no. 4, April 2018.

[7] Rongqiang Li, Yong-Xin Guo, and Guohong Du, "A conformal circularly polarized antenna for wireless capsule endoscope systems," IEEE Transactions on Antennas and Propagation, vol. 66, no. 4, pp. 2119–2124, 2018.

[8] Ruikai Zhang, Yali Zhang, and Carman CY Poon, "Polyp detection during colonoscopy using a regression based convolutional neural network with a tracker," Pattern Recognition, vol. 83, 2018.

[9] Y. Li, Y. X. Guo, and S. Xiao, "Orientation insensitive antenna with polarization diversity for wireless capsule endoscope system," Transactions on Antennas and Propagation, vol. 3743, 2017.

[10] Z. Bao, Y. X. Guo, and R. Mitra, "An ultra-wideband conformal capsule antenna with stable impedance matching," IEEE Transactions on Antennas and Propagation, vol. 65, no. 10, pp. 5086–5094, 2017.

[11] R. Das and H. Yoo, "A multiband antenna associating wireless monitoring and nonleaky wireless power transfer system for biomedical implants," IEEE Transactions Microwave Theory Technology, vol. 65, no. 7, pp. 2485–2495, 2017.

[12] R. Das and H. Yoo, "A wideband circularly polarized conformal endoscopic antenna system for high," Transactions on Antennas and Propagation, vol. 65, no. 6, p. 2826, 2017.

[13] Z. Duan and L. Xu, "Dual-band implantable antenna with circular polarization property for ingestible capsule application," Electronics Letters, vol. 53, no. 16, pp. 1090–1092, 2017.

[14] D. Nikolayev, M. Zhadobov, L. L. Coq, P. Karban, and R. Sauleau, "Robust ultra-miniature capsule antenna for ingestible and implantable applications," IEEE Transactions on Antennas and Propagation, vol. 65, no. 11, pp. 6107–6119, 2017.

[15] K. Zhang, C. Liu, X. Yang, X. Liu, and H. Guo, "An ingestible capsule system for in-body core temperature monitoring," Microwave and Optical Technology Letters, vol. 59, no. 10, pp. 2670–2675, 2017.

[16] Z. Bao, Y.-X. Guo, and R. Mittra, "Single-layer dual-/tri-band inverted antennas for conformal capsule type of applications," EEE Transactions on Antennas and Propagation, vol. 65, no. 12, pp. 7257–7265, 2017.

[17] V. T. Nguyen and C. W. Jung, "Radiation-pattern reconfigurable antenna for medical implants in Med Radio band," IEEE Antennas and Wireless Propagation Letters, vol. 15, pp. 106–109, 2016.

[18] Y. Liu, Y. Chen, H. Lin, and F. H. Juwono, "A novel differentially fed compact dual-band implantable antenna for biotelemetry applications," IEEE Antennas and Wireless Propagation Letters, vol. 15, pp. 1791–1794, 2016.

[19] Rula S. Alrawashdeh and Yi Huang, "A Broadband flexible implantable loop antenna with complementary split ring resonators," IEEE Antennas and Wireless Propagation Letters, vol. 14, 2015.

[20] L. J. Xu, Y. X. Guo, and W. Wu, "Bandwidth enhancement of an implantable antenna," IEEE Antennas and Wireless Propagation Letters, vol. 14, pp. 1510–1513, 2015.

[21] Thotahewa, Kasun MS, Jean-Michel Redoutè, and Mehmet Rasit, "Propagation, power absorption, and temperature analysis of UWB wireless capsule endoscopy devices operating in the human body," IEEE Transactions on Microwave Theory and Techniques, vol. 11, pp. 3823–3833, 2015.

[22] Carmen C. Y. Poon, Benny P. L. Lo, and Mehmet Resit Yuce, "Body sensor networks: In the era of big data and beyond," IEEE Reviews in Biomedical Engineering, vol 8, 2015.

[23] C. Liu, Y.-X. Guo, and S. Xiao, "Circularly polarized helical antenna for ism-band ingestible capsule endoscope systems," IEEE Transactions on Antennas and Propagation, vol. 62, no. 12, pp. 6027–6039, 2014.

[24] C. Schmidt, F. Casado, A. Arriola, I. Ortego, P. D. Bradley, and D. Valderas, "Broadband UHF implanted 3-D conformal antenna design and characterization for in-off body wireless links," IEEE Transactions on Antennas and Propagation, vol. 62, no. 3, pp. 1433–1444, March 2014. doi: 10.1109/TAP.2013.2295816.

[25] K. L.-L. Roman, G. Vermeeren, A. Thielens, W. Joseph, and L. Martens, "Characterization of path loss and absorption for a wireless radio frequency link between an in-body endoscopy capsule and a receiver outside the body," EURASIP Journal on Wireless Communications and Networking, vol. 2014, no. 1, p. 21, 2014.

[26] H. Rajagopalan and Y. Rahmat-Samii, "Wireless medical telemetry characterization for ingestible capsule antenna designs," IEEE Antennas and Wireless Propagation Letters, vol. 11, pp. 1679–1682, 2013.

[27] E. Y. Chow, M. M. Morris, and P. P. Irazoqui, "Implantable RF medical devices: The benefits of high-speed communication and much greater communication distances in biomedical applications," IEEE Microwave Magazine, vol. 14, no. 4, pp. 64–73, 2013.

[28] M. R. Basar et al., "The use of a human body model to determine the variation of path losses in the human body channel in wireless capsule endoscopy," Progress In Electromagnetics Research, vol. 133, pp. 495–513, 2013.

[29] E. G. Lim, J. C. Wang, Z. Wang, T. Tillo, and K. L. Man, "The UHF band in-body antennas for wireless capsule endoscopy," Engineering Letters, vol. 21, no. 2, pp. 72–80, 2013.

[30] A. Kiourti and K. S. Nikita, "A review of implantable patch antennas for biomedical telemetry: Challenges and solutions [wireless corner]," IEEE Antennas and Propagation Magazine, vol. 54, no. 3, pp. 210–228, Jun. 2012.

[31] X. Cheng, J. Wu, R. Blank, and D. E. Senior, "An omnidirectional wrappable compact patch antenna for wireless endoscope applications," IEEE Antennas and Wireless Propagation Letters, vol. 11, pp. 1667–1670, 2012.

[32] H. Rajagopalan and Y. Rahmat-Samii, "Wireless medical telemetry characterization for ingestible capsule antenna design," IEEE Antennas and Wireless Propagation Letters, vol. 11, pp. 1679–1682, 2012.

[33] Z. Duan, Y.-X. Guo, R.-F. Xue, M. Je, and D.-L. Kwong, "Differentially fed dual-band implantable antenna for biomedical applications," IEEE Transactions on Antennas and Propagation, vol. 60, no. 12, pp. 5587–5595, 2012.

[34] F. Merli et al., "Example of data telemetry for biomedical applications: An in vivo experiment," IEEE Antennas and Wireless Propagation Letters, vol. 11, pp. 1650–1654, 2012.

[35] Editorial from "Information technology in Biomedicine to biomedical and health informatics," IEEE Transactions on Information Technology in Biomedicine, vol. 16, no. 6, 2012.

[36] A. K. Ram Rakhyani, S. Mirabbasi, and M. Chiao, "Design and optimization of resonance-based efficient wireless power delivery systems for biomedical implants," IEEE Transactions on Biomedical Circuits and Systems, vol. 5, no. 1, pp. 48–63, 2011.

[37] F. Merli, L. Bolomey, J. F. Zurcher, G. Corradini, E. Meurville, and A. K. Skrivervik, "Design, realization and measurements of a miniature antenna for implantable wireless communication systems," IEEE Transactions on Antennas and Propagation, vol. 59, no. 10, pp. 3544–3555, 2011.

[38] F. J. Huang, C.-M. Lee, C.-L. Chang, L.-K. Chen, T.-C. Yo, and C.-H. Luo, "Rectenna application of miniaturized implantable antenna design for triple-band biotelemetry communication," IEEE Transactions on Antennas and Propagation, vol. 59, no. 7, pp. 2646–2653, July 2011. doi: 10.1109/TAP.2011.2152317.

[39] Francesco Merli, Léandre Bolomey, and Jean-François Zürcher, "Design, realization and measurements of a miniature antenna for implantable wireless communication systems," IEEE Transactions on Antennas and Propagation, vol. 59, no. 10, 2011.

[40] X. Cheng, D. E. Senior, C. Kim, and Y. K. Yoon, "A compact omnidirectional self-packaged patch antenna with complementary split-ring resonator loading for wireless endoscope applications," IEEE Antennas and Wireless Propagation Letters, vol. 10, pp. 1532–1535, 2011.

[41] F. Merli, "Implantable antennas for biomedical applications," PhD dissertation, Department of Electrical Engineering, EPFL University, Lausanne, 2011.

[42] M. L. Scarpello and D. Kurup, "Design of an implantable slot dipole conformal flexible antenna for biomedical applications," IEEE Transactions on Antennas and Propagation, vol. 59, no. 10, pp. 3556–3564, 2011.

[43] G. Ciuti, A. Menciassi, and P. Dario, "Capsule endoscopy: From current achievements to open challenges," IEEE Reviews in Biomedical Engineering, vol. 4, pp. 59–72, 2011.

[44] F. Merli, B. Fuchs, J. R. Mosig, and A. K. Skrivervik, "The effect of insulating layers on the performance of implanted antennas," IEEE Transactions on Antennas and Propagation, vol. 59, no. 1, pp. 21–31, 2011.

[45] S. H. Lee, J. Lee, Y. J. Yoon, S. Park, C. Cheon, K. Kim, and S. Nam, "A wide-band spiral antenna for ingestible capsule endoscope systems: Experimental results in a human phantom and a pig," IEEE Transactions on Biomedical Engineering, vol. 58, no. 6, pp. 1734–1741, 2011.

[46] C. C. Y. Poon, Q. Liu, H. Gao, W.-H. Lin, and Y.-T. Zhang, "Wearable intelligent systems for E health," Journal of Computing Science and Engineering, vol. 5, no. 3, pp. 246–256, 30 September 2011. Korean Institute of Information Scientists and Engineers.

[47] S. Yun, K. Kim, and S. Nam, "Outer-wall loop antenna for ultrawideband capsule endoscope system," Antennas and Wireless Propagation Letters, vol. 9, pp. 1135–1138, 2010.

[48] K. Kim, M. Jeon, K. Kim, J. Lee, and S. Nam, "Human body communication using chirp spread spectrum modulation," The Journal of the Korean Institute of Communication Science, vol. 35, no. 5, pp. 440–446, 2010.

[49] S. H. Lee and Y. J. Yoon, "Fat arm spiral antenna for wideband capsule endoscope systems," Radio and Wireless Symposium (RWS), pp. 579–582, 2010.

[50] S. Yun, K. Kim, and S. Nam, "Outer-wall loop antenna for ultrawideband capsule endoscope system," IEEE Antennas and Wireless Propagation Letters, vol. 9, pp. 1135–1138, 2010.

[51] K. Kim, S. Lee, E. Cho, J. Choi, and S. Nam, "Design of OOK system for wireless capsule endoscopy," IEEE ISCAS, pp. 1205–1208, 2010.

[52] F. Merli, L. Bolomey, E. Meurville, and A. K. Skrivervik, "Dual band antenna for subcutaneous telemetry applications," 2010 IEEE Antennas and Propagation Society International Symposium (APSURSI), Toronto, Canada, pp. 1–4, 2010. doi: 10.1109/APS.2010.5562054.

[53] N. Haga, K. Saito, M. Takahashi, and K. Ito, "Characteristics of cavity slot antenna for body-area networks," IEEE Transactions on Antennas and Propagation, vol. 57, no. 4, pp. 837–843, 2009.

[54] P. M. Szczypinski, R. D. Sriram, P. V. J. Sriram, and D. N. Reddy, "A model of deformable rings for interpretation of wireless capsule endoscopic videos," Medical Image Analysis, vol. 13, no. 2, pp. 312–324, 2009.

[55] Tharaka Dissanayake and Mehmet R. Yuce, "Design and evaluation of a compact antenna for implant to Air UWB communication," IEEE Antennas and Wireless Propagation Letters, vol. 8, 2009.

[56] S. Zhu and R. Langley, "Dual-band wearable textile antenna on an EBG substrate," IEEE Transactions on Antennas and Propagation, vol. 57, no. 4, pp. 926–935, 2009.

[57] P. M. Izdebski, H. Rajagopalan, and Y. Rahmat-Samii, "Conformal ingestible capsule antenna" a novel chandelier meandered design," IEEE Transactions on Antennas and Propagation, vol. 57, no. 4, pp. 900–909, 2009.

[58] Tharaka Dissanayake and R. Mehmet, "Design and evaluation of a compact antenna for implant-to-air UWB communication," IEEE Antennas and Wireless Propagation Letters, vol. 8, 2009.

[59] T. Dissanayake, K. P. Esselle, and M. R. Yuce, "Dielectric loaded impedance matching for wideband implanted antennas," IEEE Transactions on Microwave Theory and Techniques, vol. 57, no. 10, pp. 2480–2487, 2009.

[60] T. Karacolak, R. Cooper, and E. Topsakal, "Electrical properties of rat skin and design of implantable antennas for medical wireless telemetry," IEEE Transactions on Antennas and Propagation, vol. 57, no. 9, pp. 2806–2812, 2009.

[61] W. Xia, K. Saito, M. Takahashi, and K. Ito, "Performances of an implanted cavity slot antenna embedded in the human arm," IEEE Transactions on Antennas and Propagation, vol. 57, no. 4, pp. 894–899, 2009.

[62] T. Nakamura and A. Terano, "Capsule endoscopy: Past, present, and future," Journal of Gastroenterology, vol. 43, pp. 93–99, 2008.

[63] R. Warty, M. R. Tofighi, U. Kawoos, and A. Rosen, "Characterization of implantable antennas for intracranial pressure monitoring: Reflection by and transmission through a scalp phantom," IEEE Transactions on Microwave Theory and Techniques, vol. 56, no. 10, pp. 2366–2376, 2008.

[64] Xiao-Fei Teng, Yuan-Ting Zhang, Carman C. Y. Poon, and Paolo Bonato, "Wearable medical systems for P health," IEEE Reviews in Biomedical Engineering, vol. 1, 2008.

[65] T. Karacolak, A. Z. Hood, and E. Topsakal, "Design of a dual-band implantable antenna and development of skin mimicking gels for continuous glucose monitoring," IEEE Transactions on Microwave Theory and Techniques, vol. 56, no. 4, pp. 1001–1008, 2008.

[66] D. Panescu, "Emerging technologies wireless communication systems for implantable medical devices," IEEE Engineering in Medicine and Biology Magazine, vol. 27, no. 2, pp. 96–101, 2008.

[67] H. Yu, G. S. Irby, D. M. Peterson, M.-T. Nguyen, G. Flores, N. Euliano, and R. Bashirulla, "Printed capsule antenna for medication compliance monitoring," Electronics Letters, vol. 43, no. 22, pp. 1179–1181, 2007.

[68] J. Wang and D. Su, "Design of an ultra-wideband system for in-body wireless communications," in Proceedings of the 2006 4th Asia-Pacific Conference Environmental Electromagnets, Dalian, 2006, pp. 565–568.

[69] S. R. Best, "A discussion on electrically small antennas surrounded by lossy dispersive materials," Antennas and Propagation, 2006, pp. 1–7, 2006.

[70] S. Kwak, K. Chang, and Y. J. Yoon, "Small spiral antenna for wideband capsule endoscope system," Electronics Letters, vol. 42, no. 23, pp. 1328–1329, 2006.

[71] J. Kim and Y. Rahmat-Samii, "Planar inverted-f antennas on implantable medical devices: Meandered type versus spiral type," Microwave and Optical Technology Letters, vol. 48, no. 3, pp. 567–572, 2006.

[72] P. D. Bradley, "An ultra-low power, high performance medical implant communication system (MICS) transceiver for implantable devices," in Proceeding of the IEEE Biomedical Circuits and Systems Conference BioCAS 2006, November 2006, pp. 158–161.

[73] K. Gosalia, M. S. Humayun, and G. Lazzi, "Impedance matching and implementation of planar space-filling dipoles as intraocular implanted antennas in a retinal prosthesis," IEEE Transactions on Antennas and Propagation, vol. 53, no. 8, pp. 2365–2373, 2005.

[74] S. Kwak, K. Chang, and Y. J. Yoon, "Ultra-wide band spiral shaped small antenna for the biomedical telemetry," Asia-Pacific Microwave Conference Proceedings, vol. 1, 2005.

[75] S. Kwak, K. Chang, and Y. J. Yoon, "The helical antenna for the capsule endoscope," 2005 IEEE International Symposium of Antennas and Propagation, vol. 2B, pp. 804–807, 2005.

[76] P. Soontornpipit, C. M. Furse, and Y. C. Chung, "Miniaturized biocompatible microstrip antenna using genetic algorithm," IEEE Transactions on Antennas and Propagation, vol. 53, no. 6, pp. 1939–1945, 2005.

[77] W. A. Qureshi, "Current and future applications of the capsule camera," Nature Reviews Drug Discovery, vol. 3, pp. 447–450, 2004.

[78] J. Kim and Y. Rahmat-Samii, "Implanted antennas inside a human body: Simulations, designs, and characterizations," IEEE Transactions on Microwave Theory and Techniques, vol. 52, no. 8, pp. 1934–1943, 2004.

[79] P. Soontornpipit, C. M. Furse, and Y. C. Chung, "Design of implantable microstrip antenna for communication with medical implants," IEEE Transactions on Microwave Theory and Techniques, vol. 52, no. 8, pp. 1944–1951, 2004.

[80] S. Gabriel, R. W. Lau, and C. Gabriel, "The dielectric properties of biological tissues: II. Measurements in the frequency range 10 Hz to 20 GHz," Physics in Medicine and Biology, vol. 41, no. 11, pp. 2231–2293, 2004.

[81] K. Gosalia, G. Lazzi, and M. Humayun, "Investigation of a microwave data telemetry link for a retinal prosthesis," IEEE Transactions on Microwave Theory and Techniques, vol. 52, no. 8, pp. 1925–1933, 2004.

[82] L. C. Chirwa, P. A. Hammond, S. Roy, and D. R. S. Cumming, "Electromagnetic radiation from ingested sources in the human intestine between 150 MHz and 1.2 GHz," IEEE Transactions on Biomedical Engineering, vol. 50, no. 4, pp. 484–492, 2003.

[83] S. Best and J. Morrow, "On the significance of current vector alignment in establishing the resonant frequency of small space-filling wire antennas," IEEE Antennas and Wireless Propagation Letters, vol. 2, pp. 201–204, 2003.

[84] Kevin Hung and Yuan-Ting Zhang, "Implementation of WAP-based telemedicine system for patient monitoring," IEEE Transactions on Information Technology in Biomedicine, vol. 7, no. 2, 2003.

[85] W. G. Scanlon, B. Burns, and N. E. Evans, "Radio wave propagation from a tissue-implanted source at 418 MHz and 916.5 MHz," IEEE Transactions on Biomedical Engineering, vol. 47, no. 4, pp. 527–534, 2000.

[86] A. Bhattacharya, B. Roy, R. Caldeirinha, and A. Bhattacharjee, "Low-profile, extremely wideband, dual-band-notched MIMO antenna for UWB applications," International Journal of Microwave and Wireless Technologies, vol. 11, no. 7, pp. 719–728, 2019.

[87] A. De, B. Roy, A. Bhattacharya, and A. K. Bhattacharjee, "Bandwidth-enhanced ultra-wide band wearable textile antenna for various WBAN and Internet of Things (IoT) applications," Radio Science, vol. 56, 2021, p. e2021RS007315. https://doi.org/10.1029/2021RS007315.

[88] Ankan Bhattacharya, Bappaditya Roy, Santosh K. Chowdhury and Anup K. Bhattacharjee, "Computational and experimental analysis of a low-profile, isolation-enhanced, band-notch UWB-MIMO antenna," Journal of Computational Electronics, vol. 18, pp. 680–688, 2019. https://doi.org/10.1007/s10825-019-01309-3.

Chapter 10

Wearable MIMO Antenna with High Port Isolation for e-Health Monitoring Applications

Ashim Kumar Biswas, Koushik Roy, Vicky Kumar, Biswarup Neogi, and Ujjal Chakraborty

10.1 INTRODUCTION

Wearable antennas may play a very important and effective role in the purpose of e-health monitoring systems. Nowadays, wireless communication systems are very essential in the application of smart body healthcare controlling systems [1]. This type of smart healthcare management system is constructed in the vicinity of body-centric wireless communications, and it primarily consists of wearable electronic systems, which include wearable antennas [2]. In most of the cases one or more sensors or sensor networks are applied in the smart health observing systems [3]. Such a distinctive sensor device setup for a basic health intensive care application may possibly involve manifold sensor nodes, and some outside transducers are involved to collect the data in suitable form [4, 5]. These data in the appropriate arrangement can be routed to the processing centre. In these types of health monitoring systems, transceiver systems are very common where antennas may be used as very essential component. Therefore, wearable antennas are very important in these types of health monitoring systems.

The maximum of the existing strain-sensing methods like metallic foil straining instruments or fibre-type sensors involve a cable-connected sensor and acquisition system to acquire data [6, 7]. Therefore, to reduce the excessive cabling cost of these bulky assemblies and also for digitizing and communicating sensor information, wireless sensing strategies should be established [8, 9]. Therefore, integration of antennas is indispensable. When a wearable multiple-input. multiple-output (MIMO) antenna is incorporated here, a tremendous improvement is possible. This is because the MIMO structure ensures improved channel capacity with high reliability, great data rate and good network coverage [10–12]. In its place of struggling with a multipath fading problem, which is common in most wireless atmospheres, MIMO systems work on multipath reflections along with a realization of manifold virtual spatial channels [13, 14]. Wearable MIMO approaches are becoming increasingly essential in the arenas of regular functioning such as bodily fitness observations, medical care, telemedicine and exercise [15, 16]. Wearable

DOI: 10.1201/9781003459880-10

MIMO antennas can be used in any environment. The wearable antenna is designed mostly considering the body-centric consideration. It can be utilized for both off- and on-body situations. Therefore, for e-health applications, wearable MIMO antennas with good diversity performance are very important and can be utilized in wearable health monitoring systems.

Several wearable MIMO antennas are reported in the literature. But very few are designed for health-related applications. In [17], a single-layer textile MIMO antenna for wearable application is explained. It is a two-port antenna and provides isolation of above 12 dB due to the quasi-orthogonal radiations. Ref. [18] presents a sub–6 GHz ISM frequency band, flexible, portable MIMO antenna for biomedical telemetry equipment and wireless body area networks. A 2 × 2 array is used. An antenna operating at a frequency of 5–6 GHz is created using a 0.8-mm-thick Rogers RT/Duroid 5880 substrate. In [19], an idea of a flexible dual-polarized MIMO belt strap antenna with triple band application is discussed. The antenna works over 2.37–2.7, 2.8–3.22 and 5.41–6.03 GHz. A miniaturized, circularly polarized MIMO antenna for wearable biotelemetry devices is presented in [20]. The antenna covers the ISM band (2.40–2.48 GHz) with very high port isolation. Mostly, the reported MIMO configurations are for wireless body area network (WBAN) applications. Health issues are not extensively considered. Sometimes, body-centric movement is required. Therefore, a wearable MIMO antenna is important in several aspects.

In this chapter one dual-element and two port textile MIMO antenna are suggested. The antenna employs simple rectangular-shaped antenna elements and uses a jeans substrate. The antenna is designed to communicate the body-centric information, which is collected using some sensor network and converted it into suitable form using some transducer and other reliable converter. A related block diagram is illustrated in Figure 10.1. The designed MIMO antenna covers the application frequency bands of ISM (2.4–2.485 GHz) and radars, mobile phones and a commercial wireless local area network (LAN) (3.1–3.3 GHz) bands. More than 15 dB of port isolation is found over the application band. Diversity criteria such as envelope correlation co-efficient (ECC < 0.15), channel capacity loss (CCL < 0.21 BPS/Hz), diversity gain (DG), mean effective gain (MEG), etc., are investigated suitably. Radiation patterns are also explored. Ansys Electronics 2021 R1 [21] is used to simulate, analyse and adapt each of these parameters.

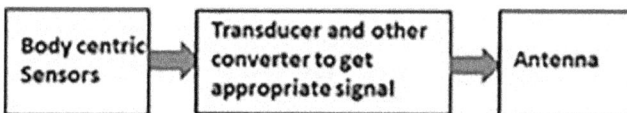

Figure 10.1 Block diagram of the body-centric transceiver system.

10.2 ARCHITECTURE AND DESIGN OF ANTENNA

The preferred wearable textile MIMO antenna's structure is described in Figure 10.2(a and b). The figure shows the antenna's top and back views, respectively. A thin copper layer is used to imprint two rectangular antenna elements. With a loss tangent of 0.02 and a relative permittivity of 1.6, we used a jeans substrate. Dielectric properties are found in line with ref. [22]. The ground plane is connected to a 'T'-shaped stub in order to increase port isolation. The antenna's elements are kept around 9 mm apart from one another. The whole volume of the antenna is about $33 \times 50 \times 1$ mm³. Two microstrip line feeds are used to excite the ports. Table 10.1 displays the suggested antenna's optimal values for each parameter. The final MIMO antenna is produced after a single element and MIMO antennas without ground stubs have been analysed. The final antenna offers very good application bands.

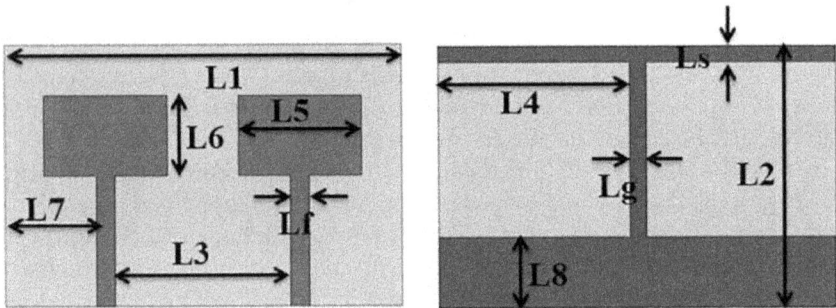

Figure 10.2 Structures of the suggested MIMO antenna in (a) top and (b) back perspectives.

Table 10.1 Dimensions of the Parameters Reported

Parameters	Dimension (mm)	Parameters	Dimension (mm)
L_s	2	L_1	50
L_g	2.1	L_2	33
L_f	2.4	L_3	22.1
S_1	1.5	L_4	23.95
S_2	6	L_5	15.5
S_3	12	L_6	10
L_8	9	L_7	11.55

10.3 DISCUSSION OF RESULTS

In the process of antenna design, one left-sided and one right-sided single element antennas are designed, as shown in Figure 10.3(a and b), respectively, for top views and in Figure 10.3(c) for back views. The second one is the mirror image of the first one. The frequency range of both single-element antennas is 2.14–3.57 GHz. A two-element MIMO antenna is created using these single-element antennas. The perspectives of it from the top and back are depicted in Figures 10.2(a) and 10.3(c), respectively. The antenna has the same frequency range coverage as the single-element antenna and has a 9 dB minimum port isolation. The S-parameters are illustrated in Figure 10.4. The antenna is further investigated to improve the isolation characteristics. The existing ground plane is combined with a 'T'-shaped stub. The new MIMO antenna has a frequency variety of 2.35–3.36 GHz and a least port field coupling of 15 dB for the application bands. In Figure 10.4, the S-parameters are displayed. According to ref. [23], a tissue model is created, as seen in Figure 10.5. On the basis of this model, the final MIMO antenna design is also examined. The S-parameters are shown in Figure 10.4 and are found to be satisfactory.

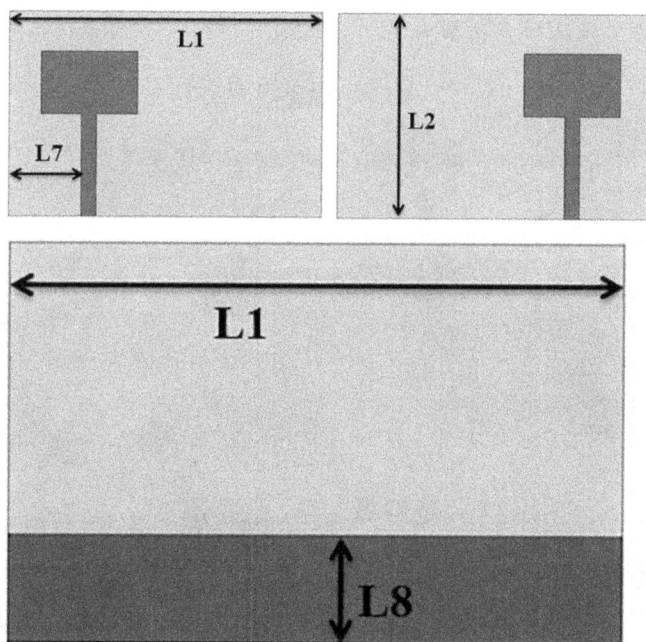

Figure 10.3 Top layer of the single element antenna: (a) left-sided element, (b) right-sided element and (c) view of the antenna's back.

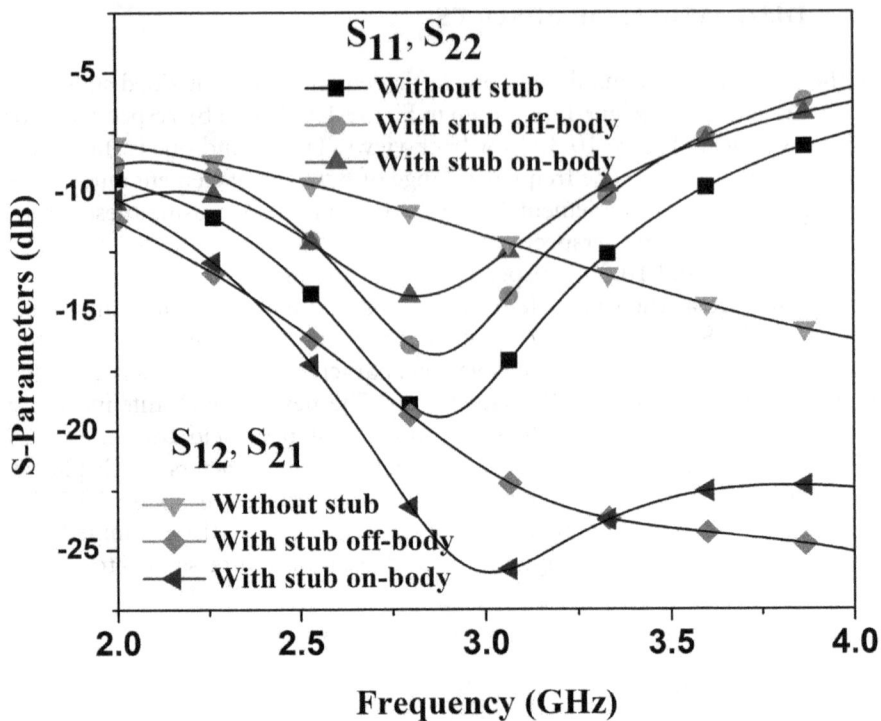

Figure 10.4 MIMO antennas' S-parameters in simulation.

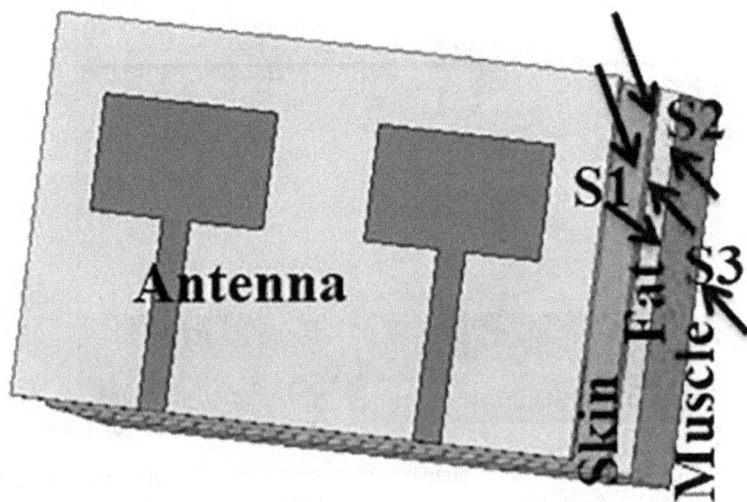

Figure 10.5 Antenna with on-body condition.

10.4 DISCUSSION OF MIMO PARAMETERS OF THE FINAL ANTENNA

Diversity factors are crucial when analysing a MIMO antenna. A MIMO constraint investigated to observe the correlation among signals anticipated by the antenna elements is called the ECC [20]. In order to achieve a MIMO antenna's best performance possible, its value must be within a tolerance of 0.5. To calculate the ECC value, we used equation 10.1 based on the S-parameters [24]. Figure 10.6 depicts the ECC variation, and its value is found as less than 0.15 over the functional band. DG is related to ECC. It is calculated using equation 10.2 [24]. Its value is found to be very high, which is good enough for MIMO operation.

$$ECC\frac{\mid S_{aa}^{*}S_{ab} + S_{ba}^{*}S_{bb}\mid^{2}}{(1-\mid S_{aa}\mid^{2} - \mid S_{ba}\mid^{2})(1-\mid S_{bb}\mid^{2} - \mid S_{ab}\mid^{2})} \tag{10.1}$$

$$DG = 10\sqrt{1-(ECC)^{2}} \tag{10.2}$$

CCL is also explored. It is calculated in line with ref. [25]. This component is susceptible to the maximum information rate, and it is regarding a suitable level beyond which message transmission will be disrupted. The CCL value is suitable if it is in the range of ≤0.4 BPS/Hz. In the current study, its value is found to be less than 0.12 BPS/Hz. The suggested MIMO antenna's MEG is computed using the equation in [24] as a further parameter. This parameter deals with the

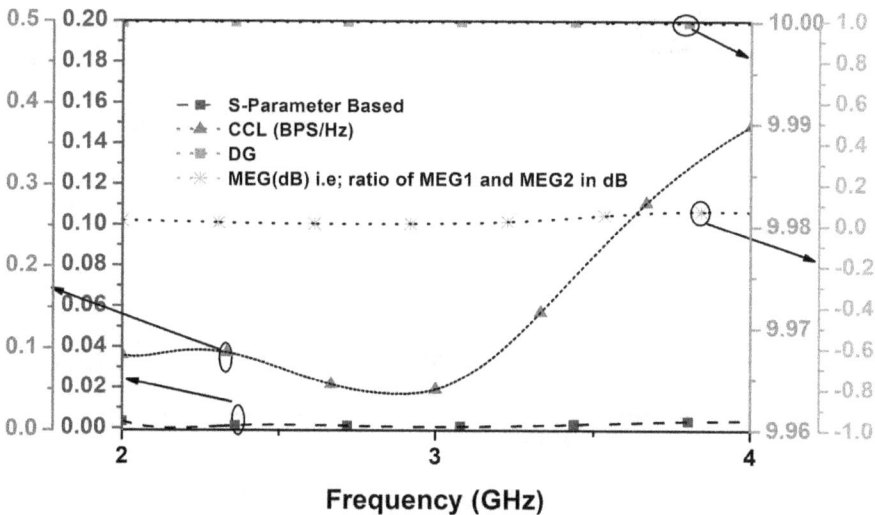

Figure 10.6 ECC, DG, CCL and MEG variations.

characterization of fading atmosphere. Figure 10.6 displays the MEG ratio for the two ports. The antenna is appropriate when the ratio is within 3 dB. The figure shows that the MEGs of the two ports are in a satisfactory range.

Another major diversity constraint is the total active reflection coefficient (TARC). TARC analysis is necessary in MIMO antennas as well to verify the impedance properties. It can be figured out by utilizing equation 10.3 [26]:

$$\Gamma = \frac{\sqrt{\left(\left|(S_{mm} + S_{mn}e^{j\theta})\right|^2 + \left|(S_{nm} + S_{nn}e^{j\theta})\right|^2\right)}}{\sqrt{2}} \tag{10.3}$$

In equation 10.3, θ implies the port's excitation phase angle. S_{mm} and S_{nn} denote the reflection coefficients, and S_{mn} and S_{nm} denote coupling characteristics. The TARC properties are investigated using some random excitation phase angles (θ). Figure 10.7 illustrates it and exhibits appropriate TARC behaviour throughout the application band.

We have analysed the current scatterings of the antenna fragments to clearly define the port isolation feature. It shows the information related to movement of current from one to another port. The integrated 'T'-shaped stub on ground plane supports to decrease the coupling of current between adjacent elements. High port isolation is achieved by producing a good band stop

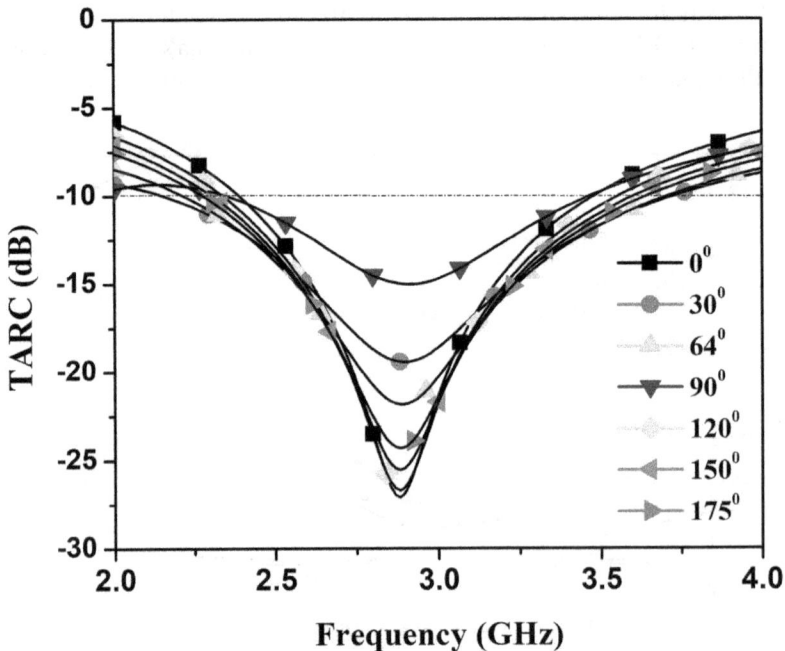

Figure 10.7 Illustration of TARC related to the specified MIMO antenna.

feature. We have looked at the 2.4 and 3.15 GHz current distributions. The scatterings of surface current are shown in Figure 10.8(a–d). From the image, it appears that using a ground stub reduces current movement from port 1 to port 2. It demonstrates that the ports' mutual coupling is quite minimal.

Figure 10.9(a–d) displays the 2.4 and 3.15 GHz E- and H-plane radiation patterns. Investigated are co-polar and cross-polarization radiation patterns. Figure 10.9 indicates very stable patterns. All the investigated parameters

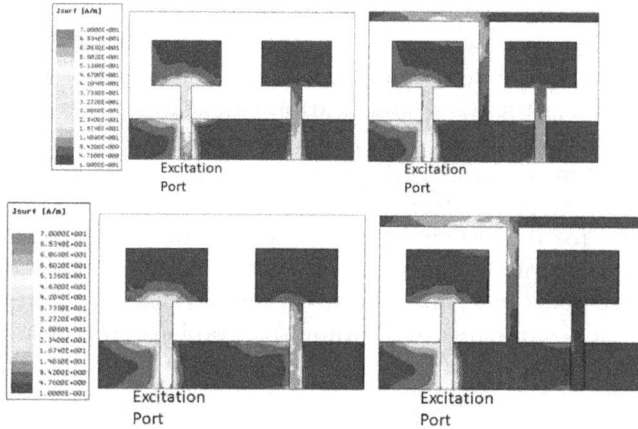

Figure 10.8 The suggested structure's surface current distributions at 2.4 GHz: (a) no stub and (b) with stub and at 3.15 GHz: (c) no stub and (d) with stub.

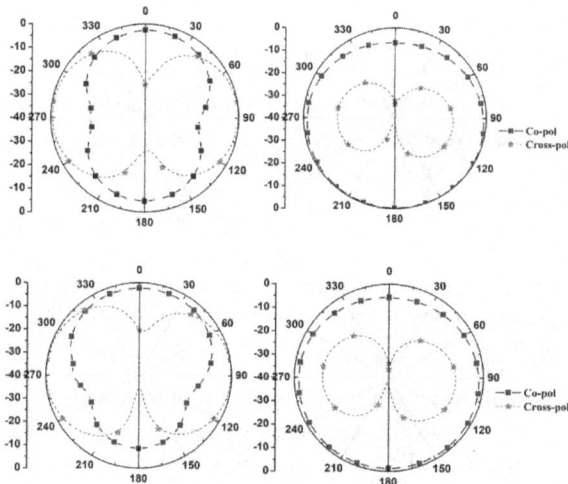

Figure 10.9 (a) E-plane, (b) H-plane, (c) E-plane and (d) H-plane, normalized simulated radiation patterns of the reported prototype at 2.4 GHz and 3.15 GHz, respectively.

are found suitable for wearable MIMO operation. Therefore, the designed prototype may meet the characteristics of the desired applications.

10.5 CONCLUSION

For wearable applications, a small textile MIMO antenna using two ports is suggested. The antenna may be used in an on-body transceiver system for e-health applications. The designed antenna deals with a wide bandwidth extending from 2.27 to 3.35 GHz ensuring the ISM (2.4–2.485 GHz) and radars, mobile phones and commercial wireless LAN (3.1–3.3 GHz) bands. Over the whole application band, it provides high port isolation of more than 15 dB. ECC, CCL, MEG and TARC characteristics are considered in the investigation, and these parameters are found very suitable. The radiation pattern also supports the antenna as an appropriate contender for the anticipated applications. In the future, the number of radiating elements of the wearable MIMO antenna will be increased to enhance the data rate more, provide added reliability, better sensitivity, user friendly, controlled radiation and good specific absorption rate (SAR) value.

REFERENCES

[1] Li, Y., and Zhang, M.: Study on a cylindrical sensor network for intelligent health monitoring and prognosis. IEEE Access 6, 69195–69201 (2018).

[2] Yuan, Q., and Ishikawa, T.: Effect of via-wheel power transfer system on human body. In Proceedings of the IEEE Wireless Power Transf. (WPT), Perugia, May 2013, 238–241 (2013).

[3] Li, Y., Yang, L., Duan, W., and Zhao, X.: An implantable antenna design for an intelligent health monitoring system considering the relative permittivity and conductivity of the human body. IEEE Access 7, 38236–38244 (2019).

[4] Kranz, M. S., English, B. A., and Michael, R. W.: RFID-inspired wireless micro sensors for structural health monitoring. 2016 IEEE Aerospace Conference, Big Sky, MT, 5–12 March 2016, 1–7 (2016).

[5] Mainwaring, Alan, Polastre, Joseph, Szewczyk, Robert, Culler, David, and Anderson, John: Wireless sensor networks for habitat monitoring. Published in ACM International Workshop on Wireless Sensor Networks and Applications (WSNA'02), (2002).

[6] Michie, W. C., Culshaw, B., Roberts, S. S. J., and Davidson, R.: Fiber optic technique for simultaneous measurement of strain and temperature variations in composite materials. In Proceedings of SPIE, Fiber Optic Smart Structures and Skins IV, Boston, MA (1991).

[7] Murray, W. M., and Miller, W. R.: The Bonded Electrical Resistance Strain Gage: An Introduction. New York, NY: Oxford University Press (1992).

[8] Liu, L., and Yuan, F.: Wireless sensors with dual-controller architecture for active diagnosis in structural health monitoring. Smart Materials and Structures 17, 025016 (2008).

[9] Kurata, N., Spencer, B. F., and Ruiz, S. M.: Risk monitoring of buildings with wireless sensor networks. Structural Control Health Monitoring 12, 315–327 (2005).

[10] Biswas, A. K., and Chakraborty, U.: Complementary meander-line-inspired dielectric resonator multiple input-multiple-output antenna for dual-band applications. International Journal of RF and Microwave Computer-Aided Engineering 29(12), e21970 (2019).

[11] Biswas, A. K., and Chakraborty, U.: A compact wide band textile MIMO antenna with very low mutual coupling for wearable applications. International Journal of RF and Microwave Computer-Aided Engineering 29(8), e21769 (2019).

[12] Biswas, A. K., and Chakraborty, U.: Reconfigurable wide band wearable multiple input multiple output antenna with hanging resonator. Microwave and Optical Technology Letters 62(3), 1352–1359 (2020).

[13] Ouyang, Y., Love, D. J., and Chappell, W. J.: Body-worn distributed MIMO system. IEEE Transactions on Vehicular Technology 58(4), 1752–1765 (2019).

[14] Biswas, A. K., and Chakraborty, U.: Compact wearable MIMO antenna with improved port isolation for ultra-wideband applications. IET Microwaves, Antennas & Propagation 13(4), 344–354 (2019).

[15] Biswas, A. K., and Chakraborty, U.: Investigation on decoupling of wide band wearable multiple-input multiple-output antenna elements using microstrip neutralization line. International Journal of RF and Microwave Computer-Aided Engineering 29(7), e21723 (2019).

[16] Yan, S., and Vandenbosch, G. A. E.: Radiation pattern-reconfigurable wearable antenna based on metamaterial structure. IEEE Antennas and Wireless Propagation Letters 15, 1715–1718 (2016).

[17] Li, H., Sun, S., Wang, B., and Wu, F.: Design of compact single-layer textile MIMO antenna for wearable applications. IEEE Transactions on Antennas and Propagation 66(6), 3136–3141 (2018). doi: 10.1109/TAP.2018.2811844.

[18] Althuwayb, Ayman A., Alibakhshikenari, Mohammad, Virdee, Bal S., Rashid, Nasr, Kaaniche, Khaled, Atitallah, Ahmed Ben, Armghan, Ammar, Elhamrawy, Osama I., See, Chan Hwang, and Falcone, Francisco: Metasurface-inspired flexible wearable MIMO antenna array for wireless body area network applications and biomedical telemetry devices. IEEE Access 11, 1039–1056 (2023). doi: 10.1109/ACCESS.2022.3233388.

[19] Yang, S., Zhang, L., Wang, W., and Zheng, Y.: Flexible tri-band dual-polarized MIMO belt strap antenna toward wearable applications in intelligent internet of medical things. IEEE Transactions on Antennas and Propagation 70(1), 197–208 (2022). doi: 10.1109/TAP.2021.3098589.

[20] Iqbal, A., Smida, A., Alazemi, A. J., Waly, M. I., Khaddaj Mallat, N., and Kim, S.: Wideband circularly polarized MIMO antenna for high data wearable biotelemetric devices. IEEE Access 8, 17935–17944 (2020). doi: 10.1109/ACCESS.2020.2967397.

[21] Ansys HFSS ver. 19, 3D High Frequency Simulation Software, Ansys Corporation. https://www.ansys.com/en-in/products/electronics/ansys-hfss.

[22] Roy, S., and Chakraborty, U.: Mutual coupling reduction in a multi-band MIMO antenna using meta-inspired decoupling network. Wireless Personal Communications 114, 3231–3246 (2020).

[23] Biswas, A. K., Pattanayak, S. S., and Chakraborty, U.: Evaluation of dielectric properties of colored resin plastic button to design a small MIMO antenna. IEEE Transactions on Instrumentation and Measurement 69(11), 9170–9177 (2020).

[24] Zhang, K., Jiang, Z. H., Hong, W., and Werner, D. H.: A low-profile and wide-band triple-mode antenna for wireless body area network concurrent on-/off-body communications. IEEE Transactions on Antennas and Propagation 68(3), 1982–1994 (2020).

[25] Biswas, A. K., Swarnakar, P. S., Pattanayak, S. S., and Chakraborty, U.: Compact MIMO antenna with high port isolation for triple band applications designed on a biomass material manufactured with coconut husk. Microwave and Optical Technology Letters 62(12), 3975–3984 (2020).

[26] Iqbal, A., Smida, A., Alazemi, A. J., et al.: Wideband circularly polarized MIMO antenna for high data wearable biotelemetric devices. IEEE Access 8, 17935–17944 (2020).

Chapter 11

Development of Multiport MIMO Antenna for C-Band Frequency Application in Wireless Communication

Kranti D. Patil and D.M. Yadav

11.1 INTRODUCTION

In today's wireless communication, the dual-port multiple-input, multiple-output (MIMO) array plays a crucial role. To execute the virtual correspondence or reconciliation of wireless local area network (WLAN) recurrence band in a single radio wire, the MIMO design is the most ideal decision. Numerous two-port MIMO antennas with various design strategies and a higher data rate that do not require additional bandwidth or power are mentioned in the existing literature. MIMO technology makes it possible to do this. The design of a MIMO array is extremely challenging, and reducing interference between antenna elements is particularly challenging.

There are a number of dual-port MIMO antennas with various methods mentioned in [1–10] of the existing literature. The design of a MIMO array is extremely challenging, and reducing interference between antenna elements is particularly challenging. As a result, various decoupling structure designs for MIMO antennas are also reported in the literature.

For the 5G NR sub-6 GHz/(Wi-Fi-5)/(DSRC)/Wi-Fi-6/Indian National Satellite, a two-port co-planar wave structure antenna [1] with a designed dimension of 3222 mm^2 is mentioned. In the ultra-wide (UW) band, the dual-band MIMO antenna described in [2] achieves isolation of 25 dB using a parasitic band between two antenna elements. The referenced MIMO radio wire was planned on a Recurrence Reach 4 substrate with an aspect of 28×16 mm^2. In [3], a dual-port, MIMO antenna with the band of frequency needed of 3.45–4.99 GHz was mentioned. The radiators' MIMO antenna was shaped by a Frequency Range-4 substrate that was 1.6 millimeters thick and had a size of $43*38$ mm^2.

According to [4], a dual-port MIMO antenna with an overall structure of 2716 mm^2 is mentioned for the wireless fidelity 5 and wireless fidelity 6 industrial standards. In [5], a dimension of 2127 mm^2 was reported for a tri-band, dual-port MIMO antenna that covered WLAN (5.3–5.7 GHz) and Worldwide Interoperability for Microwave Access (WiMAX) (2.22–2.54 GHz). In [6], a dual-port MIMO swastika antenna for wireless communication is

DOI: 10.1201/9781003459880-11

mentioned. The antenna has a total dimension of 4630 mm^2 and operates in the WLAN and WiMAX bands.

Here we have a survey on the flow research plan of MIMO antennas (Table 11.1). The review is adapted from Wiley's *International Journal of Antenna and Wave Propagation* – 2022 and 2021.

Table 11.1 Review on MIMO Antenna Design

Sr. No.	Title	Year of Publication	Findings
1	Sixteen-port numerous information different result receiving wire for 5G versatile terminal applications.	2022	Isolations of greater than 17 and 24.2 dB The efficiency is than 54% Channel capacity up to 86.6 bps/Hz ECCs are lower than 0.11
2	For 5G millimeter applications, a three-dimensional circularly polarized multibeam array with scanning capability and wideband characteristic.	2022	HPBW AR beam width of 29 3-dB is greater than HPBW ARBW — 40.5% 3-dB Covers the entire Ka band from 26.2 to 39.5 GHz G Gain of 9.9 dBi FBR of each beam is greater than 19 dB
3	A tri-band antenna combining two sub-6G and one millimeter bands with shared feeding port for 5G/B5G applications.	2022	Millimeter-wave antenna operating at 28 GHz efficiency 80%
4	Fractal loaded six element antenna configurations for super wideband operation.	2022	80 × 120 mm^2 — FR4 substrate Bandwidth 1.4–20 GHz Isolation of 20 dB is achieved
5	A wideband self-decoupled multi-input multi-output antenna with a high isolation	2022	More extensive data transmission of 3.2–5.9 GHz, 59% Exploratory effectiveness and envelope relationship coefficients are bigger than 78% and <0.04
6	A novel mushroom antenna with high gain using three dimensional printing	2022	Receiving wire works from 5.56 to 6.0 GHz for the reflection coefficient ≤−10 dB At 5.8 GHz, the maximum gain is 16.5 dBi Antenna is low cost, very cross polarization

Table 11.1 (Continued)

Sr. No.	Title	Year of Publication	Findings
7	An application of a broadband absorptive filtering antenna in a 5G multi-input, multi-output array	2022	Bandwidth between 2.55 and 5.05 GHz (65%)
			Acquire a response of 7.4 dBi
			Port seclusion in excess of 24 dB
			Due to the unique out-of-band reflection, fewer filtering characteristics of the AFA element, extremely low envelope correlation coefficient (ECC) of less than $2 * 10^{-4}$
8	Design and analysis of a Coplanar waveguide fed flexible ultra-wideband antenna for microwave imaging of breast cancer.	2022	Bandwidth from 2.26 to 13.71 GHz
			Efficiency of 85% and gain of 3.84 dBi
			Highest SAR of 0.932W/Kg at 10.6 GHz
9	A decoupling structure without sacrificing antenna-element performance for 5G smartphone.	2022	6 GHz band of 3.3–3.6 GHz is designed
			Isolation 11 dB
10	Eight-component fifth-age MIMO receiving wire planned by modular flows cancelation.	2022	Good diversity, isolation, and impedance matching performance between 3.35 and 3.71 GHz
			Isolations of 11.5 dB
			ECC of less than 0.2 of each port.

11.2 STRUCTURAL LAYOUT OF PROPOSED ANTENNA

The planned calculation and design alongside definite elements of the proposed MIMO antenna molded receiving wire is portrayed in Figure 11.1.

A 50 transmission line with a surface area of 7.5 × 1.5 mm² and a coplanar waveguide surface region are embedded on the foremost side of a Frequency Region-4 substrate with 'h', 'r', and tan(t) of 0.8 mm, 4.3 mm, and 0.025 mm, respectively, as shown in Figure 11.1. The receiving wire comprises a ring with an external span of 7 mm and internal sweep of 5.75 mm set simply over the feed point. Further a T-shaped radiator is in the middle of the hexagonal ring structures.

The upright strip of a T-shaped element measures 6 * 12 mm, while the horizontal strip measures 6 * 12 mm. Four even strips, each with an aspect of 4.5 × 1 mm², are embedded in the middle corner to corner part of the molded

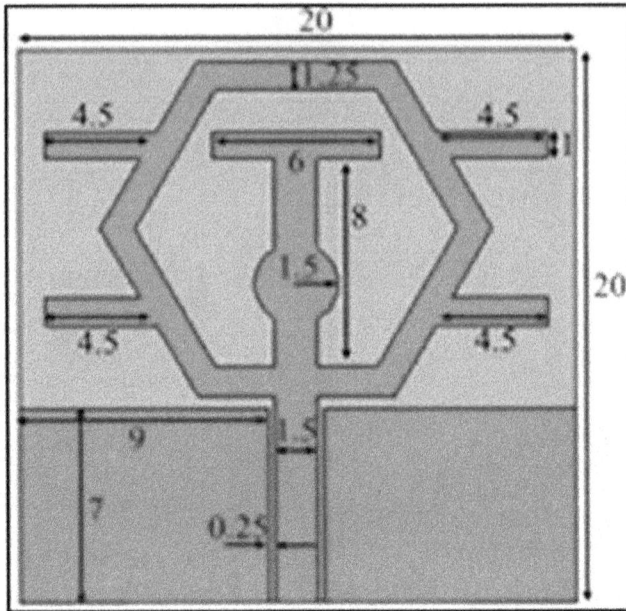

Figure 11.1 Structural design of proposed MIMO antenna (dimensions are measured in millimeters).

ring. The total construction with a planned impression of 20×20 mm^2 assists in acquiring the ideal sub-6 GHz as well as the WLAN with banding for remote application in cell phones [11].

11.3 GEOMETRY OF TWO-PORT MIMO ANTENNA

Figure 11.2 shows the step-by-step consideration for designing the array and the S_{11} parameter to provide a better understanding of the proposed hexagonal ring-shaped antenna's designed structure. The coplanar waveguide (CPW) fed method initially excited a hexagon ring, as shown in Figure 11.2 (Step 1). Stage 1 produces the tuning at 4.9 GHz with a S_{11} magnitude of –20 Db. In the second structure, a gap molded element is projected through the base of the ring that produces a recurrence range (4.42–3.59 GHz) and moves past the created recurrence range to the upper side with a recurrence scope (3.72–6.70 GHz), with a focus recurrence of 3.45 GHz and 5.65 GHz, separately. Stage 3, as shown in Figure 11.2, is designed to shift the resonance of 5.75 GHz to the lower side and also to increase the bandwidth of 3.45–3.69 GHz. Stage 3 is capable of evoking dual tuning at 3.4 GHz and 5 GHz, respectively, in frequency bands 3.23–3.52 GHz and 3.10–5.51 GHz.

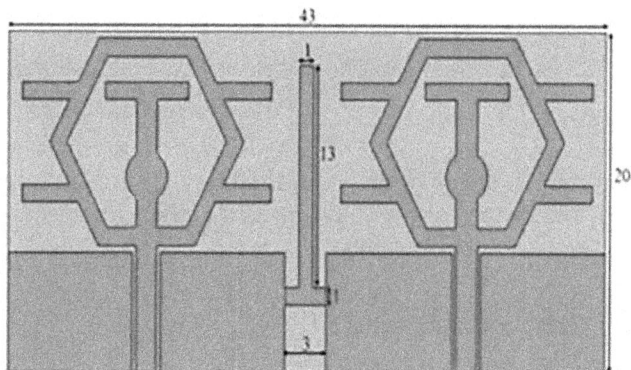

Figure 11.2 Dimensional layout of proposed dual-shaped MIMO array.

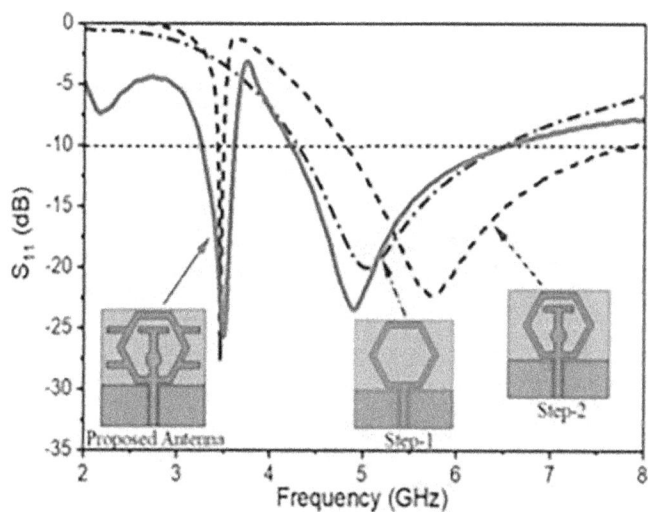

Figure 11.3 Stage-wise structure along with S_{11} (dB) of MIMO.

The step-wise plan component is additionally checked with surface current circulation examined at dual tuning at 3.49 GHz and 5 GHz, as portrayed in Figure 11.3 3, at 3.5 GHz. It very well may be seen that the higher current is moving in the T-shaped radiator and lower part of the ring, which add to produce the less (3.23–3.52 GHz) recurrence groups. Consequently, from the examination of the stage-wise plan instrument and surface current conveyance, it is noticed that the structure receiving wire is a great competitor for sub-6 GHz and WLAN.

11.4 RESULTS AND DISCUSSION OF PROPOSED MIMO ANTENNA

CST MWS software is used to conduct an analysis of the proposed dual-port MIMO antenna. For the purpose of validating the design, parameters such as the surface current distribution, gain, efficiency, envelope correlation coefficient (ECC), and diversity gain (DG), as well as the transmission coefficient (S_{12}), are taken into consideration [12].

11.4.1 S_{11} (dB) and S_{12} (dB)

The S-attributes, including S_{11} (dB)/S_{22}(dB) and S_{12} (dB)/S_{21}(dB), are outlined in Figure 11.4. It is clear from the S_{11} and S_{22} curves that there are two tunings at 4.1 GHz and 4.5 GHz.

This confirms that the two-port MIMO antenna can efficiently operate in the frequency ranges of 3.35–3.73 GHz and 4.11–5.42 GHz), which is the satisfactory sign for the impedance bandwidth requirement for mobile device applications in the sub-6 GHz and WLAN frequency bands. Here we can observe proper impedance matching in operating bands with the exact match between the S_{11} and S_{22} curves. In the lower sub-6 GHz and upper WLAN

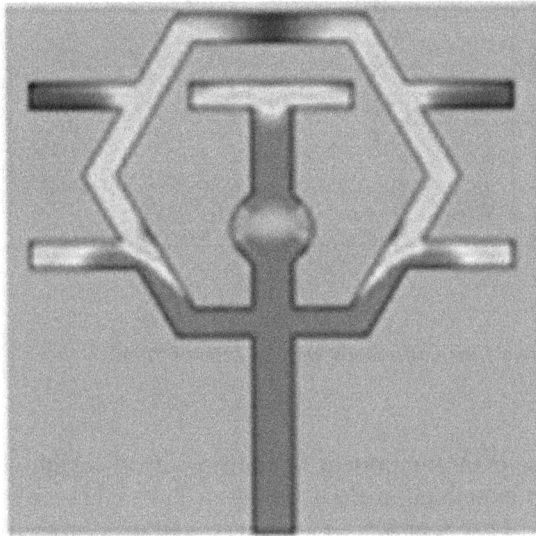

Figure 11.4 Current distribution of the planned 3.5 GHz antenna apps for mobile devices. A radiator in the form of a tap protrudes from the lower side of the ring. This radiator produces a frequency range of 5.82–7.80 GHz, with middle frequencies of 4.45 GHz and 6.15 GHz, respectively, by generating a frequency band of 3.33–4.50 GHz.

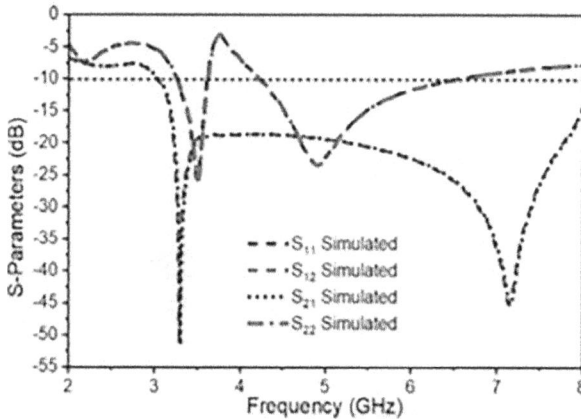

Figure 11.5 S-characteristics, including S_{11} (dB)/S_{22} (dB) and S_{12} (dB)/S_{21}(dB).

recurrence groups, the accomplished 10 dB impedance data transfer capacity is 11% and 43%, respectively. Additionally, it is noted that in both the upper WLAN frequency band and the lower frequency band, the isolation level of at least –20 dB is guaranteed by the dashed green and dotted blue lines that represent the S_{12}/S_{21} curves [7]. This demonstrates that the inverted-T-shaped isolation structure protects the two antenna elements from each other's interference while maintaining their compactness and operating simultaneously in the upper WLAN frequency band and lower frequency bands [13].

11.4.2 Distribution of Surface Current (A/m) for the Proposed MIMO Antenna

The surface current distribution at two resonating frequencies—3.49 GHz and 5 GHz—is depicted in Figure 11.6 for understanding the proposed two-port hexagonal ring molded MIMO radio wire's reduced shared coupling normally.

At 3.5 GHz, it is determined that the next array element is shielded from the initial antenna element's electric field when array elements are provided with the source. Similarly, when the next array element is excited at 4.9 GHz, the first antenna is shielded from the second antenna's electric field [6]. This is because the reversed T-shaped isolating element is inserted between the array element and serves as a stop band filter, preventing interference while simultaneously operating as the desired band.

Figure 11.7 indicates the three-dimensional (3D) radiation patterns of both array elements at 3.5 GHz and 4.9 GHz. This confirms that the proposed dual-port MIMO array radiates around 3600 KHz at 3.5 GHz, meeting the requirement for virtual communication [3]

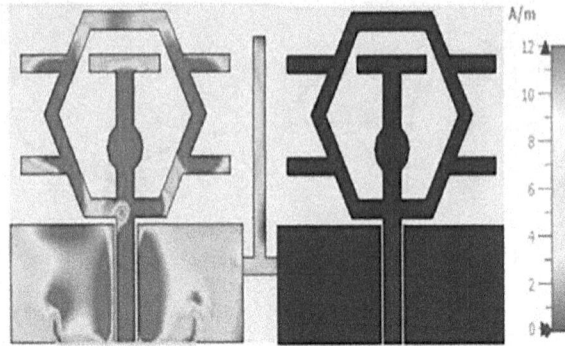

Figure 11.6 Surface current distribution of proposed two-port MIMO antenna at 3.5 GHz.

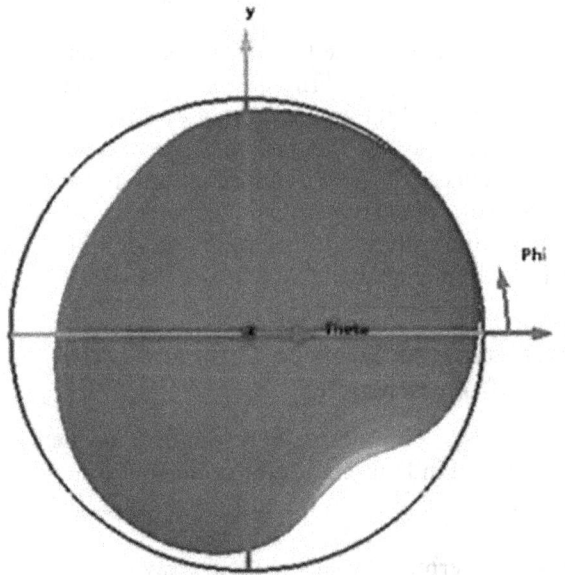

Figure 11.7 The-dimensional radiation patterns of planned dual-port MIMO antenna at 3.5 GHz.

11.4.3 Gain (Measured in dBi) and Efficiency (Measured in Percent) in Relation to Frequency (GHz)

The proposed MIMO antenna's gain (dBi) and efficiency (percent) in relation to frequency (GHz) are shown in Figure 11.7. The MIMO antenna has been observed to have a gain greater than 2 dBi and a radiation efficiency

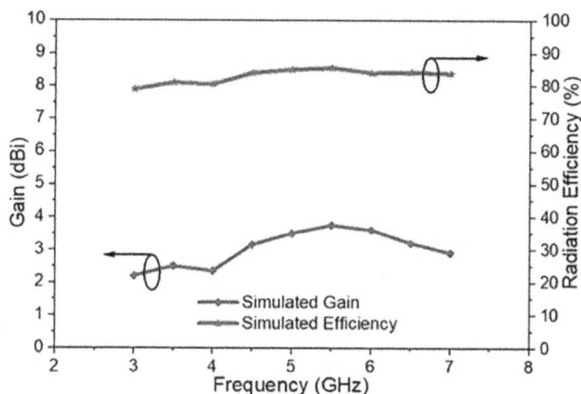

Figure 11.8 Gain and efficiency of proposed two-port MIMO antenna.

of 80.90% in the upper WLAN recurrence band and the lower recurrence band for cell phone applications. The recipe to gauge the increase is given as shown:

$$GAUT = (PR2 / PR3) \cdot GREF \qquad (11.1)$$

GAUT is the gain of the antenna under test, PR2 is the power received by the reference antenna, and PR3 is the power received by the antenna under test. GREF is the gain of the reference antenna. We can use this formula to calculate gain. First, convert dBm to dB, since dBm – 30 equals dB.

11.5 CONCLUSION

Here we have studied the design of a MIMO antenna for WLAN in handheld devices. The MIMO antenna is not difficult to configure, can be based on a modest Frequency Range-4 substrate, is little, is simple to make, and can be effortlessly mounted inside genuine cell phones to work in ideal working groups. In addition, it possesses the ECC and DG characteristics and good impedance bandwidth in the mobile device applications' preferred operating bands.

REFERENCES

[1] J. Kulkarni, A. Desai, and C.-Y. Desmond Sim, "Two port CPW-fed MIMO antenna with wide bandwidth and high isolation for future wireless applications", International Journal of RF and Microwave Computer-Aided Engineering, e22700, 2021. https://doi.org/10.1002/mmce.22700.

[2] T. Addepalli and V. R. Anitha, "Compact two-port MIMO antenna with high isolation using parasitic reflectors for UWB, X and Ku band applications," Progress in Electromagnetics Research C, vol. 102, 63–77, 2020.

[3] A. Dkiouak, A. Zakriti, M. Ouahabi, A. Zugari, and M. Khalladi, "Design of a compact MIMO antenna for wireless applications," Progress in Electromagnetics Research M, vol. 72, 115–124, 2018.

[4] J. Kulkarni, C.-Y.-D. Sim, and V. Deshpande, "Low-profile, compact, two port MIMO antenna conforming Wi-Fi-5/Wi-Fi6/V2X/DSRC/INSAT-C for wireless industrial applications," 2020 IEEE 17th India Council International Conference (INDICON), 2020, pp. 1–5. https://doi.org/10.1109/INDICON49873.2020.9342514.

[5] H. Ekrami and S. Jam, "A compact triple-band dual-element MIMO antenna with high port-to-port isolation for wireless applications," International Journal of Electronics and Communications, 2018. https://doi.org/10.1016/j.aeue.2018.09.044

[6] R. N. Tiwari, P. Singh, S. Pandey, R. Anand, D. Singh, and B. Kanaujia, "Swastika shaped slot embedded two port dual frequency band MIMO antenna for wireless applications," Analog Integrated Circuits and Signal Processing, vol. 109, 103–113, 2021.

[7] A. Iqbal, O. A Saraereh, A. Bouazizi, and A. Basir, "Metamaterial based highly isolated MIMO antenna for portable wireless applications," Electronics, vol. 7, no. 10, p. 267, 2018.

[8] J. Kulkarni, C.-Y.-D. Sim, A. Chitre, N. Kulkarni, S. Kulkarni, and R. Talware, "Design and analysis of compact 2D MIMO Sub-6 GHz 5G flexible antenna," 2021 IEEE Madras Section Conference (MASCON), 2021, pp. 1–5. https://doi.org/10.1109/MASCON51689.2021.9563492.

[9] J. Kulkarni, N. Kulkarni, C.-Y.-D. Sim, and A. Desai, "A two-port dual band microstrip feed based cylindrical dielectric resonator antenna array for sub-6 GHz 5G and super extended-C band applications," 2021 International Conference on Communication Information and Computing Technology (ICCICT), Mumbai, India, 2021, pp. 1–5. doi: 10.1109/ICCICT50803.2021.9510136.

[10] J. Kulkarni, S. Dhabre, S. Kulkarni, C.-Y. D. Sim, R. K. Gangwar, and K. Cengiz, "Six-port symmetrical CPW-fed MIMO antenna for futuristic smartphone devices," 2021 6th International Conference for Convergence in Technology (I2CT), 2021, pp. 1–5.

[11] Ankan Bhattacharya, Bappadittya Roy, Santosh K. Chowdhury and Anup K. Bhattacharjee, "Computational and experimental analysis of a low-profile, isolation-enhanced, band-notch UWB-MIMO antenna," Journal of Computational Electronics, vol. 18, 680–688, 2019. https://doi.org/10.1007/s10825-019-01309-3.

[12] A. De, B. Roy, A. Bhattacharya, and A. K. Bhattacharjee, "Bandwidth-enhanced ultra-wide band wearable textile antenna for various WBAN and Internet of Things (IoT) applications," Radio Science, vol. 56, e2021RS007315, 2021. https://doi.org/10.1029/2021RS007315.

[13] A. Bhattacharya, B. Roy, R. Caldeirinha, and A. Bhattacharjee, "Low-profile, extremely wideband, dual-band-notched MIMO antenna for UWB applications," International Journal of Microwave and Wireless Technologies, vol. 11(7), 719–728, 2019. https://doi.org/10.1017/S1759078719000266.

Chapter 12

Harmonic Suppression Triple-Band U-Slot Antenna for GPS/WLAN/5G Applications

J. Rajeshwar Goud, N. V. Koteswara Rao, and A. Mallikarjuna Prasad

12.1 INTRODUCTION

Base station antennas are used to attain the global positioning system (GPS), wireless local area network (WLAN), and 2G, 3G, 4G and 5G frequency bands. In general, numerous mobile network providers operate on distinct antennas to obtain each band. Hence, these network providers need independent base stations, which in turn shoot up maintenance costs. Multi-band antennas can be used to overcome this problem. Numerous designs are developed using multi-broadband antennas. The dual band is achieved by using folded dipole [1, 2], quasi-fractal slotted ground plane [3], loop slot [4] and open-ended slots [5]. According to reports, the triple band achieved with a printed planar monopole [6, 7], slot with L-shaped slits [8], F-shaped slots [9] and a pair of electromagnetic band gap (EBG) structures [10].

The fact that these broadband antennas pick up undesired frequencies is another barrier to employing them. There is a need to utilize a filtration device to detach these undesired frequencies. This limitation leads to an invention of designs using multi-narrowband antennas [11] which can pick up only desired frequencies in base station applications.

Dual narrow band is achieved using crossed dipole radiators [12], S-shaped slots [13], open-ended slots [14], U-slot [15], π-slot [16], two triangular rings [17] and asymmetric M-shaped patch [18]. However, the performance of the antenna degrades as a result of the harmonic frequencies that these antennas emit. In many wireless communication systems, notably in active integrated antennas (AIAs), the suppression of higher-order harmonics is a crucial design consideration.

Leading wireless communication and phased array radar systems have an extreme need for small, light transmitters. In these systems, high efficiency power amplifiers generate significant higher-order harmonics. An integrated antenna shouldn't emit harmonic radiations to meet electromagnetic interference/electromagnetic compatibility (EMI/EMC) compliance standards. The performance of the entire system is greatly diminished because of the harmonics radiation throughout the antenna. In most cases to resolve this

DOI: 10.1201/9781003459880-12

185

issue, an additional filter circuit is required to eliminate these undesired signals. Usually, this strategy is suitable for all cases because it results in an extra insertion loss and enlarges the radio frequency (RF) front end. In the literature, some typical strategies have been investigated to address this issue, including slotted ground plane structures or defective ground structures (DGSs) like photonic bandgaps (PBGs). PBG structures are among the most widely used methods for eliminating harmonic signals from the microstrip structure [19–21], along with circular, square ring slot antenna [22, 23], T-shaped wide slot antenna [24], wiggly line [25] and a circular slot loop antenna with a coplanar waveguide (CPW) feed [26].

Highly complex periodic slots etched into the ground plane are challenging to apply on a PBG structure in millimeter-wave or microwave components and increases the complexity of design. There are too many design factors that affect the bandgap property such as the relative volume proportion, lattice form and spacing. Another successful strategy is the use of DGSs below the microstrip feed line. Due to the DGS, the unit cell serves as the stopband filter in some particular frequency bands, and antenna properties differ from those of the reference antenna. A 1-D PBG structure with DGS should be used for harmonic rejection and impedance matching. The undesirable high back-radiation is the main disadvantage of combining DGS and EBG/PBG structures on the antenna's ground plane. Particularly with microwave integrated circuits, the defected physical area is a crucial factor from the design perspective. Additional components are integrated to design a compact structure with minor defects.

The impedance parameters of the patch element of an active microstrip patch antenna are severely perturbed, which results in pattern anomalies linked to the higher harmonics. To improve the performance of an antenna, it is necessary to suppress the harmonics.

The proposed antenna geometry and antenna design details are discussed in the next section. The three different design equations for three resonant frequencies are also noted. A parametric study of the antenna is also included to explain variation in resonant frequencies. The experimental and simulation results comparison is illustrated in following section.

12.2 ANTENNA DESIGN AND ANALYSIS

12.2.1 Antenna Configuration

An inset feed microstrip antenna has been designed to operate at the WLAN band. The length (L_P) and width (W_P) of the patch are calculated from standard formulae [27]. The simulation procedure has optimized the stub dimensions. Figure 12.1 shows a design of a microstrip antenna with and without stubs. Figures 12.7 and 12.8 depict the fabrication of the harmonic

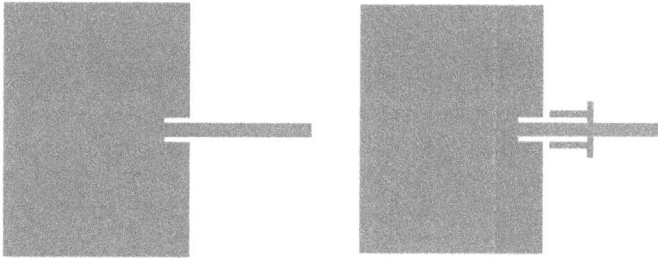

Figure 12.1 Inset feed antenna (a) without stubs and (b) with stubs.

suppressed inset-feed triple-band U-slot antenna and its geometry. The proposed antenna covers GPS, WLAN and sub-6 5G bands. The substrate size of 63 mm × 70 mm is considered for the fabrication of the antenna. To match the impedance, the proposed antenna uses a 50-strip line. The triple band is obtained by creating one symmetrical horizontal slot 'S_2' and two vertical slots 'S_1' in the form of 'U' and is shown in Figure 12.7. The return loss of the harmonic-suppressed antenna with and without the U-slot is shown in Figure 12.3. The FR4 substrate used for this design has a thickness of 1.6 mm, a loss tangent of 0.02 and a dielectric constant of 4.4. Table 12.1 displays the specific dimensions of the antenna and stubs.

In this study, we examine a better harmonic frequency suppression technique that integrates stubs with the feed line. The back-radiation present in EBG/PBG and DGS structures is not present in the proposed construction. In order to remove higher-order harmonics, we use a plane microstrip T-shaped stub with electrical length ($W_F + 2L_{S1} + 2L_{S2}$) that is approximately equal to $\lambda g / 4$ (dielectric wavelength) on either side of the feed line, where λg is the guide wavelength

Table 12.1 Design Dimensions

Parameters	Dimensions (mm)	Parameters	Dimensions (mm)
L_P	29	L_{IF}	4
W_P	40	W_{IF}	4
S_1	19.5	L_F	25
S_2	20	W_F	2.2
S_3	3	L_{S1}	3.4
S_4	1	L_{S2}	5.5
P_1	4.5	W_S	1
P_2	2	L_G	63
P_3	7	W_G	70
h	1.6	X	7

of the first harmonic. The fundamental frequency of the antenna is unaffected by the presence of stubs on the feed line and is as shown in Figure 12.2.

It can be seen that the patch antenna harmonics have been completely eliminated by comparing the harmonic properties of a microstrip antenna with and without tuning stubs. To achieve the optimized stub dimensions, parametric research was carried out. Investigations have been carried out on how the presence of stubs in the antenna design affects the antenna

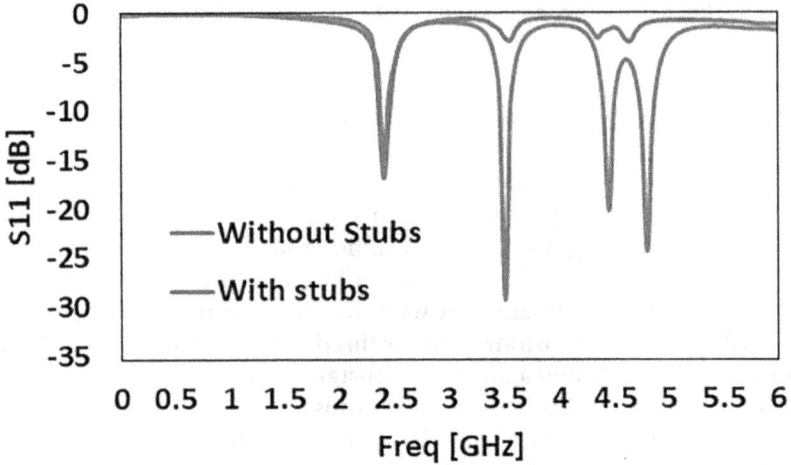

Figure 12.2 Return loss plot of inset-feed patch antenna.

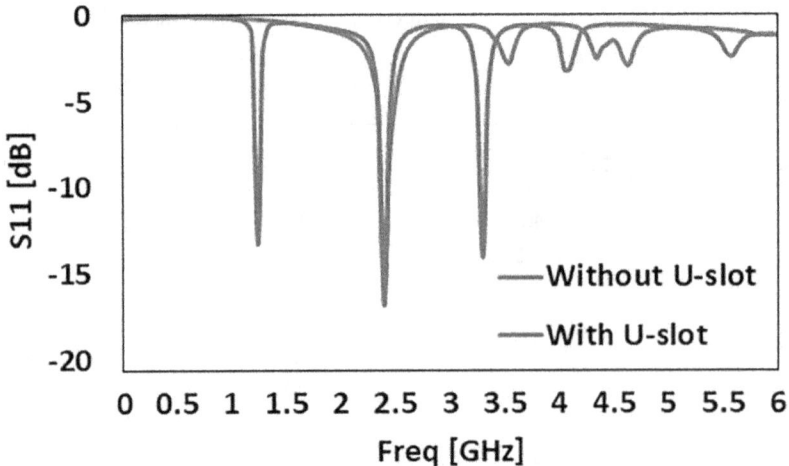

Figure 12.3 Return loss of harmonic-suppressed inset feed with and without U-slot antenna.

parameters. Ultimately, the concept of using different combinations of stubs in prototype antennas have been studied, measured, fabricated and verified.

Figure 12.2 depicts the simulated return loss of the microstrip antenna and the stub integrated patch. According to the findings, there has been a significant suppression of the higher-order modes occurring at 3.5 GHz (first harmonic), 4.45 GHz (second harmonic) and 4.8 GHz (third harmonic). Additionally, this minimizes radiation at particular frequencies, which have been thoroughly studied. As demonstrated in Figures 12.4, 12.5 and 12.6, the harmonics are minimized at X = 7 mm, L_{S1} = 3.4 mm and L_{S2} = 5.5 mm, respectively.

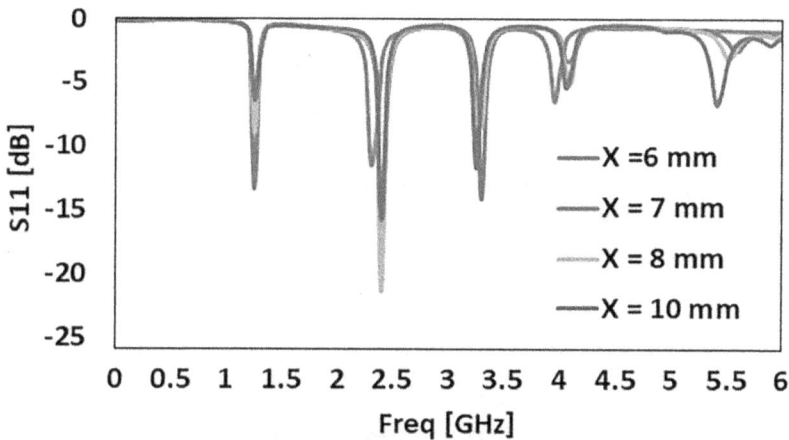

Figure 12.4 Return loss of harmonic-suppressed inset feed of various 'X' values.

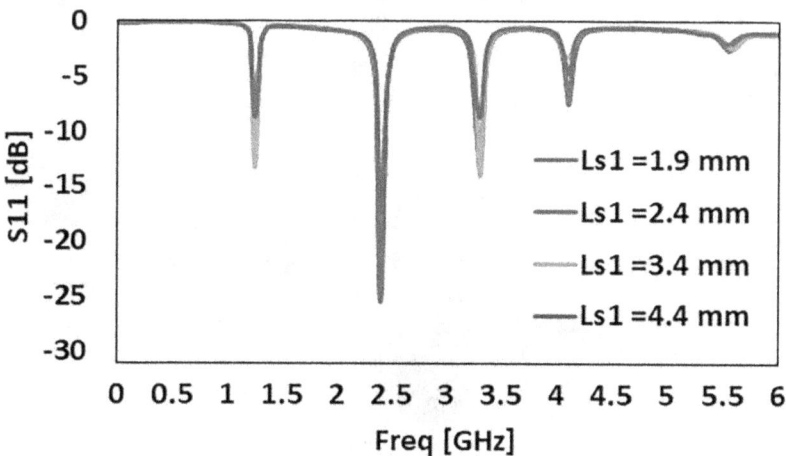

Figure 12.5 Return loss of harmonic-suppressed inset feed of various 'L_{S1}' values.

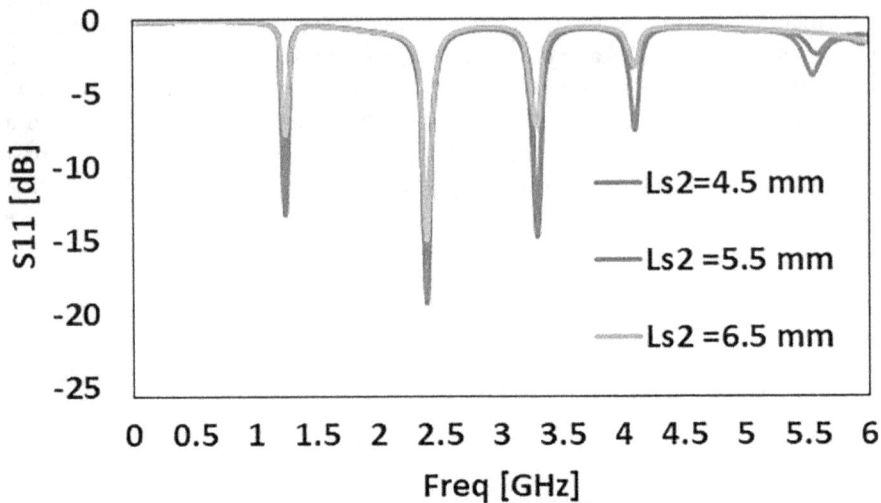

Figure 12.6 Return loss of harmonic-suppressed inset feed of various 'L_{S2}' values.

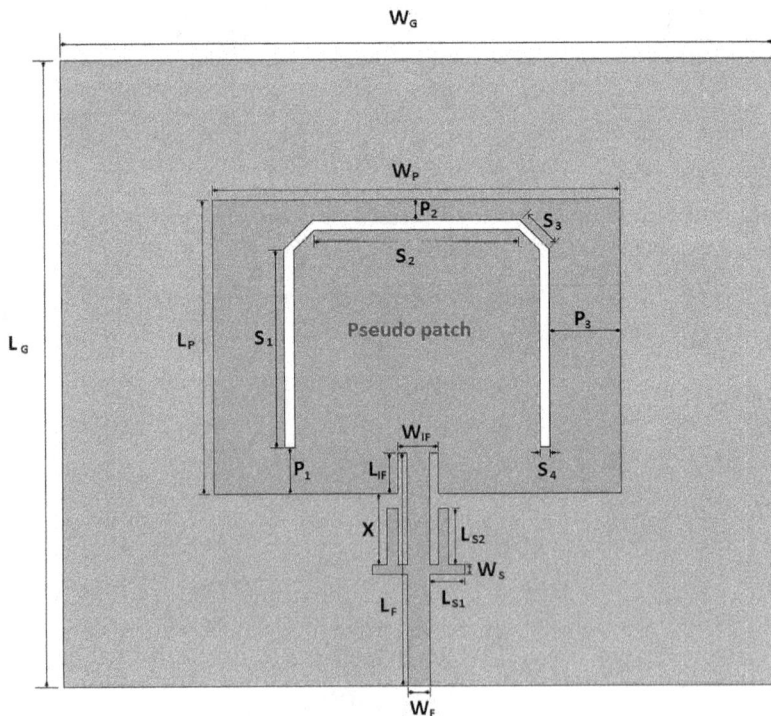

Figure 12.7 Harmonic-suppressed inset feed U-slot antenna.

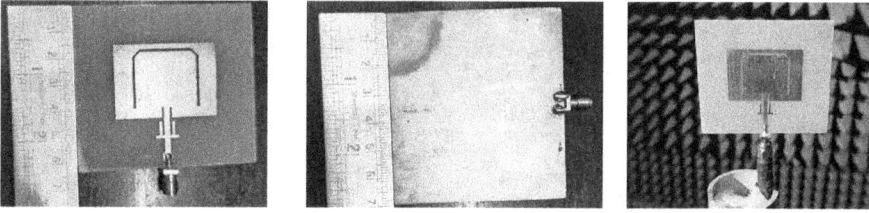

Figure 12.8 Fabricated harmonic-suppressed inset feed U-slot antenna. (a) Top view. (b) Bottom view. (c) Measurement of antenna using anechoic chamber.

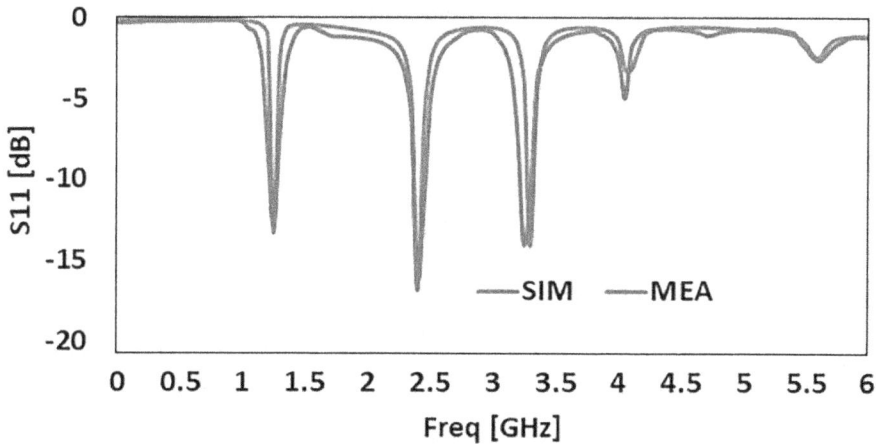

Figure 12.9 Return loss of harmonic-suppressed inset feed U-slot antenna.

12.2.2 Design and Analysis of Resonance Frequencies

The resonance frequencies are based on the patch, slot and position (i.e., 'P$_1$', 'P$_2$' and 'P$_3$') parameters. Feed widths (W$_F$) and inset feed lengths (L$_F$) have no discernible impact on resonance frequencies. Design guidelines are created for the coaxial feed rectangular U-slot microstrip antenna [28]. However, it is necessary to optimize the design expressions for inset feed structures that are suggested in [27] and [28]. This section offers significant expressions for inset feed architectures as well as design dimensions that have been optimized using the High-Frequency Structure Simulator (HFSS). The following equations (Equations 12.1–12.7) contain the analytical formulae for the resonance frequencies $f_1, f_2,$ and f_3.

First resonant frequency (f$_1$): It has been found that f_1 depends on the patch's length, the vertical slot's length (S$_1$) and the horizontal slot's length (S$_2$). The positions of 'P$_1$' and 'P$_2$', the diagonal length 'S$_3$' and the width of the vertical slot 'S$_4$' dimensions determine the vertical slot 'S$_1$'. The following equation contains the first resonance [15, 16] formula (12.1).

$$f_1 = \frac{C}{2\left(\sqrt{\epsilon_f}\right)\left(L_P + 2\Delta L_P + \frac{(S_2 + P_1)}{2} + S_1 + S_4 + P_2 - P_3\right)} \tag{12.1}$$

where C is light velocity.

$$\epsilon_f = \frac{\epsilon_r + 1}{2} + \frac{\epsilon_r - 1}{2}\left[1 + 12\frac{h}{W_P}\right]^{-\frac{1}{2}} \tag{12.2}$$

$$\Delta L = 0.412h\frac{\left[\left(\epsilon_f + 0.3\right)\left(\frac{W_P}{h} + 0.264\right)\right]}{\left(\epsilon_f - 0.258\right)\left(\frac{W_P}{h} + 0.8\right)} \tag{12.3}$$

Second resonant frequency (f$_2$): It has been found that f_2 depends on the patch length (L), positions 'P$_1$', 'P$_2$' and 'P$_3$' and the width of the vertical slot (S$_4$). The following equation contains the second resonance [15, 16] formula (12.4).

$$f_2 = \frac{C}{2\left(\sqrt{\epsilon_f}\right)\left(L_P + 2\Delta L_P + +P_3 - P_1 - P_2 - S_4\right)} \tag{12.4}$$

Third resonant frequency (f$_3$): According to the geometry in Figure 12.1, a pseudo patch (PP) is created when a U-slot is present, and the third resonance frequency relies on the dimensions of the PP. The following equation provides the third resonance's [15, 16] closest approximation (12.5).

$$f_3 = \frac{C}{2\left(\sqrt{\epsilon_{r(PP)}}\right)\left(S_1 + 2\Delta S_1 + \frac{S_3}{2} + \frac{P_1}{2} - \frac{S_1}{2P_1}\right)} \tag{12.5}$$

$$\epsilon_{f(PP)} = \frac{\epsilon_r + 1}{2} + \frac{\epsilon_r - 1}{2}\left[1 + 12\frac{h}{(S_2 + \frac{S_3}{2})}\right]^{-\frac{1}{2}} \tag{12.6}$$

$$\Delta S_1 = 0.412h \left[\frac{\left(\epsilon_{f(PP)} + 0.3 \right) \left(\frac{\left(S_2 + \frac{S_3}{2} \right)}{h} + 0.264 \right)}{\left(\epsilon_{f(PP)} - 0.258 \right) \left(\frac{\left(S_2 + \frac{S_3}{2} \right)}{h} + 0.8 \right)} \right] \tag{12.7}$$

12.2.3 Parametric Study of Patch and Slot Dimensions

As mentioned previously, the slot positions (i.e., 'P$_1$,' 'P$_2$' and 'P$_3$') and size mostly affect the resonance frequencies. Every parameter is changed with a 0.5-mm step size, and the changes to these parameters are described next.

12.2.3.1 Position P$_1$' Impact on Resonance Frequencies

All resonance frequencies will be impacted by position 'P$_1$'. The first resonance frequency has a maximum difference of 9 MHz, the second resonance frequency is 5 MHz and the third resonance frequency is 21 MHz. With each 0.5 mm, the first, second and third resonance frequencies will increase by about 9 MHz, 5 MHz and 21 MHz, respectively. Figure 12.10 displays these changes.

12.2.3.2 Position 'P$_2$' Impact on Resonance Frequencies

Position 'P$_2$' will influence the second and third resonance frequencies. For these, the proposed antenna's highest difference between analytical and simulation findings is 20 MHz and 31 MHz, respectively. The second and third

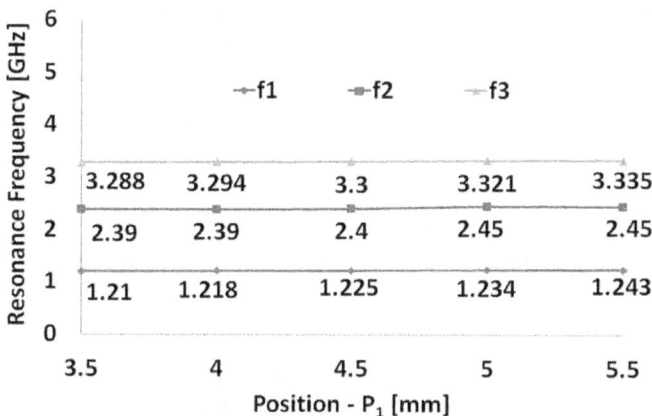

Figure 12.10 Effect of position 'P$_1$'.

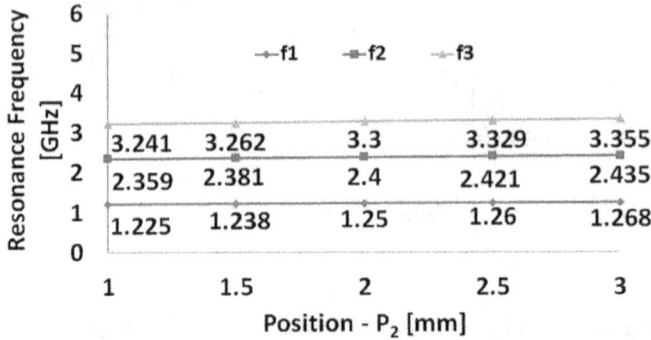

Figure 12.11 Effect of position 'P₂'.

resonance frequencies will increase by roughly 22 MHz and 39 MHz each per 0.5 mm. In Figure 12.11, these variances are displayed.

12.2.3.3 Position 'P₃' Impact on Resonance Frequencies

The first and second resonance frequencies will be impacted by position 'P₃'. The highest difference between analytical and simulation results is determined to be 11 MHz for the first resonance frequency and 15 MHz for the second resonance frequency. The first resonance frequency will increase by around 12 MHz with every 0.5 mm increase, whereas the second resonance frequency will decrease by about 21 MHz. In Figure 12.12, these variances are displayed.

12.2.3.4 Position 'S₄' Impact on Resonance Frequencies

All slots have the same slot width (S₄), and it has been found that the slot width will not affect the first resonance frequency. Increased slot width causes the second and third resonance frequencies to rise, as shown in Figure 12.9.

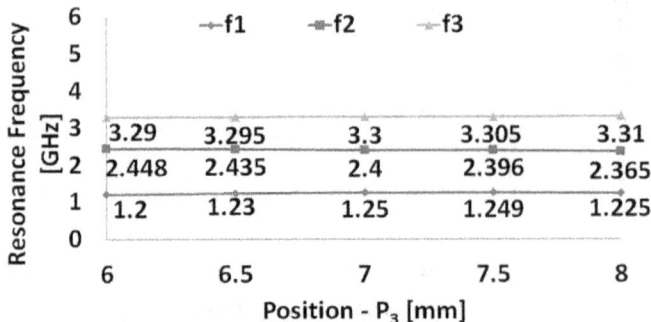

Figure 12.12 Effect of position 'P₃'.

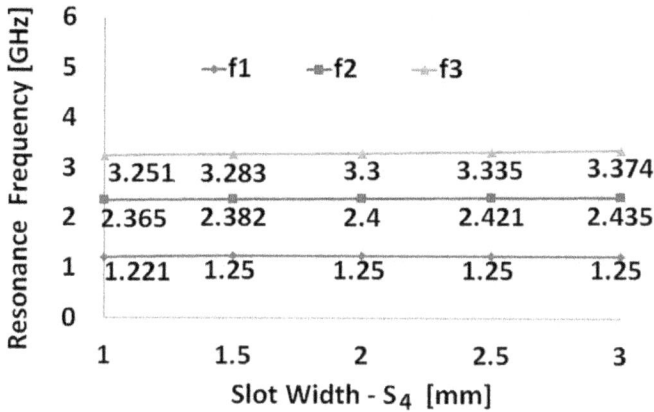

Figure 12.13 Effect of position 'S₂'.

The second and third resonance frequencies rise by approximately 21 MHz and 35 MHz each per 0.5 mm, as illustrated in Figure 12.13.

While selecting appropriate dimensions for U slots and patches to obtain the required resonance frequencies for base station applications, design expressions and parametric analysis are crucial. Table 12.1 displays the antenna dimensions for the three bands that are being suggested.

12.3 RESULTS AND DISCUSSION

The proposed antenna simulated and measured results are described in this section. Figure 12.2 shows the radiation patterns at the fundamental and harmonic frequencies with and without stubs. Figure 12.2 also demonstrates how adding stubs to the antenna reduces each harmonic frequency emission by more than 5 dBi without altering the fundamental frequency

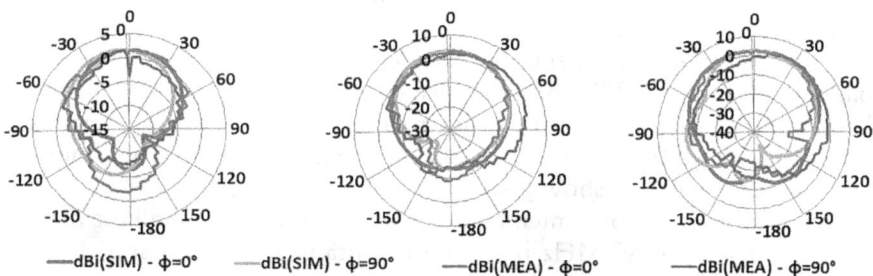

Figure 12.14 E- and H-plane patterns at (a) 1250 MHz, (b) 2400 GHz and (c) 3300 GHz.

Table 12.2 Proposed Antenna Parameters

Parameter	Band	U-slot Antenna (SIM)-[AN]	HS U-slot Antenna (MEA)
Resonance frequency (MHz)	Lower	1250-[1271]	1250
	Middle	2400-[2469]	2400
	Upper	3300-[3285]	3300
Return loss (dB)	Lower	−13.28	−12.8
	Middle	−16.84	−16.01
	Upper	−14.07	−13.93
Bandwidth (GHz)	Lower	1.21–1.26	1.22–1.27
	Middle	2.38–2.45	2.38–2.45
	Upper	3.26–3.34	3.25–3.33
VSWR	Lower	1.55	1.59
	Middle	1.33	1.37
	Upper	1.49	1.50
Peak gain (dBi)	Lower	1.53	1.48
	Middle	3.28	3.15
	Upper	2.49	2.37
Radiation efficiency (%)	Lower	38.26	37.89
	Middle	43.38	42.93
	Upper	40.71	4016

Note: SIM: simulated; AN: analytical; MEA: measured; HS: harmonic suppressed.

pattern. According to the measured results in Figure 12.9, all configurations of the stub integrated patch antennas reduce both harmonics by more than 70%.

The E- and H-plane radiation patterns of simulated and measured results of the proposed antenna are shown in Figure 12.14. In accordance with simulations and measurements, the inset feed triple-band U-slot antenna first band return loss is −1328 dB, second band return loss is −16.84 dB and third band return loss is −14.07 dB, as shown in Figure 12.9. Table 12.2 provides a summary of all further recommended antenna parameters. The simulated and measured patterns show good agreement.

Bandwidths of various antennas are listed in Table 12.3. In all the defined bands, a minimum 50 MHz bandwidth is attained, which is sufficient for base stations [27].

Table 12.3 Comparison of Proposed Antenna with Dual- and Triple-Band Antennas

Ref.	Size (mm) (L × W × H)	f_1-[BW$_1$] (MHz)	f_2-[BW$_2$] (MHz)	f_3-[BW$_3$] (MHz)	Gain (dBi)
[4]	55 × 57 × 1.6	2.45 GHz-[420 MHz]	3.5 GHz-[680 MHz]	-------	f_1:2.17 f_2:3.7
[5]	19.8 × 19.4 × 1.27	433 MHz-[154 MHz]	2.45 GH-[500MHz]	-------	f_1: 2.7 f_2: 2.2
[17]	20 × 20 ×1.57	8.33 GHz-[260 MHz]	9.6 GHz-[130 MHz]	10.78 GHz-[230 MHz]	f_1:6.2 f_2:5.3 f_3:6.0
[18]	62 × 64 × 2	2.4 GHz-[110 MHz]	3.5 GHz-[140 MHz]	5.8 GHz-[630 MHz]	f_1:1.65 f_2:3.73 f_3:6.32
[23]	100 × 91.4 × 0.51	4 GHz-[2 GHz]	-------	-------	f_1:4.5
[24]	60 × 40 × 0.51	2.4 GHz-[90 MHz]	-------	-------	f_1:1.12
Proposed	63 × 70 × 1.6	1.25 GHz-[50 MHz]	2.4 GHz-[70 MHz]	3.3 GHz-[80 MHz]	f_1:1.53 f_2:3.28 f_3:2.49

Note: f: resonance frequency; BW: bandwidth.

HFSS was used to perform numerical analysis of the antenna's characteristics, and a network analyzer and anechoic chamber are used to carry out the measurements. The measured and simulated findings from prototype antennas with stubs showed a striking degree of agreement. Moving forward, this same harmonic-suppressed triple-band design can be extended to apply to a multiple-input, multiple-output (MIMO) structure to achieve good channel capacity and a high data rate [29].

12.4 CONCLUSION

The harmonic-suppressed triple-band U-slot antenna is presented for base stations. The harmonic frequencies are minimized solely controlled by the stub. In particular, where printed antennas are implemented on a single substrate along with active components and circuitry, this architecture may find usage in microwave integrated circuits. The stubs that are suppressing harmonics have no effect on other properties like copolarized radiation over the principal planes, gain, input impedance or return loss of three specific frequency bands. The suggested method can be utilized to create better harmonic frequency control when higher-order harmonics become a significant restriction in some antennas with desirable features. The proposed antenna recommended dimensions should be chosen in accordance with the planned equations and parametric studies in order to achieve the appropriate resonance frequencies. The suggested antenna operates at first, second and third resonance frequencies of 1225 MHz, 2400 MHz and 3300 MHz, respectively, to cover the GPS, WLAN, and sub-6 5G bands. Also shown were the simulated and measured E- and H-plane patterns.

REFERENCES

1. X. Liu, Graduate Student Member, IEEE, S. Gao, Fellow, IEEE, B. Sanz-Izquierdo, Member, IEEE, H. Zhang, Lehu Wen, Member, IEEE, W. Hu, Member, IEEE, Q. Luo, Senior Member, IEEE, J. Tetuko Sri Sumantyo, Senior Member, IEEE, and X.-X. Yang, Senior Member, IEEE, 2022, December. A Mutual-Coupling-Suppressed Dual-Band Dual-Polarized Base Station Antenna Using Multiple Folded-Dipole Antenna. *IEEE Transactions on Antennas and Propagation*, vol. 70, no. 12, pp. 11582–11594. doi: 10.1109/TAP.2022.3209177.
2. Z. Wang, G.-X. Zhang, Y. Yin and J. Wu., 2014. Design of a Dual-Band High-Gain Antenna Array for WLAN and WiMAX Base Station. *IEEE Antennas and Wireless Propagation Letters*, vol. 13, pp. 1721–1724.
3. T. Hong, S.-X. Gong, Y. Liu and W. Jiang., 2010. Monopole Antenna with Quasi-Fractal Slotted Ground Plane for Dual-Band Applications. *IEEE Antennas and Wireless Propagation Letters*, vol. 9, pp. 595–598.
4. M.-T. Tan and B.-Z. Wang., 2015. A Compact Dual-Band Dual-Polarized Loop-Slot Planar Antenna. *IEEE Antennas and Wireless Propagation Letters*, vol. 14, pp. 1742–1745.
5. L.-J. Xu, Y.-X. Guo and W. Wu., 2012. Dual-Band Implantable Antenna with Open-End Slots on Ground. *IEEE Antennas and Wireless Propagation Letters*, vol. 11, pp. 1564–1567.
6. W. T. Li, X. W. Shi and Y. Q. Hei., 2009. Novel Planar UWB Monopole Antenna with Triple Band-Notched Characteristics. *IEEE Antennas and Wireless Propagation Letters*, vol. 8, pp. 1094–1098.
7. L. Li, X. Zhang, X. Yin and L. Zhou., 2016. A Compact Triple-Band Printed Monopole Antenna for WLAN/WiMAX Applications. *IEEE Antennas and Wireless Propagation Letters*, vol. 15, pp. 1853–1855.
8. J. G. Baek and K. C. Hwang., 2013. Triple-Band Unidirectional Circularly Polarized Hexagonal Slot Antenna with Multiple L-Shaped Slits. *IEEE Transactions on Antennas and Propagation*, vol. 61, no. 9, pp. 4831–4835.
9. A. K. Gautam, L. Kumar, B. K. Kanaujia and K. Rambabu., 2016. Design of Compact F-Shaped Slot Triple-Band Antenna for WLAN/WiMAX Applications. *IEEE Transactions on Antennas and Propagation*, vol. 64, no. 3, pp. 1101–1105.
10. A. Abbas, N. Hussain, J. Lee, S. G. Park and N. Kim., 2021. Triple Rectangular Notch UWB Antenna using EBG and SRR. *IEEE Access*, vol. 9, pp. 2508–2515.

11. Kai Fong Lee and Kwai Man Luk., 2011. *Microstrip Patch Antennas.* Imperial College Press.

12. Y. Chen, L. Chen, H. Wang, X.-T. Gu and X.-W. Shi., 2013. Dual-Band Crossed-Dipole Reflect array with Dual-Band Frequency Selective Surface. *IEEE Antennas and Wireless Propagation Letters*, vol. 12, pp. 1157–1160.

13. Z. N. Chen Nasimuddin and X. Qing., 2010. Dual-Band Circularly Polarized S-Shaped Slotted Patch Antenna with a Small Frequency-Ratio. *IEEE Transactions on Antennas and Propagation*, vol. 58, no. 6, pp. 2112–2115.

14. L.-J. Xu, Y.-X. Guo and W. Wu., 2012. Dual-Band Implantable Antenna with Open-End Slots on Ground. *IEEE Antennas and Wireless Propagation Letters*, vol. 11, pp. 1564–1567.

15. J. Rajeshwar Goud, N. V. Koteswara Rao and A. M. Prasad., 2020. Design of Triple Band U-Slot MIMO Antenna for Simultaneous Uplink and Downlink Communications. *Progress in Electromagnetics Research C*, vol. 106, pp. 271–283.

16. R. G. Jangampally, V. K. R. Nalam and M. P. Avala., 2022. Design of Uplink and Downlink Triple Band π: Slot Antennas for Simultaneous Communication. *Wireless Personal Communications*, vol. 124, pp. 3189–3203.

17. T. Zhang, W. Hong and K. Wu., 2015. A Low-Profile Triple-Band Triple-Polarization Antenna with Two Triangular Rings. *IEEE Antennas and Wireless Propagation Letters*, vol. 14, pp. 378–381.

18. L. Peng, C.-L. Ruan and X.-H. Wu., 2010. Design and Operation of Dual/Triple-Band Asymmetric M-Shaped Microstrip Patch Antennas. *IEEE Antennas and Wireless Propagation Letters*, vol. 9, pp. 1069–1072.

19. Y. J. Sung, M. Kim and Y. S. Kim., 2003. Harmonics reduction with defected ground structure for a microstrip patch antenna. *IEEE Antennas and Wireless Propagation Letters*, vol. 2, pp. 111–113.

20. Haiwen Liu, Zhengfan Li, Xiaowei Sun and Junfa Mao., 2005. Harmonic suppression with photonic bandgap and defected ground structure for a microstrip patch antenna. *IEEE Microwave and Wireless Components Letters*, vol. 15, no. 2, pp. 55–56.

21. Hyungrak Kim, Kwang Sun Hwang, Kihun Chang and Young Joong Yoon., 2004. Novel slot antennas for harmonic suppression. *IEEE Microwave and Wireless Components Letters*, vol. 14, no. 6, pp. 286–288,

22. W. Li, Y. Wang, B. You, Z. Shi and Q. H. Liu., 2018. Compact Ring Slot Antenna With Harmonic Suppression. *IEEE Antennas and Wireless Propagation Letters*, vol. 17, no. 12, pp. 2459–2463.

23. C.-Y.-D. Sim, M.-H. Chang and B.-Y. Chen., 2014. Microstrip-Fed Ring Slot Antenna Design With Wideband Harmonic Suppression. *IEEE Transactions on Antennas and Propagation*, vol. 62, no. 9, pp. 4828–4832.

24. N.-A. Nguyen, R. Ahmad, Y.-T. Im, Y.-S. Shin and S.-O. Park., 2007. A T-Shaped Wide-Slot Harmonic Suppression Antenna. *IEEE Antennas and Wireless Propagation Letters*, vol. 6, pp. 647–650.

25. S. I. Kwak, J. H. Kwon, D.-U. Sim, K. Chang and Y. J. Yoon., 2010. Design of the Printed Slot Antenna using Wiggly Line with Harmonic Suppression. *IEEE Antennas and Wireless Propagation Letters*, vol. 9, pp. 741–743.

26. Y.-W. Liu, Y.-J. Lu and P. Hsu., 2014. Harmonic Suppressed Slot Loop Antenna Fed by Coplanar Waveguide. *IEEE Antennas and Wireless Propagation Letters*, vol. 13, pp. 1292–1295.
27. Constantine A. Balanis., 2005. *Antenna Theory: Analysis and Design*, 4th Edition. John Wiley & Sons, Inc. https://www.wiley.com/en-in/Antenna+Theory:+Analysis+and+Design,+4th+Edition-p-9781118642061.
28. S. Weigand, G. H. Huff, K. H. Pan and J. T. Bernhard, 2003. Analysis and Design of Broad-Band Single-Layer Rectangular U-Slot Microstrip Patch Antennas. *IEEE Transactions on Antennas and Propagation*, vol. 51, no. 3, pp. 457–468.
29. A. Bhattacharya, B. Roy, R. Caldeirinha and A. Bhattacharjee., 2019. Low-Profile, Extremely Wideband, Dual-Band-Notched MIMO Antenna for UWB Applications. *International Journal of Microwave and Wireless Technologies*, vol. 11, no. 7, pp. 719–728.

Chapter 13

Mutual Coupling Reduction in a Patch Antenna Array Using a Microstrip Resonator for Wireless Communication System Applications

Santimoy Mandal and Chandan Kumar Ghosh

13.1 INTRODUCTION

Patch antennas are extensively applicable for different wireless communication systems like multi-input, multi-output (MIMO) communication systems owing to its advantageous features like low profile, low cost and simple to fabricate [1–4]. Mutual coupling arises when the patch antenna arrays are configured [5]. Because of electromagnetic coupling, the antenna performance characteristics can be tainted, for example, poorer antenna efficiency, impedance mismatching and a deviation from the radiation pattern of the array antenna [6]. Electromagnetic coupling reduction is the most recent challenge for designing an antenna array structure. One of the conformist approaches for electromagnetic coupling reduction is to maximize the distance between antenna elements. In the open literature it has been suggested that spacing between the antenna elements must be larger than 0.5 λ [7]. Numerous techniques are used to minimize the mutual coupling among closely spaced antenna radiators like electromagnetic band gap structure (EBG) [8, 9], defected ground structure (DGS) in [1, 10–12], meta-materials [6, 13, 14], resonators [15] and microstrip slots [16, 17]. Baseband algorithm and signal processing [7] are also used to resolve the issue caused by electromagnetic coupling among antenna radiators. Several other methods are also available for minimizing the mutual coupling among antenna elements like microstrip line resonator [8], inverted U-shaped microstrip resonator [9], parasitic elements [10], etc. A mutual coupling suppression of 35 dB has been observed among two E-structured antenna radiators in [8]. Furthermore, a 17 dB electromagnetic coupling suppression has been studied in [9] and 25 dB in [10]. Some other methods are also available for the reduction of electromagnetic coupling among antenna radiators with a three-dimensional antenna structure like an antenna with boxy metallic wall in [11], antenna with observer [12], etc. There are certain disadvantages of these three-dimensional structures like complex design and large structure. Some of the most commonly used antenna structures like DGS [13, 14] and EBG [15, 16] are also available for electromagnetic coupling supersession among antenna radiators. Nevertheless, it suffers from high back-radiation which degrades the radiation pattern. A noteworthy

DOI: 10.1201/9781003459880-13

quantity of electromagnetic coupling reduction is observed in [17]. This structure also suffers from high back-radiation. A 43 dB electromagnetic coupling reduction was observed in [18] using a simple U-shaped resonator where 0.6λ spacing is used among antenna arrays. A significant amount of back-radiation has been observed in this technique. On the other hand, electromagnetic coupling affects the radiation characteristics of the designed antenna. The problem of isolation is a very crucial factor in MIMO antennas, as discussed in articles in [19–22]. Numerous techniques are also presented in the open literature for the reduction of electromagnetic coupling like DGS [23–25], EBG [26–32] and parasitic resonator among antenna elements [33–37]. All these structures are designed on the substrate among antenna radiators to disrupt the current circulation in the radiators. Due to that, a high line capacitance and inductance have been developed in the microstrip line. Consequently, it achieves an ample stop band as well as miniaturization of antenna size, which is suitable for the applications of different wireless communications systems. In [34], a structure consisting of two parasitic elements is used, though it suffers from design complexity. A complicated structure has been implemented in [35] for the reduction of mutual coupling. It also suffers from high back-radiation. An electromagnetic coupling reduction of 18 dB is observed in [36], using a complex split ring resonator structure. Another technique is used for the suppression of electromagnetic coupling among antenna radiators [37] where the current vectors are canceled out at the center of the resonator due to its $\lambda/2$ line resonator. MIMO antenna systems comprise diverse applications in wireless communication systems owing to several radiators deployed at both transmitter and receiver ends. To overcome multipath fading, these radiators are engaged for both sending and receiving superior signals. In contrast with the single-input, single-output (SISO), the MIMO antenna system has certain advantageous features like enhanced data rate, reliability without any extra power and bandwidth requirement. Some of the major disadvantages for a MIMO antenna is mutual coupling, as multiple antennas have been placed in the same substrate and the radiated energy is absorbed by the adjacent antennas. Due to that, electromagnetic coupling arises between adjacent antenna radiators.

The potential differences vs. current association for the attached antenna structure is expressed as:

$$V - ZI$$

$$
\begin{pmatrix} E_1 \\ E_2 \\ \cdot \\ \cdot \\ \cdot \\ E_m \end{pmatrix}_{m \times 1}
=
\begin{pmatrix}
Z_{11} & Z_{12} & Z_{13} & \cdots & Z_{1n} \\
Z_{21} & Z_{22} & Z_{23} & \cdots & Z_{2n} \\
\cdot & \cdot & \cdot & \cdots & \cdot \\
\cdot & \cdot & \cdot & \cdots & \cdot \\
\cdot & \cdot & \cdot & \cdots & \cdot \\
Z_{m1} & Z_{m2} & Z_{m3} & \cdots & Z_{mn}
\end{pmatrix}_{m \times n}
\begin{pmatrix} I_1 \\ I_2 \\ \cdot \\ \cdot \\ \cdot \\ I_n \end{pmatrix}_{n \times 1}
$$

where $E_1, E_2, E_3 \ldots E_m \equiv$ Voltages of antenna (1, 2 . . . m).

$I_1, I_2, I_3 \ldots I_n \equiv$ Current flow through antennas $(1, 2 \ldots n)$.

$Z_{11}, Z_{22} \ldots Z_{mn} \equiv$ Auto-impedance of antenna $(1, 2 \ldots n)$.

$Z_{ij} (i \neq j) \equiv$ Mutual impedance of all antennas $(1, 2, 3, \ldots .. n)$.

The appearances of mutual coupling of various antenna $Z_{ij} (i \neq j)$ are different, so the near-field scatterer (NFS) is propagated among all the antennas and the mutual coupling configuration is transformed. In the occurrence of NFS, the previous matrix transforms to:

$$
\begin{pmatrix} e_1 \\ e_2 \\ \cdot \\ e_n \end{pmatrix} = \begin{pmatrix} Z_{11} + Z_{11}' & Z_{12} + Z_{12}' & \cdots & Z_{1n} + Z_{1n}' \\ Z_{21} + Z_{21}' & Z_{2n} + Z_{2n}' & \cdots & Z_{2n} + Z_{2n}' \\ \cdot & \cdot & \cdots & \cdot \\ Z_{m1} + Z_{m1}' & & \cdots & Z_{mn} + Z_{mn}' \end{pmatrix} \begin{pmatrix} i_1 \\ i_2 \\ \cdot \\ i_n \end{pmatrix}
$$

where $e_1, e_2 \ldots e_n \equiv$ Transformed voltages, $i_1, i_2 \ldots i_n \equiv$ Transformed currents and $Z_{mn} + Z_{mn}' \equiv$ Transformed impedances.

Numerous other procedures have been discussed by different authors for the minimization of electromagnetic coupling in [38–72]. Some of the techniques are like DGS [38–43], EBG [44–46], parasitic elements [47–49], microstrip resonator [50–56], electromagnetic soft surface structure (EMSS), frequency selective surface structure (FSSS), metamaterial structure [63], coupling matrix-based band stop filter [64], mender line resonator [65–66], etc. (Table 13.1).

Here, a straightforward technique is projected to remove the occurrences of mutual coupling among antenna radiators. A resonator structure is being considered and is placed among the microstrip antennas and results in a mutual coupling reduction of more than 43 dB. The anticipated design is being simulated by the end-to-end procedure of a numerical model by means of the method of moment (MOM) supported by the IE3D electromagnetic simulator and the S-parametric analysis of the same has been presented. The planned antenna diagram is shown in Figure 13.1.

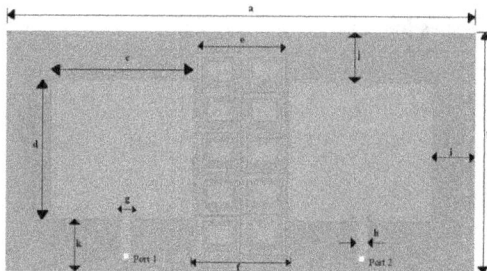

Figure 13.1 Schematic diagram of patch antennas with a resonator sandwiched between two antenna elements (all dimensions are in mm).

13.2 ANTENNA DESIGN AND ITS CONFIGURATION

The aim of the proposed design was to minimize the effect of electromagnetic coupling on the array antennas. Here, in this projected design we have developed a microstrip resonator for the minimization of the electromagnetic coupling among neighboring radiators and require a trouble-free fabrication procedure. Consequently, once the resonator structure and the antenna are placed in the same layer, this helps to shrink the electromagnetic coupling so that it also improves the radiation pattern of the anticipated antenna. In this design we planned an original procedure, where a coaxial probe feed is used among two identical rectangular patches. A unit cell representation of the microstrip resonator is presented in Figure 13.2.

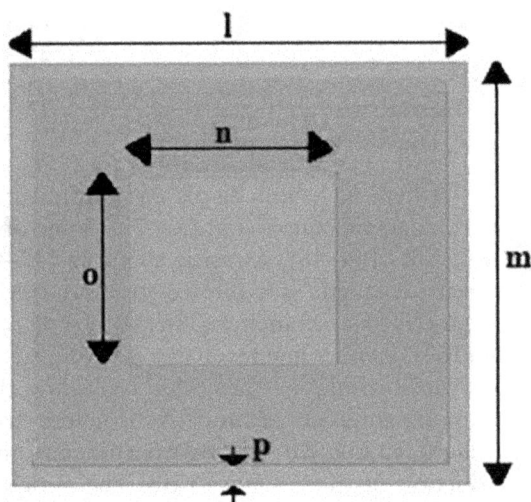

Figure 13.2 Unit cell representation of the microstrip resonator.

Table 13.1 Device Parameters of the Proposed Antenna

Parameters	Values (mm)	Parameters	Values (mm)	
a	70	f	20	
b	60	g	2	
c	20	h	1.6	
d	20	i	5	
e	19	j	15	
k	15	l		9
m	8	n		6
o	5.5	p		0.2

13.3 SIMULATION AND MEASUREMENT RESULTS

Here, a 70×60 mm^2 FR4 substrate has been considered with permittivity (ε_r) 4.4, width 1.6 mm and loss tangent 0.02. Figure 13.3 illustrates the deviation of computer-generated S-parameters with frequency of the planned antenna with and without a resonator. Here we come across an apparent mark of approximately 48 dB electromagnetic coupling suppression at 3.8 GHz with the existence of the resonator structure. The electrical distinctiveness of the projected antenna is depicted in Figure 13.4. The E-plane radiation pattern revealed in Figure 13.5 depicts the existence of a resonator structure that prevents degradation of the pattern in the broadside direction and no back-radiation.

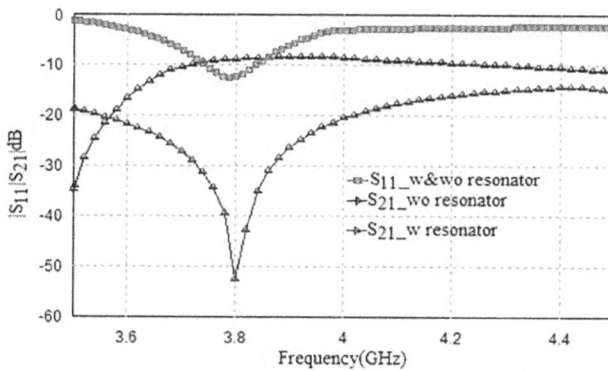

Figure 13.3 Simulated S-parameter variations with frequency with and without a resonator.

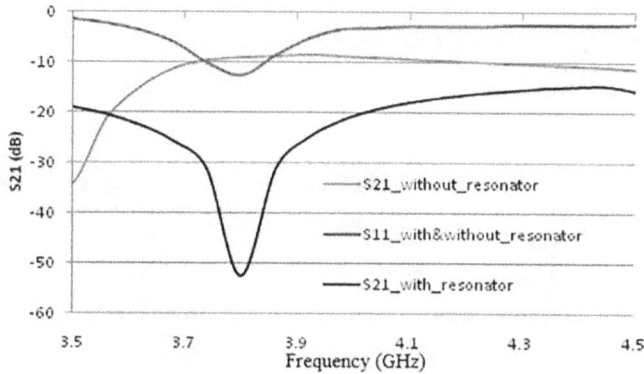

Figure 13.4 Study of S_{11} and S_{21} parameters with and without a resonator.

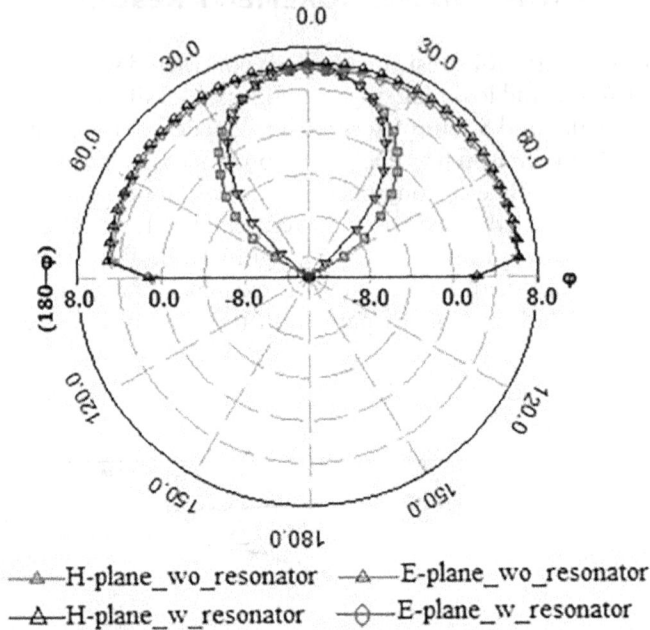

Figure 13.5 Elevation pattern gain with and without a resonator structure.

The simulated current circulation pattern of the proposed antenna with a resonator is shown in Figure 13.6(a). The current vectors are un-evenly distributed in the radiators and in the resonator structure at 3.8 GHz (resonance frequency). Because of that, an electrical wall is being established among nearby antenna radiators to disallow the signal from impending on another radiating element. Furthermore, Figure 13.6(b) shows the occurrence of an electromagnetic coupling with no resonator resulting in one inactive radiating element with no resonator structure.

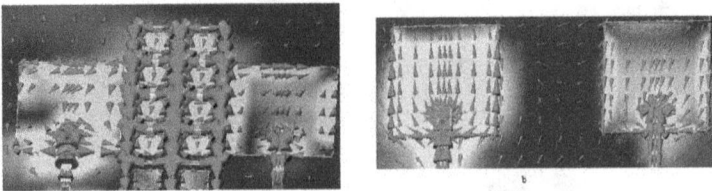

Figure 13.6 (a) Simulated current distribution with resonator structure and (b) without resonator structure.

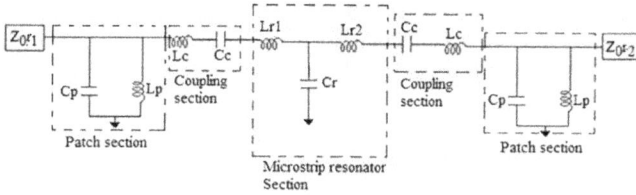

Figure 13.7 Equivalent circuit diagram of the proposed model.

Figure 13.7 indicates the theoretical equivalent model representation for proper perspective where three different circuit sections are connected together like patch section, coupling section and decoupling section. For each patch section Lp is the patch inductance and Cp is the patch capacitance, which are connected in parallel. For the coupling section, a series connected inductance (L_c) and capacitance (C_c) are used. The decoupling structure consists of parallel inductance (L_{r1}) and (L_{r2}) with a capacitor (C_r). The coupling unit is used to couple all these three sections together. By varying the length and gap among the two successive parallel wings of the resonator section, a strong decoupling phenomenon exits.

13.4 CONCLUSION

Our purpose is to study the prospective outcome of electromagnetic coupling among two contiguous antenna elements as the microstrip resonator structure was positioned among the radiators. A MOM-based IE3D simulator is used to simulate the projected design and optimize it all the way through the progression of a numerical simulation. At the resonance frequency of 3.8 GHz a significant amount of electromagnetic coupling suppression has been achieved as depicted in the S_{11}, S_{21} vs. frequency characteristics of the planned structure; in addition, it does not generate any back-radiation and also it does not alter the other antenna characteristics. The anticipated antenna has relevance in satellite communication, mobile communication, Wi-Fi technology, etc.

REFERENCES

1. Wei, K., Li, J. Y., Wang, L., Xing, Z. J., and Xu., R., 2016. Mutual coupling reduction by novel fractal defected ground structure bandgap filter. *IEEE Transactions on Antennas and Propagation*, vol. 64, no. 10, pp. 4328–4335.
2. Hwangbo, S., Yang, H. Y., and Yoon, Y. K., 2017. Mutual coupling reduction using micro-machined complementary meander line slots for a patch array antenna. *IEEE Antennas and Wireless Propagation Letters*, vol. 16, pp. 1667–1670.

3. Emadeddin, A., Shad, S., Rahimian, Z., and Hassani, H. R., 2017. High mutual coupling reduction between microstrip patch antennas using novel structure. *AEU—International Journal of Electronics and Communications*, vol. 71, pp. 152–156.

4. Wu, G.-C., Wang, G.-M., Liang, J.-G. Gao, X.-J., and Zhu, L., 2015. Novel ultra-compact two-dimensional waveguide-based metasurface for electromagnetic coupling reduction of microstrip antenna array. *International Journal of RF and Microwave Computer-Aided Engineering*, vol. 25, no. 9, pp. 789–794.

5. Murch, R. D., and Letaief, K. B., 2002. Antenna systems for broadband wireless access. *IEEE Communications Magazine*, vol. 40, no. 4, pp. 76–83.

6. Yang, X. M., Liu, X. G., Zhou, X. Y., and Cui, T. J., 2012. Reduction of mutual coupling between closely packed patch antennas using wave guided metamaterials. *IEEE Antennas and Wireless Propagation Letters*, vol. 11, pp. 389–391.

7. Mohammadian, A. H., Martin, N. M., and Griffin, D. W., 1989. A theoretical and experimental study of mutual coupling in microstrip antenna arrays. *IEEE Transactions on Antennas and Propagation*, vol. 37, no. 10, pp. 1217–1223.

8. Rajo-Iglesias, E., Quevedo-Teruel, Ó., and Inclan-Sanchez, L., 2008. Mutual coupling reduction in patch antenna arrays by using a planar EBG structure and a multilayer dielectric substrate. *IEEE Transactions on Antennas and Propagation*, vol. 56, no. 6, pp. 1648–1655.

9. Lee, J. Y., Kim, S. H., and Jang, J. H., 2015. Reduction of mutual coupling in planar multiple antenna by using 1-D EBG and SRR structures. *IEEE Transactions on Antennas and Propagation*, vol. 63, no. 9, pp. 4194–4198.

10. Zhu, F. G., Xu, J. D., and Xu, Q., 2009. Reduction of mutual coupling between closely-packed antenna elements using defected ground structure. *Electronics Letters*, vol. 45, no. 12, pp. 601–602.

11. Salehi, M., and Tavakoli, A., 2006. A novel low mutual coupling microstrip antenna array design using defected ground structure. *AEU—International Journal of Electronics and communications*, vol. 60, no. 10, pp. 718–723.

12. Mahmoudian, A., Rashed-Mohassel, J., and Kong, J. A., 2008. Reduction of EMI and mutual coupling in array antennas by using DGS and AMC structure. *Progress in Electromagnetics Research Symposium, Vols I and II, Proceedings, Electromagnetics Academy*, p. 106.

13. Bait-Suwailam, M. M., Siddiqui, O. F., and Ramahi, O. M., 2010. Mutual coupling reduction between microstrip patch antennas using slotted-complementary split-ring resonators. *IEEE Antennas and Wireless Propagation Letters*, vol. 9, pp. 876–878.

14. Qamar, Z., and Park, H.-C., 2014. Compact waveguided metamaterials for suppression of mutual coupling in microstrip array. *Progress in Electromagnetics Research*, vol. 149, pp. 183–192.

15. Ghosh, J., Ghosal, S., Mitra, D., and Bhadra Chaudhuri, S. R., 2016. Mutual coupling reduction between closely placed microstrip patch antenna using meander line resonator. *Progress in Electromagnetics Research Letters*, vol. 59, pp. 115–122.

16. Ou Yang, J., Yang, F., and Wang, Z. M., 2011. Reducing mutual coupling of closely spaced microstrip MIMO antennas for WLAN application. *IEEE Antennas and Wireless Propagation Letters*, vol. 10, pp. 310–313.

17. Sonkki, M., and Salonen, E., 2010. Low mutual coupling between monopole antennas by using two λ/2 slots. *IEEE Antennas and Wireless Propagation Letters*, vol. 9, pp. 138–141.

18. Mandal, S., and Ghosh, C. K., 2021. Low mutual coupling of microstrip antenna array integrated with dollar shaped resonator. *Wireless Personal Communications*, vol. 119, pp. 777–789. https://doi.org/10.1007/s11277-021-08237-1.

19. Ouyang, J., Yang, F., and Wang, Z. M., 2011. Reducing mutual coupling of closely spaced microstrip MIMO antennas for WLAN application. *IEEE Antennas and Wireless Propagation Letters*, vol. 10, pp. 310–312.

20. Sato, H., Koyanagi, Y., Ogawa, K., and Takahashi, M., 2013. A method of dual-frequency decoupling for two-element MIMO antenna. *PIERS Proceedings*, Stockholm, 12–15 August, pp. 1853–1857.

21. Chen, S. C., Wang, Y. S., and Chung, S. J., 2008. A decoupling technique for increasing the port isolation between two strongly coupled antennas. *IEEE Transactions on Antennas and Propagation*, vol. 56, no. 12, pp. 3650–3658.

22. Sato, H., Koyanagi, Y., Ogawa, K., and Takahashi, M., 2011. A method of dual-frequency decoupling for closely spaced two small antennas. *IEICE Transactions B*, vol. J94-B-II, no. 9, pp. 1104–1113.

23. Zhu, F. G., Xu, J. D., and Xu, Q., 2009. Reduction of mutual coupling between closely-packed antenna elements using defected ground structure. *Electronics Letters*, vol. 45, no. 12, pp. 601–602.

24. Mandal, M. K., and Sanyal, S., 2006. A novel defected ground structure for planar circuits. *IEEE Microwave and Wireless Components Letters*, vol. 16, no. 2, 93–95.

25. Ahn, D., Park, J. S., Kim, C. S., Kim, J., Qian, Y., and Itoh, T., 2001. A design of the low-pass filter using the novel microstrip defected ground structure. *IEEE Transactions on Microwave Theory and Techniques*, vol. 49, no. 1, pp. 86–93.

26. Yang, F., and Samii, Y. R., 2003. Microstrip antennas integrated with electromagnetic band-gap (EBG) structures: A low mutual coupling design for array applications. *IEEE Transactions on Antennas and Propagation*, vol. 51, no. 10, pp. 2936–2946.

27. Zhang, L., Castaneda, J. A., and Alexopoulos, N. G., 2004. Scan blindness free phased array design using PBG materials. *IEEE Transactions on Antennas and Propagation*, vol. 52, no. 8, pp. 2000–2007.

28. Iluz, Z., Shavit, R., and Bauer, R., 2004. Microstrip antenna phased array with electromagnetic band gap substrate. *IEEE Transactions on Antennas and Propagation*, vol. 52, no. 6, pp. 1446–1453.

29. Farahani, H. S., Veysi, M., Kamyab, M., and Tadjalli, A., 2010. Mutual coupling reduction in patch antenna arrays using a UC-EBG superstrate. *IEEE Antennas and Wireless Propagation Letters*, vol. 9, pp. 57–59.

30. Fu, Y., and Yuan, N., 2004. Elimination of scan blindness in phased array of microstrip patches using electromagnetic band gap materials. *IEEE Antennas and Wireless Propagation Letters*, vol. 3, no. 1, pp. 63–65.

31. Yu, A., and Zhang, X., 2003. A novel method to improve the performance of microstrip antenna arrays using a dumbbell EBG structure. *IEEE Antennas and Wireless Propagation Letters*, vol. 2, no. 1, pp. 170–172.

32. Rajo-Iglesias, E., Quevedo-Teruel, O., and Inclan-Sanchez, L., 2008. Mutual coupling reduction in patch antenna arrays by using a planar EBG structure and a multilayer dielectric substrate. *IEEE Transactions on Antennas and Propagation*, vol. 56, no. 6, pp. 1648–1655.

33. Nikolic, M., Djordjevic, A., and Nehorai, A., 2005. Microstrip antennas with suppressed radiation in horizontal directions and reduced coupling. *IEEE Transactions on Antennas and Propagation*, vol. 53, no. 11, pp. 3469–3476.

34. Habashi, A., Nourinia, J., and Ghobadi, C., 2011. Mutual coupling reduction between very closely spaced patch antennas using low-profile folded split-ring resonators. *IEEE Antennas and Wireless Propagation Letters*, vol. 10, pp. 862–865.

35. Farsi, S., Aliakbarian, H., Nauwelaers, B., and Vandenbosch, G. A. E., 2012. Mutual coupling reduction between planar antenna by using a simple microstrip U-section. *IEEE Antennas and Wireless Propagation Letters*, vol. 11, pp. 1501–1503.

36. Bait-Suwailam, M. M., Siddiqui, O. F., and Ramahi, O. M., 2010. Mutual coupling reduction between microstrip patch antennas using slotted-complementary split-ring resonators. *IEEE Antennas and Wireless Propagation Letters*, vol. 9, pp. 876–878.

37. Ghosh, C. K., and Parui, S. K., 2013. Reduction of mutual coupling between E-shaped microstrip antenna array by using a simple microstrip I-section. *Microwave and Optical Technology Letter*, vol. 55, no. 11, pp. 2544–2549.

38. Ghosh, C. K., 2016. A compact 4-channel microstrip MIMO antenna with reduced mutual coupling. *International Journal of Electronics and Communications (AEÜ)*, vol. 70, pp. 873–879.

39. Teng, L., Jingrui, P., and Shoulin, Y., 2016. Mutual coupling optimization of compact microstrip array antenna. *Indonesian Journal of Electrical Engineering and Computer Science*, vol. 4, no. 3, pp. 538–541.

40. Acharjee, J., Mandal, K., and Sujit Kumar, M., 2018. Reduction of mutual coupling and cross polarization of a MIMO/diversity antenna using a string of H-shaped DGS. *AEU-International Journal of Electronics and Communications*. doi:10.1016/j.aeue.2018.09.037.

41. Kun, W., Li, J.-Y., Ling, Wang, Z.-J., Xing, R. X., 2016. Mutual coupling reduction of microstrip antenna array by periodic defected ground structures. *IEEE 5th Asia-Pacific Conference on Antennas and Propagation (APCAP)*. doi:10.1109/apcap.

42. Zhang, Z., Wei, K., Xie, J., Li, J., and Wang, L., 2018. The MIMO antenna array with mutual coupling reduction and cross-polarization suppression by defected ground structures. *Radio Engineering*, vol. 27, no. 4. doi:10.13164/re.2018.0969.

43. Ghosh, C. K., Bappaditya, M., and Susanta, K. P., 2014. Mutual coupling reduction of a dual-frequency microstrip antenna array by using U-shaped DGS and inverted U-shaped microstrip resonator. *Progress in Electromagnetic Research C*, vol. 48, pp. 61–68.

44. Ou Yang, J., Yang, F., and Wang, Z. M., 2011. Reducing mutual coupling of closely spaced microstrip MIMO antennas for WLAN applications. *IEEE Antennas and Wireless Propagation Letters*, vol. 10.

45. Sajjad, H., Khan, S., and Arvas, E., 2017. Mutual coupling reduction in array elements using EBG structures. *International Applied Computational Electromagnetics Society Symposium—Italy*. doi:10.23919/ropaces.2017.7916410.

46. Sarbandi, H., Mehdi, F., Kamyab, V. M., and Tadjalli, A. 2010. Mutual coupling reduction in patch antenna arrays using a UC-EBG superstrate. *IEEE Antennas and Wireless Propagation Letters*, vol. 9.

47. Faraz, F., Li, Q., Chen, X., Abdullah, M., Zhang, S., and Zhang, A., 2019. Mutual coupling reduction for linearly arranged MIMO antenna cross strait quad-regional. *Radio Science and Wireless Technology Conference (CSQRWC-2019)*. doi:10.1109/csqrwc.2019.8799266.

48. Emadeddin, A., Shad, S., Rahimian, Z., and Hassani, H. R., 2017. High mutual coupling reduction between microstrip patch antennas using novel structure. *AEU—International Journal of Electronics and Communications*, vol. 71, pp. 152–156. doi:10.1016/j.aeue.2016.10.017.

49. Sun, X.-B., and Cao, M.-Y., 2017. Mutual coupling reduction in an antenna array by using two parasitic microstrips. *AEU—International Journal of Electronics and Communications*, vol. 74, pp. 1–4. doi:10.1016/j.aeue.2017.01.013.

50. Saadallah, A., Raghad, G., and Yetkin, G. Ö., 2018. Mutual coupling reduction of E-shaped MIMO antenna with matrix of C-shaped resonators. *International Journal of Antennas and Propagation*, p. 4814176. doi:10.1155/2018/4814176.

51. Suwailam-Bait, M. M., Siddiqui, O. F., and Ramahi, O. M., 2010. Mutual coupling reduction between microstrip patch antennas using slotted-complementary split-ring resonators. *IEEE Antennas and Wireless Propagation Letters*, vol. 9, pp. 876–878.

52. Bhattacharjee, S., and Ghosh, C. K., 2020. Reduction of mutual coupling between two adjacent microstrip antennas using I-shaped resonators. *2020 IEEE International Conference on Advent Trends in Multidisciplinary Research and Innovation (ICATMRI-2020)*, Buldhana, India, pp. 1–5. doi: 10.1109/ICATMRI51801.2020.9398327.

53. Ghosh, C. K., Hazra, R., Biswas, A., Bhattachrjee, A. K., and Parui, S. K., 2014. Suppression of cross-polarization and mutual coupling between dual trace dual column coaxial microstrip array using dumbbell-shaped resonator. *Microwave and Optical Technology Letters*, vol. 56, no. 9, pp. 2182–2186. doi:10.1002/mop.28531.

54. Mandal, S., and Ghosh, C. K., 2019. Reduction of the mutual coupling in patch antenna arrays based on dollar shaped electrical resonator. *Lecture notes on Electrical Engineering 602, proceeding of 2nd international conference on Communication, Devices and Computing ICCDC 2019*, Springer, pp. 71–76.

55. Ghosh, C. K., and Parui, S. K., 2013. Reduction of mutual coupling between E-shaped microstrip antennas by using a simple microstrip I-section. *Microwave and Optical Technology Letters*, vol. 55, no. 11, pp. 2544–2549. doi: 10.1002/mop.27928.

56. Ghosh, C. K., Pratap, M., Kumar, R., and Pratap, S., 2020. Mutual coupling reduction of microstrip MIMO antenna using microstrip resonator. *Wireless Personal Communications*. doi: 10.1007/s11277–020–07138-z.

57. Mandal, S., and Ghosh, C. K., 2022. Mutual coupling reduction in a patch antenna array based on planar frequency selective surface structure. *Radio Science*, vol. 57, no. 2, p. e2021RS007392. https://doi.org/10.1029/2021RS007392.

58. Abushamleh, S., Al-Rizzo, H., Abbosh, A., and Kishk, A. A., 2013. Mutual coupling reduction between two patch antennas using a new miniaturized soft surface structure. *IEEE AP-S Dig*, pp. 1822–1823.

59. Islam, M. T., and Alam, S., 2013. Design of high impedance electromagnetic surfaces for mutual coupling reduction in patch antenna array. *Materials*, vol. 6, pp. 143–155. doi:10.3390/ma6010143.

60. Quevedo-Teruel, Ó., Inclán-Sánchez, L., and Rajo-Iglesias, E., 2010. Soft surfaces for reducing mutual coupling between loaded PIFA antennas. *IEEE Antennas and Wireless Propagation Letters*, vol. 9, pp. 91–94. doi:10.1109/lawp.2010.2043632.

61. Akbari, M., Ali, M. M., Farahani, M., Sebak, A. R., and Denidni, T., 2017. Spatially mutual coupling reduction between CP-MIMO antennas using FSS superstrate. *Electronics Letters*, vol. 53, no. 8, pp. 516–518.

62. Beiranvand, E., Afsahy, M., and Sharbati, V., 2017. Reduction of the mutual coupling in patch antenna arrays based on EBG by using a planar frequency-selective surface structure. *International Journal of Microwave and Wireless Technologies*. doi: 10.1017/S1759078715001440.

63. Bhattacharjee, S., Mandal, S., Ghosh, C. K., and Banerjee, S., 2022. Reduction of mutual coupling and cross-polarization of microstrip MIMO antenna using electromagnetic soft surface (EMSS). *Radio Science*, vol. 57, no. 5, pp. 1–10. doi: 10.1029/2021RS007377.

64. Yu, K., Leo Liu, X., and Li, Y., 2017. Mutual coupling reduction of microstrip patch antenna array using modified split ring resonator metamaterial structures. *IEEE International Symposium on Antennas and Propagation & USNC/URSI National Radio Science Meeting*. doi:10.1109/apusncursinrsm.2017.8073186.

65. Bilal, T., 2020. Mutual coupling reduction using coupling matrix based band stop filter. *International Journal of Electronics and Communications*. doi: 10.1016/j.aeue.2020.153342.

66. Biswas, S., Ghosh, C. K., Banerjee, S., Mandal, S., and Mandal, D., 2020. High port isolation of a dual polarized microstrip antenna array using DGS. *Journal of Electromagnetic Waves and Applications*, vol. 34, no. 6, pp. 683–696. doi: 10.1080/09205071.2020.1736647.

67. Ghosh, J., Ghosal, S., Mitra, D., and Chaudhuri, S. R. B., 2016. Mutual coupling reduction between closely placed microstrip patch antenna using meander line resonator. *Progress in Electromagnetics Research Letters*, vol. 59, pp. 115–122.

68. James, J. R., and Hall, P. S. Handbook of microstrip and printed antennas. Wiley, New York, 1997.

69. Ghosh, C. K., Rana, B., and Parui, S. K., 2013. Reduction of cross polarization of slotted microstrip antenna array using spiral-ring resonator. *Microwave and Optical Technology Letters*, vol. 55, no. 9. doi:10.1002/mop27778.

70. De, A., Roy, B., Bhattacharya, A., and Bhattacharjee, A. K., 2021. Bandwidth-enhanced ultra-wide band wearable textile antenna for various WBAN and Internet of Things (IoT) applications. *Radio Science*, vol. 56, p. e2021RS007315. https://doi.org/10.1029/2021RS007315

71. Bhattacharya, A., Roy, B., Caldeirinha, R., and Bhattacharjee, A., 2019. Low-profile, extremely wideband, dual-band-notched MIMO antenna for UWB applications. *International Journal of Microwave and Wireless Technologies*, vol. 11, no. 7, pp. 79–728. doi:10.1017/S1759078719000266

72. Bhattacharya, Ankan, Roy, Bappadittya, Chowdhury, Santosh K., and Bhattacharjee, Anup K., 2019. Computational and experimental analysis of a low-profile, isolation-enhanced, band-notch UWB-MIMO antenna. *Journal of Computational Electronics*, vol. 18, pp. 680–688. https://doi.org/10.1007/s10825-019-01309-3

Chapter 14

Filter Synthesis–Based Compact Dual-Band Filtenna for C-Band Applications

Pravesh Pal, Rashmi Sinha, Ranjeet Kumar,
Praveen Kumar, and Santosh Kumar Mahto

14.1 INTRODUCTION

With the exponential growth of wireless communication services such as satellite communication, cellular communication, Wi-Fi, infrared communication, and Bluetooth, the integrated design of circuit elements is becoming increasingly important in the development of miniaturised devices. Traditionally, the filter and antenna of a wireless system are made separately and connected by a 50-ohm line, thus resulting in large system volume. Lately, researchers have investigated the possibility of merging a filter and an antenna in one component, referred to as filtenna. In addition to the filtenna having filtering and radiating characteristics, they also have features such as closely packed, inexpensive, and low crosstalk.

In the past, antenna input impedance is optimised by introducing a small alteration in the antenna, such as vias or slits into the feedline or the radiating structure to get filter-like characteristics. This technique resulted in the development of filtennas [1–3]. However, this approach involves a complicated structure. As an alternative, the bandpass filter's final resonator can be replaced with an antenna element using the filter synthesis technique. The filtenna in [4] gives a restricted gain bandwidth response but with a good narrowband selectivity. In [5], an microstrip antenna is integrated with a defected ground structure (DGS), utilising a complex impedance inverter circuit to realise a wideband filtenna. An impedance matching structure is inserted into [6] to enhance antenna-filter integration while also enhancing filtering performance. This increases the overall dimension, but also enhances antenna-filter integration. To decrease loss but increase overall size, the synthesis process for a vertically integrated filtenna is shown in [7]. Similarly, despite the different properties of ultra-wideband (UWB) antennas in terms of bandwidth, data rate, and low cost, a band notch function is used to eliminate interference from neighbouring communication systems [8–11], thereby acting as a class of filtenna. A very wideband, dual-band notch, multiple-input, multiple-output (MIMO) antenna in [8] provides a dual notch band at 3.5 GHz and 5.5 GHz with the use of a complementary

DOI: 10.1201/9781003459880-14

split ring resonator and twin L-shaped slits. In [9] a band-notched UWB antenna made up of a MIMO antenna and a split ring resonator provides a 3.5 GHz center frequency notch band. A highly selective dual-notched band filtenna is realised by implementing the filter portion antenna feed line in [10]. In [11] a coplanar wavelength (CPW)–fed antenna with a DGS made of denim fabric for UWB communication is investigated for wireless body area network application.

Moreover, many dual-band filtennas have been investigated in the literature [12–15]. In [12], two orthogonal meander-line slot dipoles are drilled on a metallic sheet of the horn antenna to get two orthogonal polarisations (RHCP and LHCP) at the two bands. An annular ring printed on the top and bottom substrates generates two operating bands in [13]. In [14], the rectangular patch generates two orthogonal polarisations using orthogonal coupling slots by introducing a step impedance resonator (SIR) to realize dual bands. The patch's TM10 and TM30 modes were used to get dual-band characteristics in [15]. Furthermore, peak gains within the two bands are low due to the transmission loss caused by the feed. The multistub feeding network is used in [16] to get a dual-band filtering response. By introducing a complementary split ring resonator (CSRR), two notch bands are obtained in planar filtenna [17]. Further, short-circuited stubs are used to achieve a better cut-off frequency.

This chapter describes a novel method for developing a dual-band microstrip filtenna. It is made up of a stub-loaded open loop resonator (SLOLR) coupled with an L-shaped antenna through an admittance inverter. By proper tuning and placement of the tapped feed position, a dual operating mode is realised. It has an omnidirectional antenna pattern, with a fractional bandwidth (FBW) of 4.14% and 3.83%, in the lower and higher operating frequency band, respectively.

This chapter is categorised into five sections: Section 14.2 illustrates the filter synthesis process, Section 14.3 explains the filtenna design process, Section 14.4 which summarises the results and discussion, and Section 14.5 concludes the work.

14.2 FILTER SYNTHESIS

A fourth-order Chebyshev dual-band band pass filter (BPF) is designed by introducing a shunt stub in a conventional open-loop resonator with an FBW of 5.38% and 3.3%, at a center frequency of 2.4 GHz and 3.6 GHz in the lower and the upper band, respectively. For microstrip implementation, a Rogers/RT Duroid substrate with an ε_r of 10.2, $\tan\delta = 0.0023$ with a height of 1.6 mm is utilised. The dual-band SLOLR BPF dimensions illustrated in Figure 14.1 are

W = 19 mm, L = 27 mm, Lf = 5 mm, Wf = 1.4 mm, Wr = 0.5 mm, Lr = 7 mm, g = 0.8 mm, t = 0.7 mm, $Ls1$ = 4.4 mm, $Ws1$ = 4.4 mm, $Ls2$ = 3.1 mm, and S = 1.5 mm. The BPF simulated response is shown in Figure 14.2.

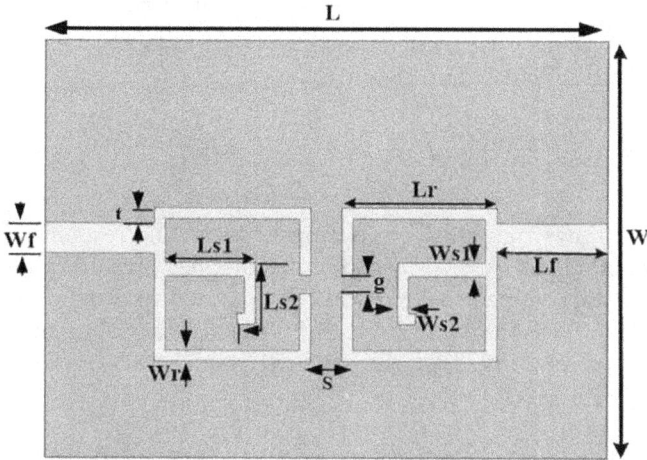

Figure 14.1 SLOLR dual-band BPF.

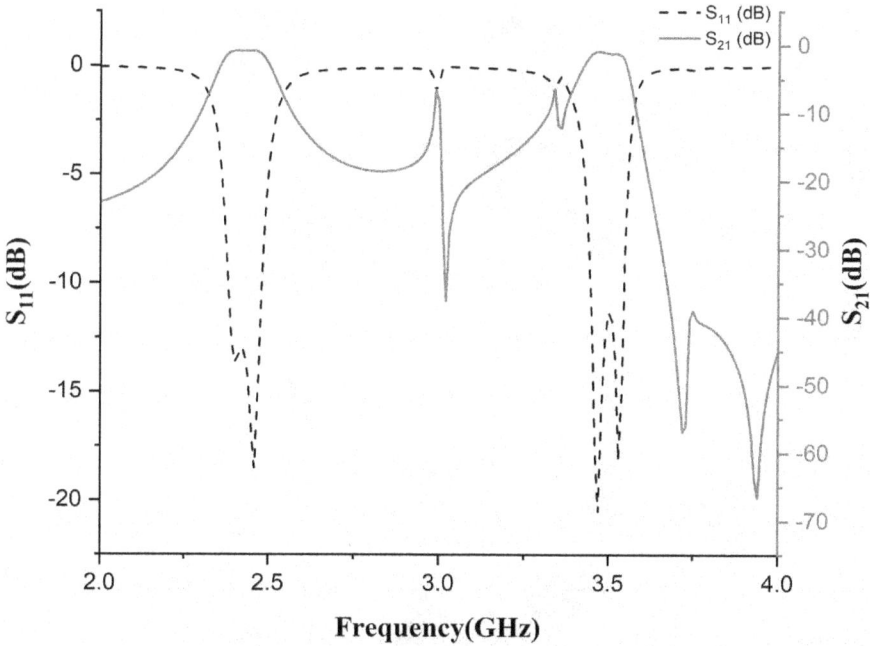

Figure 14.2 Frequency response of SLOLR dual-band BPF.

External quality factor Q_{e1} and coupling coefficients $K_{i,i+1}$ are obtained using electromagnetic simulation software Ansys HFSS. The SLOLRs are only weakly excited by ports in order to get a coupling coefficient. Two characteristic frequencies are seen when the distance "S" between them is changed. Equation (9.1) is used to obtain the coupling coefficient [18] using characteristic frequencies f_1 and f_2,

$$k_{12} = \frac{f_2^2 - f_1^2}{f_2^2 + f_1^2} \qquad (14.1)$$

Figure 14.3 shows the effect of spacing between the resonator design vs. coupling coefficients $K_{i,i+1}$. Once the spacing between SLOLRs is determined, a suitable tapped feed position is obtained to get the desired dual-band function. Both passbands of the SLOLR filter are obtained by altering the length and placement of the stub on the SLOLRs, as illustrated in Figure 14.4.

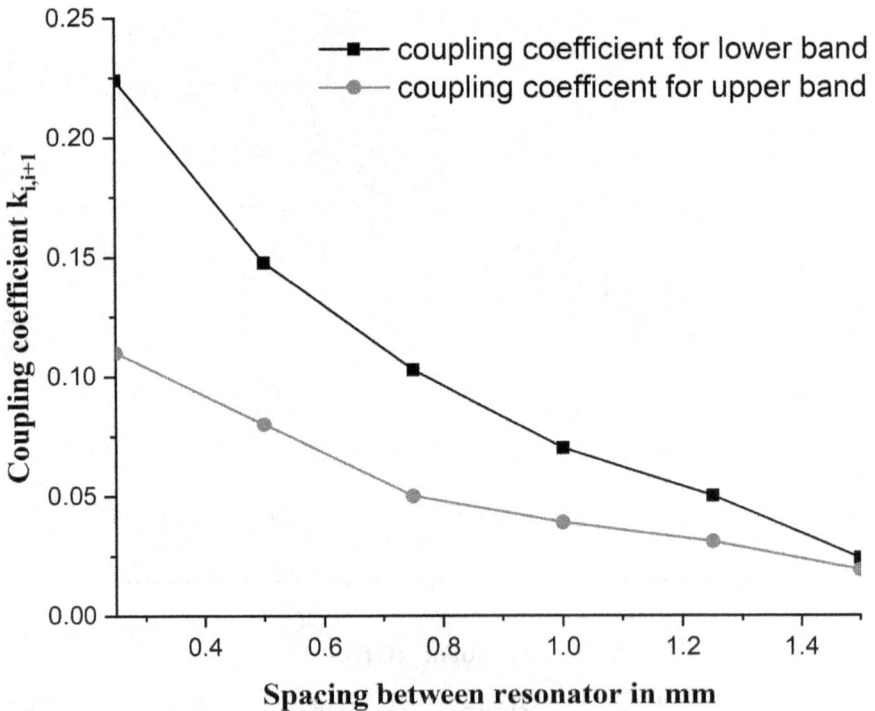

Figure 14.3 Coupling coefficient with respect to spacing between the SLOLR.

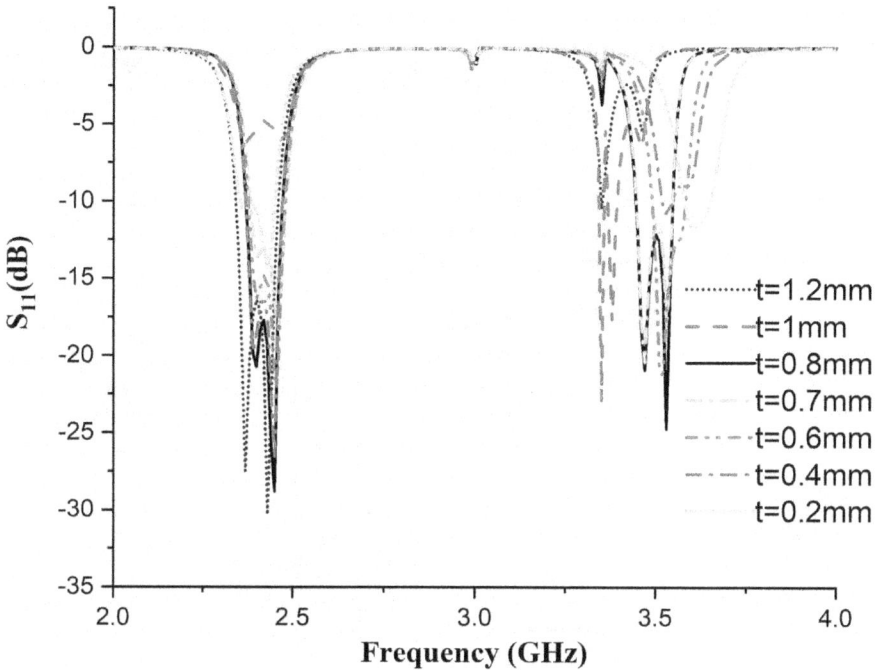

Figure 14.4 Variation of reflection coefficient with tapped feed position in SLOLR.

14.3 FILTENNA DESIGN

The antenna characteristics are obtained by altering the last resonator of the SLOLR BPF with an L-shaped antenna. The 10-dB antenna bandwidth is from 2.6 to 2.88 GHz as shown in Figure 14.5. It is implemented using the RT Duroid substrate having the same permittivity as the filter. The antenna dimensions are Wa = 19 mm, La = 27 mm, Wfa = 1.4 mm, Wa1 = 1 mm, Wa2 = 1.4 mm, La1 = 23 mm, La2 = 5 mm, and Lg = 13.2 mm. The reflection coefficient S_{11} of an L-shaped radiating element is illustrated in Figure 14.6.

The L-shaped resonator and SLOLR BPF are properly integrated using a coupled line admittance inverter. The spacing and width of the coupled line and admittance inverter with even (Z_{0e}) and odd mode (Z_{0o}) impedance are obtained using [19]. The values of Z_0 and filter elements g_1, g_2 are 50, 1.1088, and 1.30621, respectively, and chosen for design implementation. This translates to $J_{12}Z_0 = 0.24$, $Z_{0e} = 64.88$ Ω, and $Z_{0o} = 40.88$ Ω, with coupled line spacing and width as 0.7 mm and 0.5 mm, respectively. The coupled line spacing and width [20] are determined using characteristic impedance (Z_0) and are illustrated in Figure 14.7(a) and (b). The filtenna design in which the L-shaped antenna is replaced with second SLOLR is illustrated in Figure 14.8.

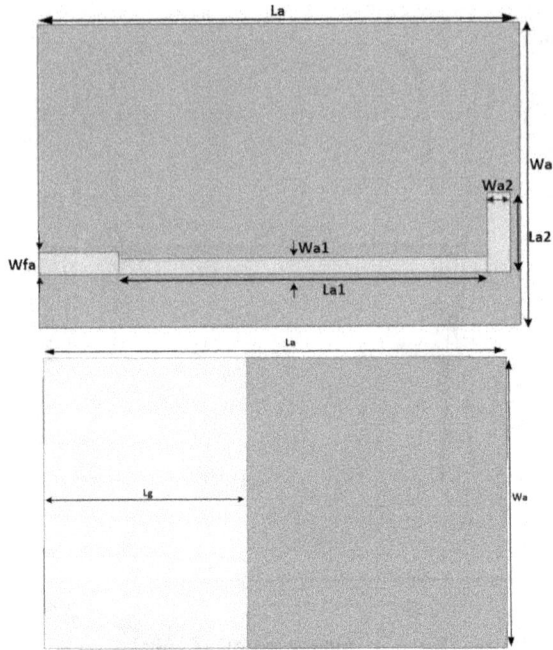

Figure 14.5 (a) Front and (b) back view of an L-shaped antenna.

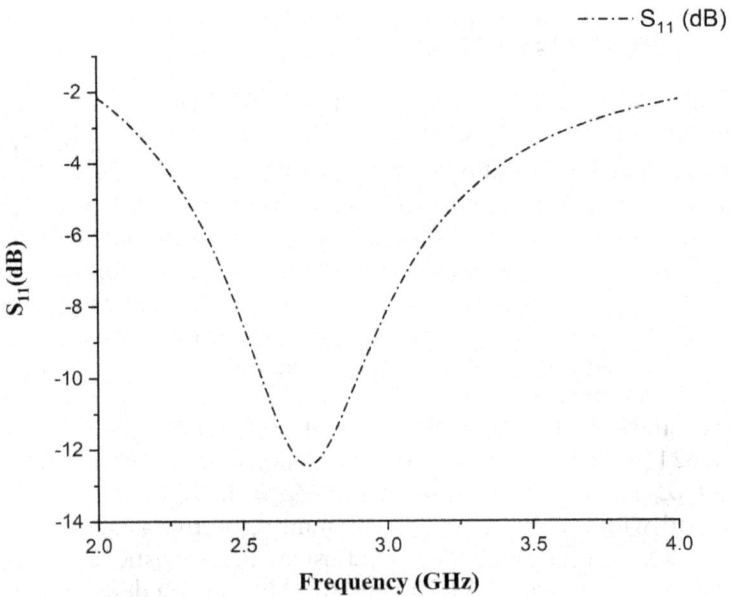

Figure 14.6 L-shaped antenna reflection coefficient (S_{11}) in dB.

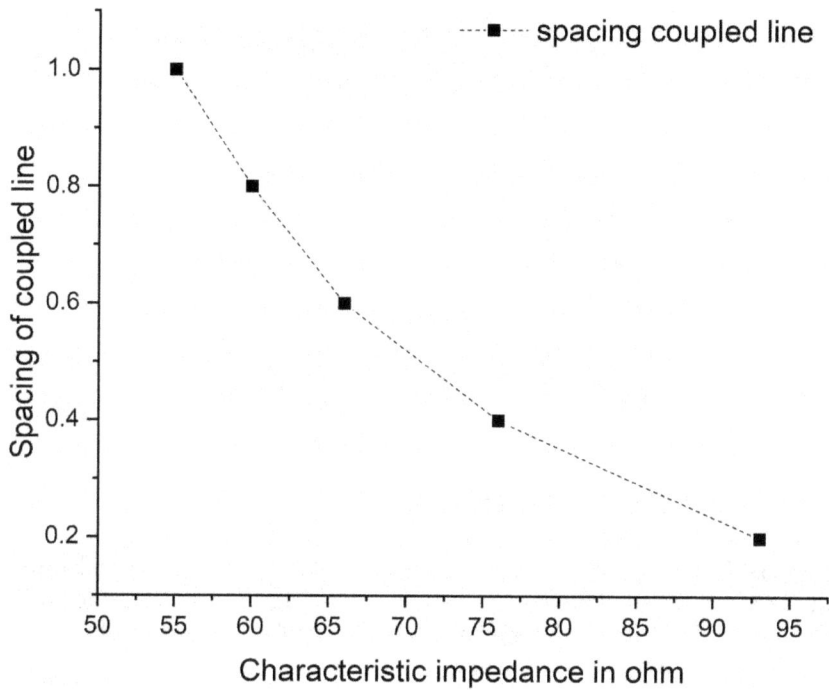

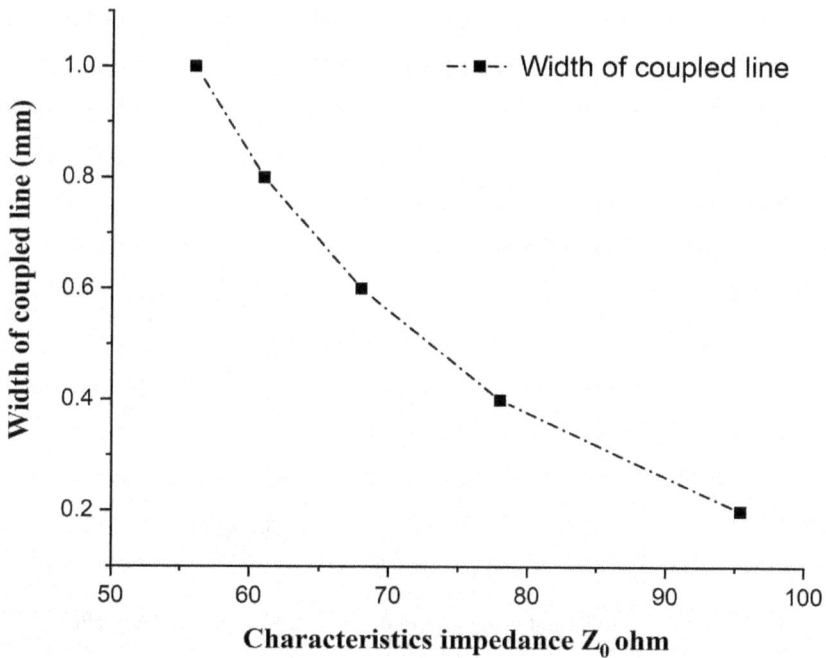

Figure 14.7 (a) Spacing vs. Z_0 (b) width vs. Z_0 of coupled line.

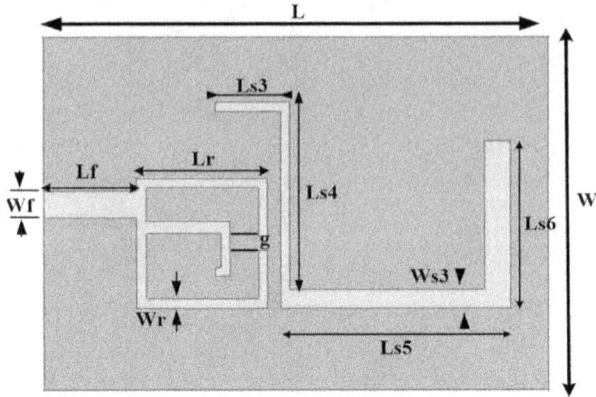

Figure 14.8 The layout of the proposed dual-band filtenna.

14.4 RESULTS AND DISCUSSION

Figure 14.9 depicts the reflection coefficient S_{11} of the dual-band filtenna and the SLOLR BPF. The filtenna possess a dual-band characteristic with an FBW of 4.14% (2.36–2.46 GHz) and 3.83% (3.54–3.68 GHz), which agrees

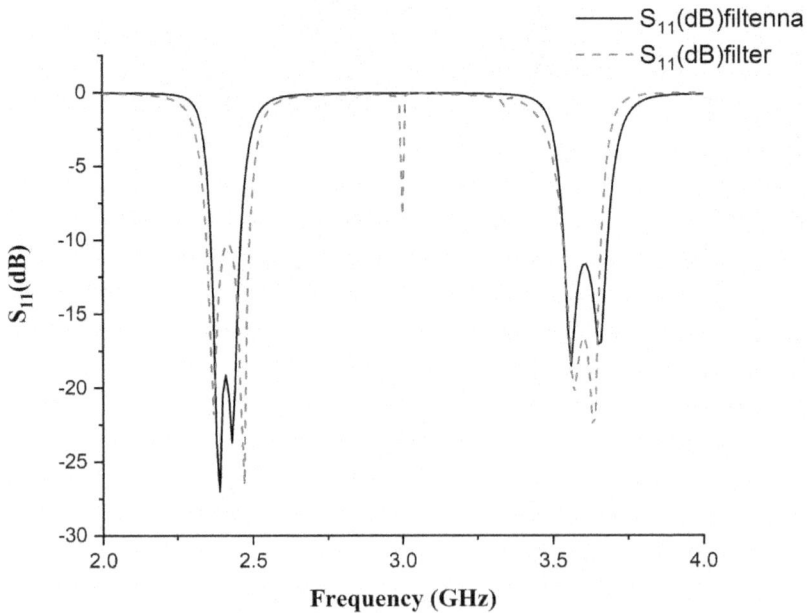

Figure 14.9 Reflection coefficient of dual-band filtenna along with dual-band BPF.

well with the response of the SLOLR dual-band BPF in the lower and higher band, respectively.

Figure 14.10 depicts the gain response of the proposed dual-band SLOLR-based filtenna. The filtenna achieves a maximum gain of 1.1 dBi in the lower band and –0.36 dBi the in upper band, which shows a good impedance match between the SLOLR filter and the L-shaped antenna.

Figure 14.11(a), (b), (c), and (d) depict the co-polarisation and cross-polarisation antenna patterns at 2.41 GHz and 3.6 GHz. From the graph it has been investigated that the filtenna possess an omnidirectional pattern in the YZ plane in both the lower band and upper band. It possesses a low value of cross-polarisation in the lower band but not in the upper band operating frequency range.

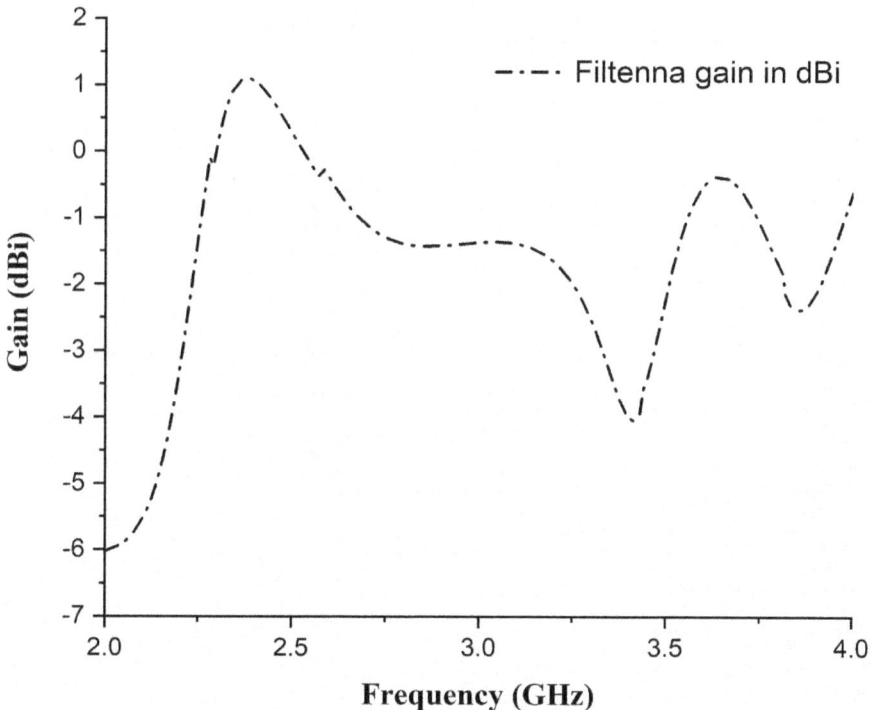

Figure 14.10 Gain response of SLOLR-based dual-band filtenna.

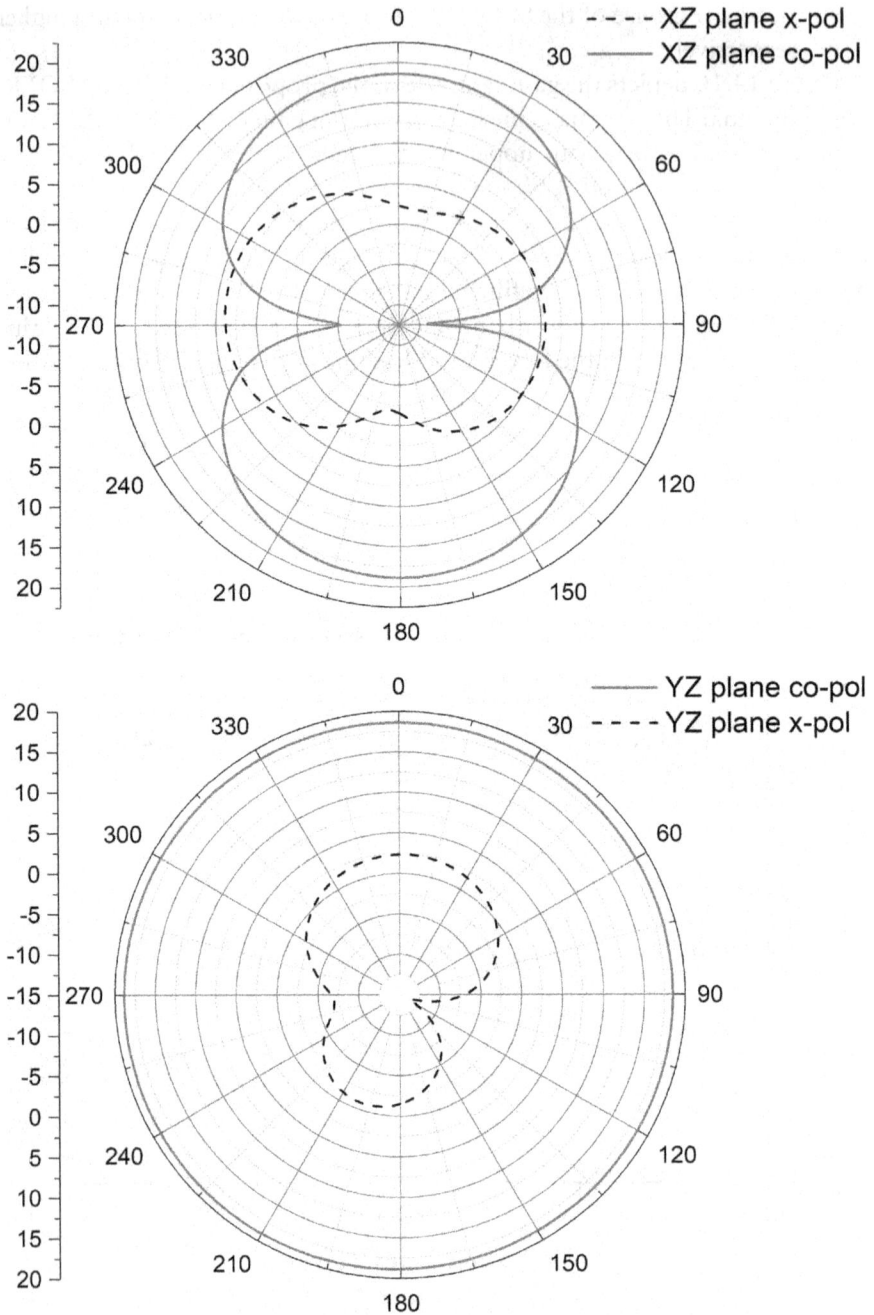

Figure 14.11 Simulated radiation pattern: (a) XZ plane, (b) YZ plane at 2.4 GHz, (c) XZ plane, and (d) YZ plane at 3.6 GHz.

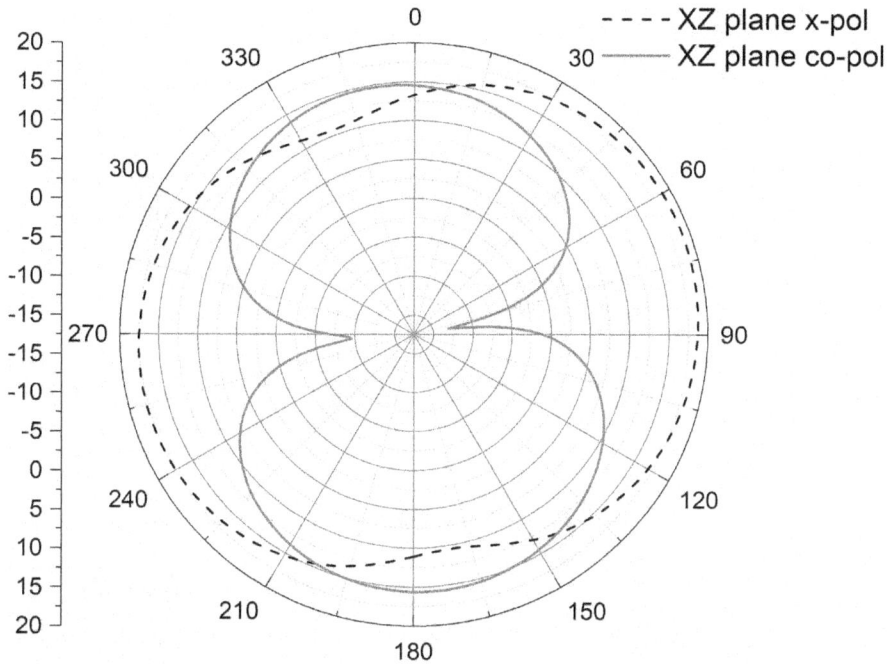

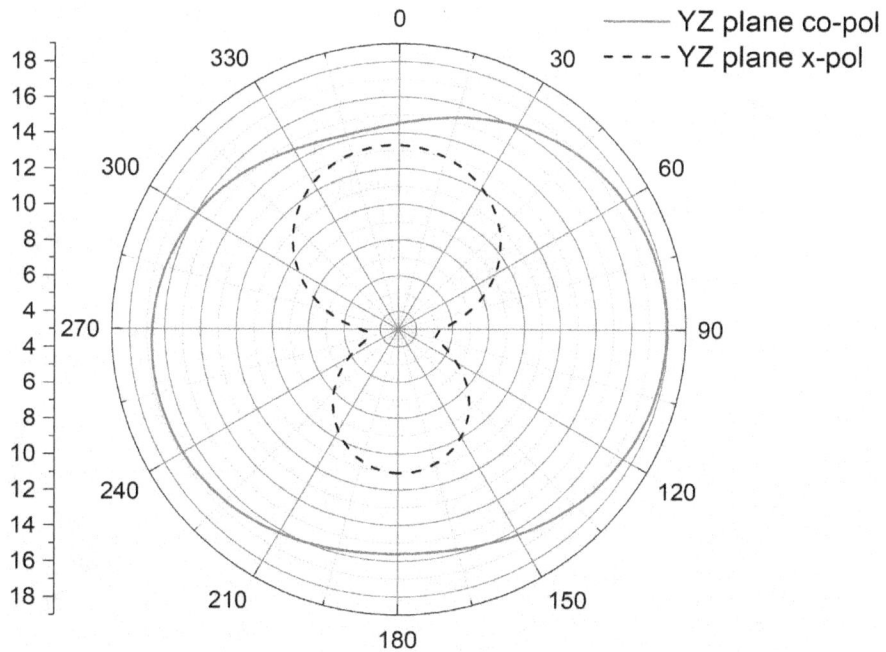

Figure 14.11 (Continued)

14.5 CONCLUSION

In this chapter, a miniaturised planar dual-band filtenna comprising an SLOLR has been proposed. A tapped feed SLOLR is coupled to an L-shaped antenna through an admittance inverter to achieve compact integration. The desired external quality factor is obtained by varying the tapped feed position to get the dual-band filtenna characteristics. Compared to the traditional dual-band filtenna, the proposed filtenna obtained dual-band characteristics without the need for extra space. The results validate that the filtenna is well suited for C-band wireless services applications.

REFERENCES

[1] Z. H. Jiang and D. H. Werner, "A Compact, Wideband Circularly Polarized Co-designed Filtering Antenna and Its Application for Wearable Devices with Low SAR," *IEEE Trans. Antennas Propag.*, vol. 63, no. 9, pp. 3808–3818, 2015, doi: 10.1109/TAP.2015.2452942.

[2] I. Bulu and H. Caglayan, "Designing Materials with Desired," *Microw. Opt.*, vol. 48, no. 12, pp. 2611–2615, 2006, doi: 10.1002/mop.

[3] P. F. Hu, Y. M. Pan, X. Y. Zhang, and B. J. Hu, "A Compact Quasi-Isotropic Dielectric Resonator Antenna with Filtering Response," *IEEE Trans. Antennas Propag.*, vol. 67, no. 2, pp. 1294–1299, Feb. 2019, doi: 10.1109/TAP.2018.2883611.

[4] X. Min and H. Zhang, "Compact Filtering Antenna Based on Dumbbell-Shaped Resonator," *Prog. Electromagn. Res. Lett.*, vol. 69, pp. 51–57, 2017, doi: 10.2528/PIERL17042803.

[5] M. Li and K.-M. Luk, "A Differential-Fed UWB Antenna Element With Uni-directional Radiation," *IEEE Trans. Antennas and Propag.*, vol. 64, no. 8, pp. 3651–3656, Aug. 2016, doi: 10.1109/TAP.2016.2565726.

[6] C. T. Chuang and S. J. Chung, "A Compact Printed Filtering Antenna using a Ground-Intruded Coupled Line Resonator," *IEEE Trans. Antennas Propag.*, vol. 59, no. 10, pp. 3630–3637, 2011, doi: 10.1109/TAP.2011.2163777.

[7] L. Li and G. Liu, "A Differential Microstrip Antenna with Filtering Response," *IEEE Antennas Wirel. Propag. Lett.*, vol. 15, pp. 1983–1986, 2016, doi: 10.1109/LAWP.2016.2547884.

[8] A. Bhattacharya, B. Roy, R. F. S. Caldeirinha, and A. K. Bhattacharjee, "Low-Profile, Extremely Wideband, Dual-Band-Notched MIMO Antenna for UWB Applications," *Int. J. Microwave Wirel. Technol.*, vol. 11, no. 7, pp. 719–728, 2019, doi: 10.1017/S1759078719000266.

[9] A. Bhattacharya, B. Roy, S. K. Chowdhury, and A. K. Bhattacharjee, "Computational and Experimental Analysis of a Low-Profile, Isolation-Enhanced, Band-Notch UWB-MIMO Antenna," *J. Comput. Electron.*, vol. 18, no. 2, pp. 680–688, 2019, doi: 10.1007/s10825-019-01309-3.

[10] A. K. Bhattacharya, A., A. De, B. Roy, and A. K. Bhattacharjee, "Investigations on a Low-Profile, Filter Backed, Printed Monopole Physics," *Indian J. Pure Appl. Phys.*, vol. 58, no. 2, pp. 106–112, 2020.

[11] A. K. De, R. Bappadittya, A. Bhattacharya, and A. K. Bhattachaqee, "Bandwidth-Enhanced Ultra-Wide Band Wearable Textile Antenna for Various WBAN and Internet of Things (IoT) Applications," *Radio Sci.*, vol. 56, no. 11, pp. 1–16, 2021.

[12] M. Barbuto, F. Trotta, F. Bilotti, and A. Toscano, "Design and Experimental Validation of Dual-Band Circularly Polarised Horn Filtenna," *Electron. Lett.*, vol. 53, no. 10, pp. 641–642, 2017, doi: 10.1049/el.2017.0145.

[13] D. S. La, J. H. Zhao, S. M. Chen, C. X. Zhang, M. J. Qu, and J. W. Guo, "Dual-Band Omnidirectional Coupled-Fed Monopolar Filtering Antenna," *Eng. Sci. Technol. an Int. J.*, vol. 35, 2022, doi: 10.1016/j.jestch.2022.101188.

[14] C. Y. Hsieh, C. H. Wu, and T. G. Ma, "A Compact Dual-Band Filtering Patch Antenna using Step Impedance Resonators," *IEEE Antennas Wirel. Propag. Lett.*, vol. 14, pp. 1056–1059, 2015, doi: 10.1109/LAWP.2015.2390033.

[15] Y. J. Lee, G. W. Cao, and S. J. Chung, "A Compact Dual-Band Filtering Microstrip Antenna with the Same Polarization Planes," *Asia-Pacific Microw. Conf. Proc., APMC*, pp. 1178–1180, 2012, doi: 10.1109/APMC.2012.6421862.

[16] X. Y. Zhang, Y. Zhang, Y. M. Pan, and W. Duan, "Low-Profile Dual-Band Filtering Patch Antenna and Its Application to LTE MIMO System," *IEEE Trans. Antennas Propag.*, vol. 65, no. 1, pp. 103–113, 2017, doi: 10.1109/TAP.2016.2631218.

[17] W. T. Li, Y. Q. Hei, H. Subbaraman, X. W. Shi, and R. T. Chen, "Novel Printed Filtenna with Dual Notches and Good Out-of-Band Characteristics for UWB-MIMO Applications," *IEEE Microw. Wirel. Components Lett.*, vol. 26, no. 10, pp. 765–767, 2016, doi: 10.1109/LMWC.2016.2601298.

[18] P. Pal, R. Sinha, and S. K. Mahto, "Synthesis Approach to Design a Compact Printed Monopole Filtenna for 2.4 GHz Wi-Fi Application," *Int. J. RF Microw. Comput. Eng.*, vol. 31, no. 5, pp. 2–9, 2021, doi: 10.1002/mmce.22619.

[19] P. Pal, R. Sinha, and S. Kumar Mahto, "A Compact Wideband Circularly Polarized Planar Filtenna using Synthesis Technique for 5 GHz WLAN Application," *AEU Int. J. Electron. Commun.*, vol. 148, p. 154180, 2022, doi: 10.1016/j.aeue.2022.154180.

[20] W. J. Wu, Y. Z. Yin, S. L. Zuo, Z. Y. Zhang, and J. J. Xie, "A New Compact Filter-Antenna for Modern Wireless Communication Systems," *IEEE Antennas Wirel. Propag. Lett.*, vol. 10, pp. 1131–1134, 2011, doi: 10.1109/LAWP.2011.2171469.

Chapter 15

Reviews on Electromagnetic Interference/Compatibilities

Himadri Sekhar Das, Santanu Mishra,
Gourisankar Roy Mahapatra, and
Takashiro Akitsu

15.1 INTRODUCTION

It was first noticed by the US Federal Aviation Agency in a routine flight where an aircraft altimeter showed the wrong measurement of altitudes. It was then identified as undesired electromagnetic interference (EMI) generated by the ignition unit of the aircraft. In general, transmitters are known as a source of radio frequency signals for propagation through the air media [1, 2]. During such transition, some undesired signals known as spurious signals also are generated by the instruments guiding the antenna to transmit, which are then called EMI. These kinds of undesired signals are not seen only for transmitters but also for other electronic instruments such as computers, precious research instruments etc. So, it is a common problem in almost all scientific signal processing which we need to eliminate. At first, we need to identify the presence of any undesired signal in our electromagnetic signal generator units or transmission line (data cables) or control units by suitable measurements [3, 4]. Next, we need to expose our instrument in specific frequency enabled electromagnetic environments and observe whether our instrument works undisturbed or not. This is called immunity testing.

In essence, electromagnetic fields are created when electrical and electronic equipment interferes with other equipment. When the electrical and electronic fields are present simultaneously, this interference is either produced or is already existent. When two or more signals travel in parallel, EMI may happen [5, 6]. Electromagnetic waves of one signal may interfere with those of another during that time. Noise or unwanted signal/interference that originates from an external source is referred to as EMI. Radio frequency (RF) interference is another name for this. Any electronic gadget could stop functioning due to these types of interference. The electrical circuit may become damaged or cease to function as a result of external noise sources, which may also have an impact on the data signal. It can result in a rise in error rates and complete data loss. Several types of equipment, including AM radios, FM radios, mobile phones, televisions, and

DOI: 10.1201/9781003459880-15

radio astronomy, can be impacted by EMI. Natural or artificial sources can produce EMI. The use of contemporary error correcting systems, electrical shielding, and high-quality electronics can lower interference levels. When a cell phone is placed next to an audio speaker, noise is made, and a beeping sound can be heard. This is an example of EMI. During a storm, RF waves are produced in the atmosphere, which have an impact on the operational antenna in this setting. These radio frequencies from the storms interfere with the initial communication signals, resulting in information loss and signal distortion [7, 8].

15.1.1 Fundamentals of EMI

It is common knowledge that electromagnetic waves involved in EMI have both electric and magnetic field components and that these components oscillate at right angles to one another, as seen in Figure 15 1. It is quite difficult to understand the nature of EMI when it responds to different parameters like frequency, voltage, distance, and current.

High-conductivity materials have the potential to improve EMI attenuation; however, materials with increased permeability have the opposite effect, reducing EMI attenuation while improving it for the magnetic field component. As a result, greater permeability in a system with an E-field–dominated EMI will cause attenuation to decrease, but an H-field–dominated EMI will cause attenuation to increase. However, the E-field is typically the main source of the interference because of recent developments in the technologies used to make electronic components.

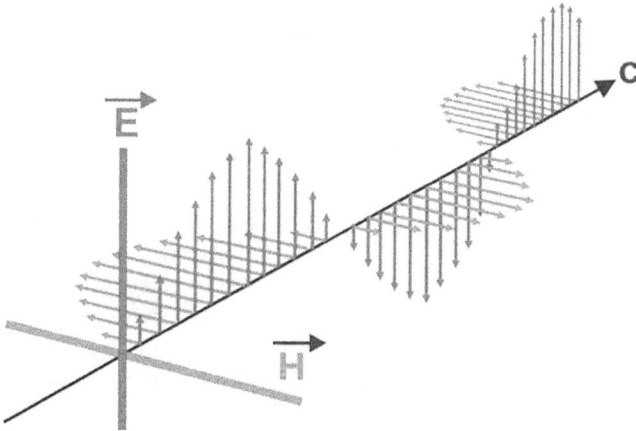

Figure 1. Nature of EMI.

Figure 15.1 Electromagnetic wave propagation.

15.1.2 What Causes Electromagnetic Interference?

When an undesired signal reaches the receiver side, EMI happens. The fact that EMI happens when electrical signals flow and produce a magnetic field, or the opposite, when a magnetic field causes an electrical current, was also widely known. Electricity and magnetism are closely related to one another. Unwanted signals are typically generated by powerful electrical and radio sources [9–14].

The EMI of today affects a variety of systems, including those used for communication, transportation, healthcare, and mobile devices. Sensitive electronic equipment nearby is impacted by ambient electromagnetic radiation. All electrical and electronic equipment nearby can be completely destroyed by a powerful electromagnetic pulse source.

Any equipment subjected to fluctuation is considered to be a source of EMI and reaches victim equipment either by conduction or radiation, which results in changes in the signal or damages the instrument. The effect of EMI on any instrument is different depending on the applicability area or environments of the intended instruments. It can also treat as EM pollution in EM world.

15.2 TYPES OF ELECTROMAGNETIC INTERFERENCE

In recent years, as technology has continued to advance, so has the industrial environment. Circuits, power wirings, signal processing units, and other electrical components frequently interact with one another in electronic devices [15]. This device contact causes some signal noises, or EMI, in the circuit, which can impair the devices' functionality and measurement capabilities. The following sections discuss a few examples of EMI effects.

15.2.1 Manmade EMI

Manmade EMI can be subdivided as shown in Figure 15.2:

Voluntary Sources: In this case EMI is used to disable or destroy enemy arsenals or communications in a hostile situation. Different EMI systems are made to disrupt space communications, telecommunications, and power.

Involuntary Sources: A circuit's component or components may have an impact on how another component of the same circuit or any other surrounding devices operate through conduction (through conducting lines) or radiation. Involuntary effects are observed for the small manmade electronic gadget, automobiles, medical equipment, microwave oven, and so on. Though small gadgets are not that much affected by

Conducted ⮞
      Intentioanl
      Unintentional

⮞ Narrowband ⮞

Radiated ⮞
      Intentioanl
      Unintentional
      Restricted

Man made

Coherent ⮞
      Conducted
      Radiated

⮞ Broadband ⮞

Incoherent ⮞
      Conducted
      Radiated

Figure 2 Different kinds of manmade EM.

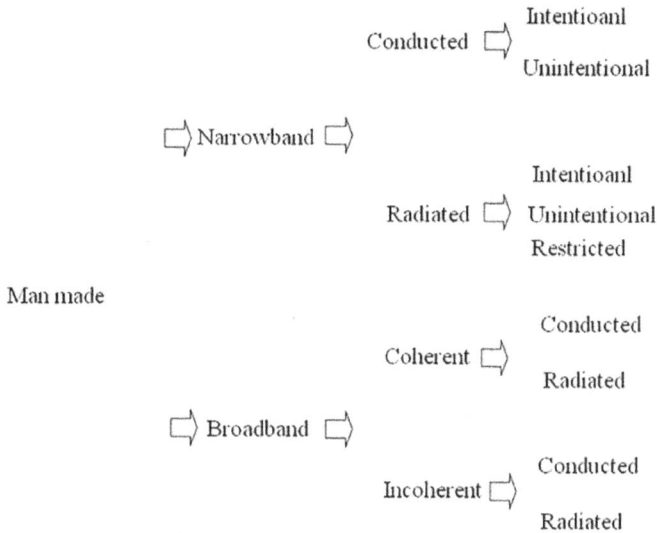

Figure 15.2 Different kinds of manmade EMI.

EMI as large electronic instruments are, still interference can also be seen due to the presence of a neighbouring electronic gadget or different parts of a large instrument.

15.2.2 Natural EMI

Some natural sources of EMI are solar flares, lightning, auroras, and cosmic radiations which can produce strong electric as well as magnetic fields. These natural phenomena can affect the operation of small electronic devices, satellite communications, etc. [16].

15.2.3 Narrowband EMI

Narrowband EMI is frequently from a radio transmitter and only affects a certain radio frequency.

15.2.4 Broadband EMI

Several frequencies of the radio spectrum are affected by EMI, which is frequently brought on by broken equipment. An illustration of radiated EMI is when wireless phones cause Wi-Fi to drop.

15.2.5 Radiated EMI

When a high-power transmitter or electrical device emits a radio frequency that is picked up and has an unintended impact on another device, this is known as radiated EMI. It is most likely radiated EMI if there is EMI and the source and receptor are far apart. Examples of this include a malfunctioning kitchen microwave that prompts a PC to restart or outdated wireless phones that cause Wi-Fi to fail [17–19].

The antenna is a component of the major source of EMI in communication systems. Since it is an essential part of an instrument which takes part in transmitting and receiving signal, it will lead to the source of EMI.

Analog modulation techniques, which are utilized in broadcasting and other communication systems, have high signal-to-noise ratio (SNR) values between transmitter and receiver. In analogue modulation communication, SNR represents the presence of an undesired signal received in the desired signal strength. SNR measurement is used to gauge the effectiveness of communication along the communication trajectory, which in turns defines EMI. Hence, it might be categorized as additive white gaussian noise (AWGN). It is an estimation of statistically random noise with sources of consistent spectral densities over a wide frequency range. This could fall under the thermal class of noise, which is caused by obstructions to electron transport in a conductor depending on temperature, or shot noise, which is caused by fluctuations in current flow.

Thermal noise mathematically can be expressed as:

$$P = K_B TB$$

where P is the noise power (W), K_B is the Boltzmann constant (1.38×10^{-23} m^2 Kg s-2 K-1), T is the temperature in [K], and B is the frequency bandwidth in hertz (Hz).

Shot noise is sometimes expressed as:

In Ampere [A] it is defined as:

$$I_n = \sqrt{2qI_{DC}B}$$

Where I_{DC} is a DC bias current in the electric circuit, B is the frequency bandwidth in [Hz], and q is the electron's charge, which is -1.6×10^{-19}C.

15.2.6 Conducted EMI

When there is a direct connection between the source and the receptor by wires or cables, this is the type of EMI produced. It frequently occurs near electrical transmission lines or cables. One or more sources could be a sizable motor or power source. It is observed that when a treadmill or clothes dryer is turned on, a computer connected to the same electrical mains circuit may restart automatically. This is an example of conducted EMI.

15.2.7 Coupled EMI

How EMI travels from the source to the receiver is described by an EMI coupling mechanism (affected devices). The solution to the issue depends on an understanding of the pattern of the interferences as well as the way it is being coupled between the receiver and the transmitter. There are four main types of EMI coupling: conductive, radiative, capacitive, and inductive based on the mechanism of coupling of EMIs between a source and a receiver. These coupling methods are generated by either the H-field or E-field or both. Let's examine each of the coupling mechanisms in turn.

15.2.7.1 Conduction Coupling

Conduction coupling is related to wires and cables which happens when EMI emissions are transmitted to link the source and the receiver. A power supply cable in general exhibits this kind of coupling of EMI and depends significantly on the H-field part. With conduction on power lines, coupling can take place in the form of either differential mode conduction or common mode conduction type. In differential mode, the interference appears on two conductors in and out of phase. In common mode, the interference appears between the +ve and −ve line and is in-phase in nature. The application of filters and shields over wires is the most often used remedy for conduction coupled interference.

15.2.7.2 Radiation Coupling

The most prevalent and well-known type of EMI coupling is radiation coupling. It does not require a physical connection between the source and the receiver, like conduction requires, since the interference is sent (radiated) to the receiver across space.

15.2.7.3 Capacitive Coupling

This happens when two connected devices interact. When a source's voltage changes and transmits a charge to the victim capacitively, this is known as capacitive coupling. The conductors must be extremely close to one another for capacitively linked EMI to occur, which is most frequently found on electronic circuit boards or in collections of densely packed wires travelling great distances.

15.2.7.4 Inductive/Magnetic Coupling

According to the principles of electromagnetic induction, this type of EMI results from a conductor coil's magnetic-inducing influence on another conductor nearby. Inductive or magnetic coupling EMI happens when the magnetic field produced due to the flow of current in any conductor induces an unwanted current in another neighbouring conductor. It can be detected when a power cable and an audio cable are near to each other and a hum is heard on the audio output.

15.3 ELIMINATION METHODS OF EMI

To reduce the noise and enhance the performance of the device, it is required to protect the system by taking proper precautions against these noises [20]. Signal drop can be reduced by using proper grounding and shielding techniques. One can eliminate the noise source/EMIs or improve the receiving equipment that is impacted by the noise signals, such as capacitive coupled interference, by ensuring proper compatibility between the source and receptor. Twisted-pair technology and inductive coupling both help to reduce magnetic interference.

There are several methods for reducing EMI in switched mode power supply (SMPS) circuits, such as using larger power supplies that can generate far less EMI than linear power supplies. Grounding systems are essential for not only returning current but also for several factors such as frequencies, impedances, and wiring length depending on the setup. A mix of grounding processes is much more effective in handling any interference. A metal shield is used as electromagnetic shielding in electronic circuits, which creates a physical barrier between the source and destination of the interference and results in weakening or attenuation of the interference. To reduce the EMI, a combination of inductor–capacitor filter (LC filters) is also used. Proper design is the most effective way to avoid EMI damage. So, it is required to place the noisy nodes as far away as possible from the sensitive nodes and design the ground plane below the EMI source. Also bundle the wires to reduce the loop area and use the EMI filter for conducted EMI.

Today's electronic appliances are made up of tens of thousands of passive and/or active parts and employ small, compact technologies. It is difficult to fit a lot of components onto a device only a few centimetres in size. It is also difficult for electronics engineers to create high-speed, multipurpose devices that are both compact and electromagnetically compatible. The term "electromagnetic compatibility" (EMC) refers to a parameter used to assess a device's resistance to unwanted interferences present in its electromagnetic environment. It is necessary to lessen the interference up to a specific level in order for the device to work better. The frequency range of EMI spans between 2 kHz and 150 kHz. Both lower-frequency and higher-frequency regions experience EMI. Hence, it is necessary to minimize EMI across the full frequency range. As an example, electrical power transmission lines, nuclear submarines, low-frequency brain waves, FM and AM communication, TV, radar, cell phones, and infrared satellite dishes are all operated in different frequency ranges [21].

EMC is an important step in designing an electronic instrument which addresses unwanted signal effects in the instruments. It eliminates generation, propagation, and reception of undesired EM signals (EMI), which may cause damage to electronic instruments. EMC ensures efficient operation of electronic instruments in an EM environment by employing countermeasures against EMI.

EMC is the best possible engineering which can protect any victim instrument from such kind of EMI by providing better identification, filtration, and control over the potentially affected instrument.

Any appliance (television, computer, remote, mobile, etc.) needs to be protected from different kinds of interference such as RFI, EMI, etc. There are several methods to minimize or EMI. Some of the prevention steps are discussed next.

15.3.1 Shielding

The most common way to prevent EMI is to ensure the devices only use shielded or insulated electric circuits. Though complete prevention of all types of EMI is impossible, insulating, shielding, and filtering circuits and cables can help appreciably. Shielding is the suggested approach for containing radiation or coupling in source or victim devices and requires putting the circuit in a completely sealed enclosure, such as a metallic box. It is crucial for electronic equipment because electromagnetic waves are reflected into the enclosure and waves that aren't reflected are absorbed by the shielding material. If the barrier is not thick enough, some radiation normally manages to get through. The materials for EMI shielding include metals, carbons, ceramics, cement or concrete, conductive polymers, and associated composites. Almost any common metal, including copper, steel, and aluminium, can be used as a shielding material. Carbon nano-fibre mats made from extremely thin (0.16 μm diameter) carbon filament by the paper-making (thin film) process is low-cost. The lightweight binder-less mats provide high shielding effectiveness. With the rapid advancement in smart and wearable electronic devices, the manufacture of flexible EMI-shielding materials with high performance becomes an imperative requirement. Synthesis of flexible polyaniline (PANI) paper with high bending durability is a substantial challenge because of the integrally inflexible nature of its polymeric chains. To overcome the problem, PANI particles are decorated on top of a polyethylene terephthalate (PET) thin film and are tightly wrapped by the PANI coating layer and thus, the PET/PANI composite thin films become a reliable and effective EMI shielding material. The process is very simple and economical [22].

Emulating a Faraday cage is the most effective approach to protect a device. A conductor-enclosed Faraday cage serves to prevent or significantly reduce the amount of electromagnetic signals that enter the object. Mesh can be employed if its pores are smaller than the wavelength of the signals being blocked, even if solid conductors are preferable. The Faraday case is a significant tool used for many years in preventing EMI. This device was invented in the nineteenth century to block the spreading of electromagnetic waves in a specific area. The effect resembling a Faraday cage is produced by using a metal chassis as the device enclosure. Openings in the cage are needed for cables and connectors; however, noise may still enter. Combining the use of metal connections with cable shielding can be beneficial. There are various kinds of cable shields and methods for attaching them to the connectors on either end. While not as structurally robust, foil shield performs better at high frequencies than braided shield. To gain the benefits of both, some cables combine braided and foil shielding. Common-mode noise, which affects both conductors, can be lessened by using twisted pairs. In order to combat the Faraday effect, cables should, whenever possible, have their shields connected to ground at both ends. The method used can potentially have an impact on EMI.

A microwave cage is also considered as a potential tool for shielding against EMI. The operation is opposite to that of a Faraday cage, where electromagnetic microwaves are kept inside a microwave appliance and are not allowed to escape from it.

A fibreoptic cable is one significant shielding material that is virtually unsusceptible to EMI. These types of cables are now widely used in place of copper cables for data communication.

Additionally, EMI limits the signal's range and produces erroneous signal decoding. With the right filtering, shielding, and grounding, EMI can be reduced significantly. Because fibreoptic cable uses light as a signal, magnetic and electric field related EMI could not affect the signal and is fully eliminated. Table 15.1 shows the types of equipment malfunction due to EMI.

Table 15.1 Malfunctions of Equipment Owing to EMI

Kind of Equipment	Equipment Mode	Degradation
Communication	Analog, digital, and voice data	Diminished intelligence, a higher bit error rate, and distortion
Radar	Search	Range reduction, processor overload, higher fire alarm frequency, and lower detection likelihood
	Tracking	Mistakes in tracking and break-lock
Navigational aids		Overloads, range/analog errors
Identification, friend, or foe (IFF) system		Overloads and false decoding

15.3.2 Filtering

Filters can be used to minimize transmitted EMI when shielding techniques are insufficient. A low-pass filter composed of an inductor and capacitor may muffle high frequencies while maintaining DC voltage stability. Because it may reduce noise in a range while keeping the frequency of the remaining frequencies, a band-stop filter may be more useful in some situations. Other filter configurations abound, and most of them have online calculators that can be used to filter a particular problem frequency.

One of the most popular ways to get rid of EMI is by filtering out unwanted signals and noise from a particular signal. In this scenario, passive filters – which are now found in most types of modern equipment – are employed to accomplish filtration. As the initial phase in the filtering process, an AC line filter is installed to prevent unwanted signals from entering the power supply or powered circuits. It stops internal signals from being connected to the AC line. Filtering is widely utilized with cables and connections on lines into and out of a circuit. The major purpose of some special connectors' built-in low-pass filters is to mellow digital waveforms by allowing for longer rise and fall times and reduced harmonic production. Low-voltage analogue signals frequently need to be amplified and then filtered to eliminate background noise before being translated to digital form. Filtering is widely used to reduce noise outside of a specific frequency range. For instance, the magnetic components product line from TT Electronics uses common-mode chokes to prevent (choke) unwanted EMI noise while allowing the required signals to pass through inductive filters, thus reducing EMI.

By introducing undesired currents and voltages into an electrical device's circuitry, EMI causes performance to be impeded or degraded. Such interferences, which can be lethal for the system as a whole, are typically caused by a variety of factors.

15.3.3 Ground

Distance and utilization based routing protocol (DAU) defines grounding as an electrical connection to the Earth. It entails creating a reference point or plane that is referenced to ground between an electrical or electronic system and a link that is electrically conductive. Several grounding locations on a big ground plane are necessary for better results. To create the best possible ground, keep leads from internal circuits or other components as far away as you can in order to minimize inductance. Try to separate circuits from the ground if there is no other way to control ground loop voltages.

15.3.4 Transmission Mode

The two main EMI sources that are present are conducted and radiated. Although transmitted noise travels through a conductor into or out of a device, radiated noise is received or transmitted wirelessly. Any electronic

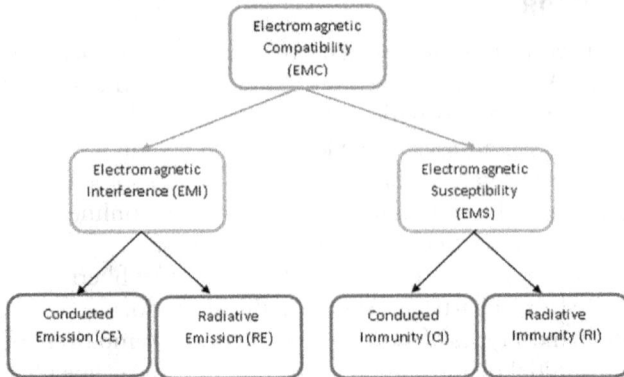

Figure 15.3 Compatible electromagnetic field.

system is known to be subject to a variety of sounds, which necessitates the use of numerous mitigation techniques. The various varieties are depicted in Figure 15.3 [23].

At higher frequencies, noise is often emitted through the air, whereas at lower frequencies, it usually travels through conducted means. The frequencies that a device transmits on are typically also sensitive to them. Another possibility is that what appears to be conducted noise is actually radiation-related, such as cable cross-talk in cable bundles or a poor printed circuit board (PCB) layout.

15.3.4.1 Conducted Susceptibility/Immunity

Conducted susceptibility (CS) refers to a system's vulnerability to noise that enters via a conductive route, such as power inputs or returns. A device may share a power rail with numerous other devices, and each one could introduce noise to the rail, so it is crucial to ensure that all devices can continue to function normally. Most integrated circuit (IC) manufacturers advise decoupling capacitors to be placed close to the power pins on the PCB to stop high-frequency noise from entering the IC. Depending on the noise frequency involved, this can also be successfully applied to the entire system. Incorporating a low-pass filter that reduces the noise's frequency may also be advantageous.

15.3.4.2 Conducted Emissions

Conducted emissions (CE) are the noise that a system generates onto its conductive connections that could disrupt equipment. To stop noise from entering the power rails, the best course of action is to locate and remove the noise source. If this is not possible, a filter can be used to reduce the noise. Switching power sources and regulators are known to produce this type of noise. Furthermore, conducted noise can mimic cable cross-talk. Using distinct, twisted, insulated pairs in a cable helps mitigate the consequences.

15.3.4.3 Radiated Susceptibility/Immunity

Radiated susceptibility (RS) is the term used to describe a system's vulnerability to wireless background noise. A device must always be able to withstand outside noise to some extent while continuing to function as intended. Both properly shielded cables with twisted shielded pairs and a chassis similar to a Faraday cage can reduce radiated noise. Filters and internal shielding around sensitive components might be necessary if it is not possible to prevent radiated noise from entering a device.

15.3.4.4 Radiated Emissions

Radiated emissions (RE) are the sounds a system makes when operating. To prevent disrupting surrounding systems or equipment, this noise must be kept at a moderate volume. Design considerations like the choice of cable and case material can have a significant impact on RE. A properly protected system will emit very little noise. If protecting the entire arrangement is not possible, locating the radiation source inside the device and finding a way to attenuate it there, such as localized shielding or using a separate element operating at a different frequency, may be helpful.

In addition to being sensitive to these erroneous noises, electronic equipment is also capable of emitting or creating these noise signals into the environment, which may interfere with the functionality of neighbouring electronic systems. The EMI connected to a device must be measured in order to determine its immunity and the volume of interference it might conceivably produce in its surroundings.

An antenna is a crucial part of the machinery used to send or receive electromagnetic radiation. By using proper antenna design, installation, and other considerations, EMI in communications systems can be reduced. As a result, antenna systems in broadcasting stations come in a variety of sizes and configurations, including monopole, dipole, corner reflector, yagi, log dipole, and folded dipole antenna systems. A typical antenna system is made up of collinear arrays, which are typically a stack of vertical dipoles spaced equally apart and powered by currents of the same amplitude and phase. The transmitted signal is applied across two identical cylindrical conductors that make up a dipole antenna [24]. Antenna characteristics must be adjusted to meet the requirements for EMI mitigation. The following are the properties that can be changed.

15.3.4.4.1 Antenna Radiation

The variation in power radiation of an antenna with distance from the antenna [24] defines the quality of an antenna itself.

15.3.4.4.2 Directivity

Another characteristic of an antenna called directivity helps to reduce EMI. It is the ability of an antenna of concentrated radiation power towards a specific direction. It is defined as the ratio of the antenna's radiation along a specific direction to its average radiation. It is also known as directivity. The directivity can be expressed mathematically as:

$$D(\theta, \varphi) = \frac{E(\theta, \varphi)}{E_{avg}} = \frac{4\pi E(\theta, \varphi)}{P_{rad}}$$

where D is directivity, E is power radiated in specific direction, P_{rad} is power radiated from the antenna, and E_{avg} is average radiated power.

15.3.4.4.3 Antenna Gain

An electronic system's antenna gain should be taken into account while doing an EMI study. It is described as the difference between the amount of power an antenna radiates at a specific point in relation to the amount of power there when an isotropic antenna is utilized. A reference antenna that transmits power evenly in all directions is said to be isotropic. Antenna gain describes how an antenna converts electrical power input into radio waves that are radiated and go in a certain direction for transmission, and vice versa for a reception antenna. Increasing or reducing antenna gain can be utilized as an EMI mitigation strategy to boost or lower effective isotropic radiated power (EIRP), which facilitates a changeable electric field strength [25].

15.3.4.4.4 Polarization

The orientation of the electric field wave with regard to the ground is referred to as an antenna polarization. Vertical, horizontal, circular, and mixed polarization are only a few of the different types of antenna polarization. The physical makeup of some antennas, including dipole antennas, determines their polarization. Depending on how it is positioned on its support, a dipole antenna can provide both horizontal and vertical polarization [25].

15.3.4.4.5 Impedance and Voltage Standing Wave Ratio

The ratio of voltage to current passing through antenna terminals is known as antenna impedance. Antenna impedance plays a great role in the transmission and receiving of electromagnetic energy signals through the antenna. This impedance must be matched between the transmitter and antenna for smooth transfer of the electromagnetic energy signal to the antenna. Input impedance ranges from 50 ohms to 150 ohms are usually used in communication systems. Any kind of mismatch in input impedance of different radiation sources is bridged by using an additional circuit called a Balun impedance matching circuit [26]. Voltage standing wave ratio (VSWR) quantifies the mismatch in a link when the communication system's impedances are not compatible. The VSWR is a measurement based on the reflection coefficient mechanism of the antenna i.e., how well the antenna is dissipating power from a transmitter. It can be expressed as:

$$VSWR = \frac{1+\Gamma}{1-\Gamma}$$

where Γ is the reflection coefficient, which is calculated by the equation:

$$\Gamma = \frac{Z - Z_0}{Z + Z_0}$$

where Z_0 represents the line's characteristic impedance and Z represents the load impedance.

15.4 MEASURING METHODS FOR EMI

According to accepted standards, EMI testing can be carried out as pre-compliance testing or compliance testing. In terms of hardware, software, and methodology, the pre-compliance setup test beds must closely mimic the compliance test setup [27]. The origin of most of the unwanted interferences on the receiver or susceptor is the emitter. These interferences are transmitted from the origin point to the receiving end by the coupling channel. There are three main elements that influence the development of EMI in an electrical system. If the communicating channel is a conductor, then EMI is thought to have occurred as a result of conducted emission. Radiated emission occurs if the coupling channel is of the radiating kind. These components may be several systems that result in intrasystem EMI, or they may be various subsystems of a larger system that results in intersystem EMI. EMI measurements and testing are grouped as illustrated in Figure 15.4.

EMI measurement techniques

Emission Testing   Immunity Testing

Radiated Emission	Conducted Emission	Continuous Sources	Transient Sources
Open Air Test Anechoic Chamber	Transverse Electromagnetic Cell Line Impedence	Magnetic Field Radiation Conducted	Electrostatic Discharge
Reverberation Chamber	Current Probe		Surge
	Voltage Probe		Electrical Fast Transient/Burst (EFT/B)

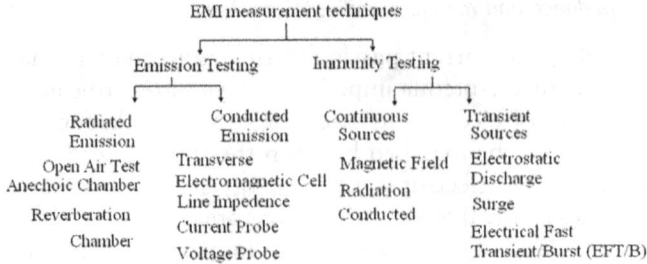

Summery chart of different EMI measurement technique

Figure 15.4 Summary chart of different EMI measurement techniques.

15.4.1 Emission Testing

Almost every electronic device emits conducted or radiated emissions, which deliberately or unknowingly contribute to electromagnetic pollution. These unwanted emissions, which may travel up to GHz ranges, can be either carried directly through antennas or indirectly through AC power systems. They come from power wires, resistors, capacitors, Op-Amps, and other components (in the case of radiated emission). In order to maintain a secure electromagnetic environment that can be utilized for other permitted applications, every electronic device must undergo emission testing. In these tests, much as in the cases of conducted or radiated emissions, the instrument being tested acts as the emitter (EUT) [28].

15.4.2 Radiated Emission Testing

It is a description of a specific measuring mechanism for radiated emission EMI testing. Standard radiated emission measurement is generally done in the frequency range between 30 MHz to 1 GHz or wavelength ranges from 10 m to 0.3 m. This type of setup normally consists of a receiving antenna connected via cables to an EMI receiver or spectrum analyzer [29], a potentially infinite metallic ground plane, and the EUT, which is typically maintained 3–10 metres away from the receiver (unless otherwise stated).

15.4.3 Conducted Emission Testing

A sudden fluctuation in voltage or current in the circuits of an electronic instrument causes noise on peripheral or loaded devices, which is measured or quantified by emission testing [30]. The instrument noise could damage the connected devices and cause the equipment to malfunction. There are several popular methods for doing emission testing, including:

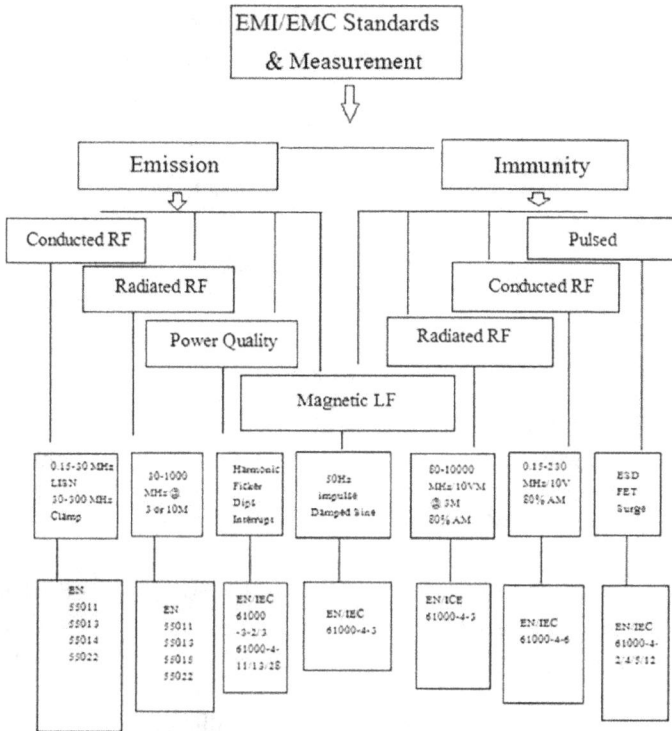

Figure 15.5 Summary chart of standard EMI/EMC measurement processes.

Network for line impedance stabilization (LISN)

1. X technique
2. Current and voltage probes
3. Transverse electromagnetic (TEM) cell

15.4.4 Immunity Testing

A completely opposite of emission testing is called immunity testing. Emission testing requires detecting the noise the EUT emitted, whereas immunity testing involves exposing the EUT to an electromagnetically hostile environment and assessing whether any changes in the EUT's performance are present or absent. If there is a change in how the EUT functions, it is measured and, if necessary, compared to national or international standards. The instrument is unable to function effectively in the actual world if it fails to pass these tests [31]. This subsection explains the continuous and transient types of immunity measuring procedures used in pre-compliance or compliance EMI testing.

15.4.4.1 Continuous-Source Immunity Testing

Continuous immunity testing is used to determine whether the device under test (DUT) or equipment under test (EUT) will function as intended under continuous exposure to a noise environment or not. The typical examples of such noise sources are cosmic microwave background, solar radiation, broadcast stations, moving vehicles, and magnetic fields. This kind of immunity testing is carried out for a few to several minutes [32]. Immunity testing in the magnetic field, radiated, and conducted category are other testing options also available.

15.4.4.2 Transient-Source Immunity Testing

Lightning, electromagnetic pulses (EMPs), electrostatic discharge (ESD), voltage changes, rapid switching, and relaying are a few examples of transient electromagnetic interference that, despite occurring for extremely brief intervals, can have disastrous impacts on system performance. Electronic systems of an aeroplane or spacecraft, as well as equipment used in meteorology to monitor storms, tornadoes, etc., are few cases of expected recipients of transient EMI. Therefore, it is crucial to assess the system's tolerance for such circumstances [33]. Since transient sources emit a considerable amount of EM radiation in a brief period of time, immunity testing against them is done in the temporal domain (less than a few milliseconds).

15.5 CONCLUSION

Since we don't live in a flawless world, EMI has the potential to cause harm to our electronic systems and equipment. We have made significant progress in our knowledge of how EMI affects our electronic equipment and how we deal the matter to completely avoid it. The development of regulations for allowable levels of EMI and EMC is one option. We may make sure we're effectively combating interference from artificial and naturally occurring sources by adhering to best practices when determining which method(s) fit our electrical system. This chapter covers types of EMI, measuring techniques, and reduction approaches. Consequently, it might be advantageous for both experienced and novice researchers to gain a thorough understanding of EMI. This review focuses on both the fundamental methods for reducing EMI produced by electrical equipment and the aforementioned EMI measurement methods. Some of these techniques include the use of electromagnetic shields, EMI filters, altered circuit topologies, and spread spectrum technology. Special attention is paid to electromagnetic shielding because it is most likely the popular EMI reduction technique.

REFERENCES

[1] P. Mathur and S. Raman, 2020, Electromagnetic Interference (EMI): Measurement and Reduction Techniques, Journal of Electronic Materials, 49, pp. 2975–2998.

[2] J. L. N. Violette, D. R. J. White, and M. F. Violette, 1987, *Electromagnetic Compatibility Handbook*, New York: Springer.

[3] M. J. Horst, W. A. Serdijn, and A. C. Linnenbank, 2013, *EMI-Resilient Amplifier Circuits*, Berlin: Springer.

[4] M. Shalaby, W. Saad, and M. Shokair, 2017, Evaluation of Electromagnetic Interference in Wireless Broadband Systems. Wireless Personal Communications, 96, pp. 2223–2237.

[5] D. Morgan, 1994, *A Handbook for EMC Testing and Measurement*, London: IET.

[6] B. R. Archambeault and J. Drewniak, 2013, *PCB Design for Real-World EMI Control*, Berlin: Springer.

[7] S. H. Voldman and E. S. D. Testing, 2016, *From Components to Systems*, New York: Wiley.

[8] M. I. Montrose and E. M. Nakauchi, 2004, *Testing for EMC Compliance: Approaches and Techniques*, New York: Wiley.

[9] D. G. Baker, 2015, *Electromagnetic Compatibility: Analysis and Case Studies in Transportation*, New York: Wiley.

[10] A. Raveendran, M. T. Sebastian, and S. Raman, 2019, Applications of Microwave Materials: A Review, Journal of Electronic Materials, 48, pp. 2601–2634.

[11] X. C. Tong, 2009, *Advanced Material and Design for Electromagnetic Interference Shielding*, Boca Raton: CRC Press.

[12] G. Dong, Q. Li, X. Jiang, S. Chen, and X. Chen, 2022, The Analysis and Suppression of Electromagnetic Interference on Satellite Automatic Identification System, *2022 Asia-Pacific International Symposium on Electromagnetic Compatibility (APEMC)*, Beijing, 2022, pp. 470–472.

[13] S. K. Srivastava and K. Manna, 2022, Recent Advancements in the Electromagnetic Interference Shielding Performance of Nanostructured Materials and Their Nanocomposites: A Review, Journal of Materials Chemistry A, 10, pp. 7431–7496.

[14] V. Shukla, 2020, Observation of Critical Magnetic Behavior in 2D Carbon Based Composites, Nanoscale Advances, 2, p. 962.

[15] Sabu Thomas and Suji Mary Zachariah, 2021, *Nanostructured Materials for Electromagnetic Interference Shielding*, Boca Raton: CRC Press. https://www.routledge.com/Nanostructured-Materials-for-Electromagnetic-Interference-Shielding/Thomas-Zachariah/p/book/9781032108360.

[16] L. Tihanyi, 1995, *Electromagnetic Compatibility in Power Electronics*, New York: IEEE Press; Sarasota, Fla J.K. Eckert; Oxford: Butterworth-Heinemann.

[17] P. S. Crovetti, 2021, *Electromagnetic Interference and Compatibility*, Switzerland: MDPI AG. https://www.mdpi.com/books/book/3843-electromagnetic-interference-and-compatibility.

[18] W. Prasad Kodali, V. Prasad Kodali, 2001, *Engineering Electromagnetic Compatibility Principles, Measurements, Technologies, and Computer Models*, US: Wiley. https://www.wiley.com/en-us/Engineering+Electromagnetic+Compatibility%3A+Principles%2C+Measurements%2C+Technologies%2C+and+Computer+Models%2C+2nd+Edition-p-9780780347434.

[19] J. Carr, 2000, *The Technician's EMI Handbook Clues and Solutions*, Netherlands: Elsevier Science. https://www.sciencedirect.com/book/9780750672337/the-technicians-emi-handbook.

[20] R. Perez, 2013, *Handbook of Electromagnetic Compatibility*, Netherlands: Elsevier Science. https://shop.elsevier.com/books/handbook-of-electromagnetic-compatibility/perez/978-0-12-550710-3.

[21] J. R. Mediavilla, C. A. Espinoza, A. P. Espinosa, and F. Salaza, 2022, Design of a Passive Radar Based on RTL-SDR Technology with Coherent Dual Channel, Journal of Physics: Conference Series, 2199, p. 012023.

[22] Y. Zhang, T. Pan, and Z. Yang, 2020, Flexible Polyethylene Terephthalate/Polyaniline Composite Paper with Bending Durability and Effective Electromagnetic Shielding Performance, Chemical Engineering Journal, 389, p. 124433.

[23] H. M. Schlicke, 2020, *Electromagnetic Compossibility*, 2nd edition, US: CRC Press. https://www.routledge.com/Electromagnetic-Compossibility-Second-Edition/Schlicke/p/book/9780824718879.

[24] K. Malaric, J. Bartolic, and R. Malaric, 2005, Immunity Measurements of TV and FM/AM Receiver in GTEM-Cell, Measurement, 38(3), pp. 219–229.

[25] F. M. Rea, N. Tovar, M. F. Barba and J. C. Aviles, 2018, Approximate antenna radiation pattern determination using a low cost measuring system at 2.4 GHz, *2018 IEEE 10th Latin-American Conference on Communications (LATINCOM)*, Guadalajara, 2018, pp. 1–5. doi: 10.1109/LATINCOM.2018.8613212.

[26] T. Wang, C. Ruf, S. Gleason, A. O'Brien, D. McKague, B. Block, and A. Russel, 2021, Dynamic Calibration of GPS Effective Isotropic Radiated Power for GNSS-Reflectometry Earth Remote Sensing, IEEE Transactions on Geoscience and Remote Sensing, pp. 1–12. doi:10.1109/TGRS.2021.3070238.

[27] P. Kumar and J. L. Masa, 2016. Dual Polarized Monopole Patch Antennas for UWB Applications with Elimination of WLAN Signals. Advanced Electromagnetics, 5, pp. 46–52. doi:10.7716/aem.v5i1.305.

[28] S. Sankaran, K. Deshmukh, M. Basheer Ahamed, and S. K. Khadheer Pasha, 2018, Recent Advances in Electromagnetic Interference Shielding Properties of Metal and Carbon Filler Reinforced Flexible Polymer Composites: A Review, Composites Part A: Applied Science and Manufacturing, 114, pp. 49–71.

[29] P. F. Wilson, M. T. Ma, and J. W. Adams, 1988, Techniques for Measuring the Electromagnetic Shielding Effectiveness of Materials. I. Far-Field Source Simulation, IEEE Transactions on Electromagnetic Compatibility, 30(3), pp. 239–250.

[30] M. Rudd, T. C. Baum, and K. Ghorbani, 2020, Determining High-Frequency Conductivity Based on Shielding Effectiveness Measurement using Rectangular Waveguides, *IEEE* Transactions on Instrumentation and Measurement, 69(1), pp. 155–162.

[31] X. C. Tong, 2008, *Advanced Materials and Design for Electromagnetic Interference Shielding*, Boca Raton: CRC Press. https://www.taylorfrancis.com/books/mono/10.1201/9781420073591/advanced-materials-design-electromagnetic-interference-shielding-xingcun-colin-tong.

[32] J. D. Kraus, 1992, *Electromagnetics*, New York: McGraw-Hill.

[33] International Electrotechnical Commission, 2005, Electromagnetic compatibility (EMC) –Part 4–5: Testing and Measurement Techniques – Surge Immunity Test, Document IEC 61000-4-5:2014. https://webstore.iec.ch/publication/4223.

Chapter 16

Application of a Frequency Selective Surface in the Modern Medical Field

M. Shobana and S. Raghavan

16.1 INTRODUCTION

The history of the frequency selective surface (FSS) dates back to 1786 when David Rittenhouse observed the suppression of colors from a lamp through silk [1, 2]. The evolution of the FSS is from diffraction grating, and earlier intensive studies on FSS have been dealt with in [3–5]. FSS is considered the counterpart of filters in transmission line theory, and it has become a fascinating research topic in the electromagnetic field. Metamaterial saw unprecedented market growth due to rapid industrialization and urbanization with the increase in the use of electrically small antennas in applications of telecommunication networks, television networks, and satellites [6]. Over the years FSS has been widely used in applications such as multifrequency reflector antenna [7–10], absorber [11–12] radar cross-section (RCS) reduction, [13] electromagnetic interference (EMI) reduction and shielding in rooms [14, 15], and millimeter and terahertz wave applications [16]. FSS use in radomes has reached new heights. A radome is a bandpass FSS that is integrated to minimize the RCS of the antenna [17–21]. Recently researchers have demonstrated the successful implementation of FSS in 5G [22, 23]. Though FSS has been applied in several applications, we aim at reviewing how FSS could be used in medical applications. In today's scenario where unknown viruses cause a pandemic, there is a huge demand for wireless medical devices that could be implanted in the human body for medical surveillance. Though a large number of implantable antennas are available in the market, the race is on to increase the performance of the antenna. Biomedical engineering requires twofold requirements from an antenna: 1. conformal structures and 2. efficient performance when planted in human tissues. For these requirements, the best candidate would be filters called FSS. When an antenna is implanted in the human body, the specific absorption rate (SAR) should be taken into consideration. Direct exposure to radiation from antennas causes health hazards. Hence we need to isolate the antenna from the human body. FSS-based high impedance

DOI: 10.1201/9781003459880-16

surfaces (HISs) are used as a ground where perfect isolation from a human body is provided. With the help of an HIS, numerous lightweight antennas and thinner antennas are investigated and realized with high gain [24–27]. The chapter explores the different FSS-based electromagnetic structures for application in the medical field.

The chapter is structured as follows. The introduction section introduces the readers to the theoretical perspective of FSS, the types of FSS, and the simulation setup to measure FSS. The second section discusses how FSS is used to improve the performance of antennae used in the industrial scientific medical (ISM) band, FSS as an absorber, FSS for biomedical sensing, artificial magnetic conductor (AMC) structures, electronic band gap (EBG) structures, FSS in magnetic resonance imaging (MRI), and FSS as a superstrate. The last section of the chapter discusses the summary of FSS structures used for biomedical applications.

16.2 THEORETICAL PERSPECTIVE: HOW DOES FSS TRANSMIT OR BLOCK INCOMING WAVES?

The FSS can be an aperture or patch, and the resonant element i.e. inductance and capacitance determine the operation mechanism of the FSS. When an electric field strikes an electron, as in Figure 16.1(a), the electron will oscillate and some energy is transmitted to nearby electrons as kinetic energy. The remaining energy will be absorbed by the electron. The transmission will be zero if all energy is absorbed. If, however, electron movement is restricted along the horizontal direction, the electron is imperceptible to the incident wave and the wave will pass through the filter. Hence, if the energy is not absorbed by electrons, then the transmission

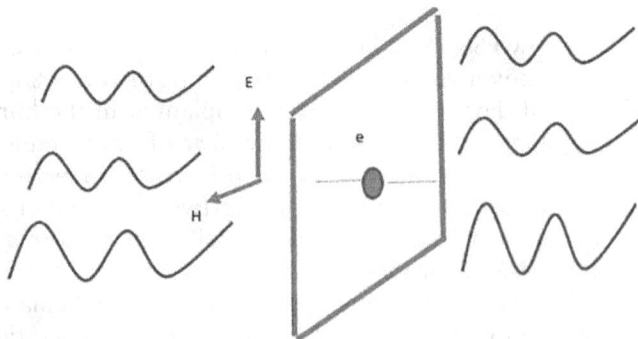

Figure 16.1(a) FSS transmitting waves [1].

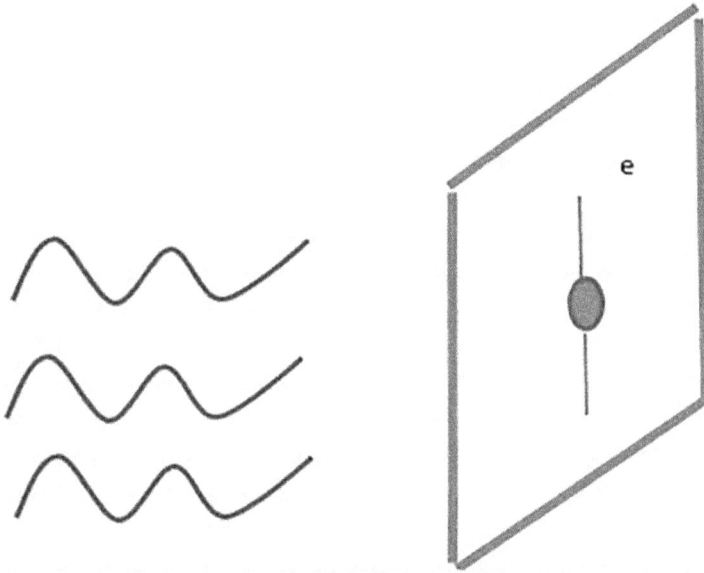

Figure 16.1(b) FSS blocking incoming waves [1].

is high. On the other hand, the transmission is zero when the emitted radiation on the right side destructively interferes with the wave coming from
the left as shown in Figure 16.1(b) [28]. The required filtered response can
be produced by designing elements. The parameters that have a significant
influence on the filtering behavior are the parameters of the substrate and
the interelement spacing. Interelement spacing affects the bandwidth, and
smaller interspacing produces a wide bandwidth, which is the preferred
choice. FSS is either a capacitive or inductive type similar to a filter in circuit theory. FSSs are periodic structures, and Floquet's theorem is applied
to design an FSS [29].

The resonance frequency of an FSS unit cell is given by equation (16.1),
in which L and C represent the equivalent inductance and capacitance,
respectively. FSSs come in different shapes and are classified as groups.
Figure 16.2, Figure 16.3, Figure 16.4, and Figure 16.5 show types of FSSs
and their structures.

$$f = \frac{1}{2\pi LC} \qquad (16.1)$$

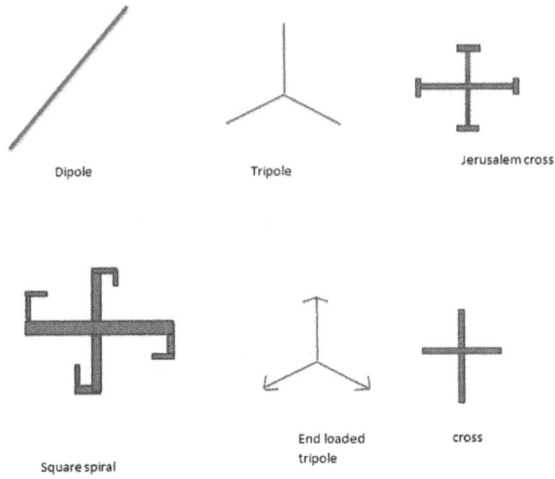

Figure 16.2 Group 1 center connected or n-pole type of FSS [1].

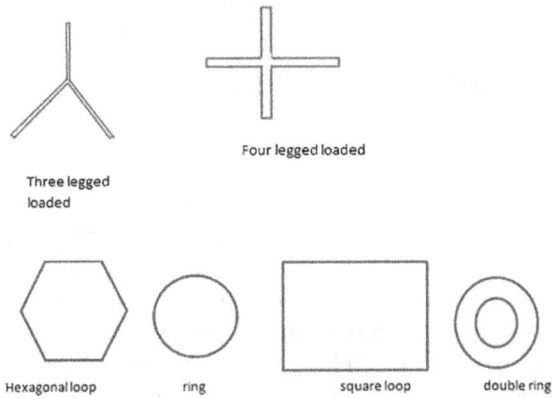

Figure 16.3 Group 2 loop type of FSS [1].

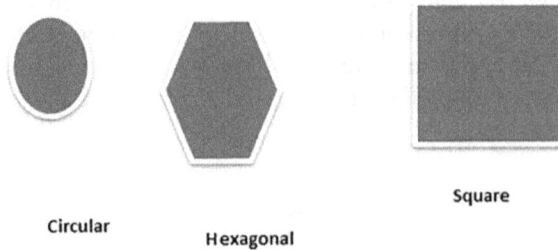

Figure 16.4 Group 3 plate type of FSS [1].

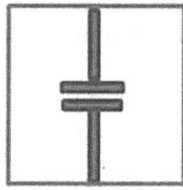

| Square loop with capacitor | split ring |

Figure 16.5 Group 4 split type of FSS [1].

16.3 SIMULATION AND MEASUREMENT

After designing the required FSS for medical applications, the design is simulated for verification. Numerous simulation tools are available like High Frequency Structure Simulator (HFSS), Computer Simulation Tool (CST), and COMSOL PHYSICS. Though FSS is implemented in the infinite periodic array, it is enough to simulate a one-unit cell with a perfect boundary condition (PBC) thereby reducing the simulation time.

16.4 SIGNIFICANCE OF FSS ON ANTENNAS

The FSS structure is incorporated into microwave antennas to increase the radiation aperture and thereby achieve directivity. Research on the combination of FSS with an antenna has progressed over the years, and new designs are being published from time to time. FSSs are used as a superstrate and substrate with the antennas. They are used in conjunction with antennas primarily for miniaturization, multiband operation and gain improvement. FSSs can also be incorporated into antennas together with switches and varactor diodes to achieve reconfigurability and tunability. EBG structures based on FSS technology are used as a superstrate (i.e. suspended above the patch antenna) to enhance the gain of the parent antenna. Surface wave suppression is accomplished by designing EBG structures that mimic mushroom-like structures and that are integrated with an antenna. Mutual coupling between antennas is greatly reduced with the help of FSS.

16.5 APPLICATION OF FSS IN THE MEDICAL FIELD

16.5.1 Enhancing the Performance of the ISM Antenna

The microstrip antenna is a crucial element in a biomedical communication system. When doctors treat patients with a history of contagious disease, the need for telemedicine becomes important. The doctors can remotely monitor the clinical conditions of the patients. This not only saves the lives of doctors and patients but also prevents the spread of contagious diseases. Hence, the wireless medical system is in the demand.

As the antenna is an important component in the wireless system, the performance characteristics of this antenna must be increased. A metamaterial is integrated with a microstrip antenna to improve performance. Metamaterials are artificial materials that exhibit negative permeability and permittivity [30]. These artificial structures in the form of rectangular complementary split ring resonators are successfully implemented in multiple-input, multiple-output (MIMO) antennas to enhance isolation [31, 32]. Hence, FSSs, a class of metamaterial, are incorporated with an antenna in the form of EBG, AMC to improve the gain, radiation characteristics, and front to back ratio (FBR). The FSS is compact and facilitates easier integration with an antenna. Hence, researchers have shown interest in designing microwave and millimeter-wave components. Radiation characteristics of the planar antenna can be increased by using FSSs [33–40]. In [41] a dual-layer frequency selective surface (DLFSS) is employed to improve the gain of a microstrip patch antenna operating at 2.4 GHz. The identical anchor-shaped FSS elements as shown in Figure 16.6 are arranged periodically on an inexpensive FR4 substrate. The DLFSS is positioned 10 mm from the antenna to achieve the best gain. When the DLFSS is brought in front of the antenna, there is a good increase in return loss. In [42] a microstrip patch antenna for 5.8 GHz is proposed. The novelty of the idea is that the patch is printed on organic magnetic material. The magnetic material not only reduces the size of the microstrip antenna but improves impedance bandwidth significantly. FSS structures have been

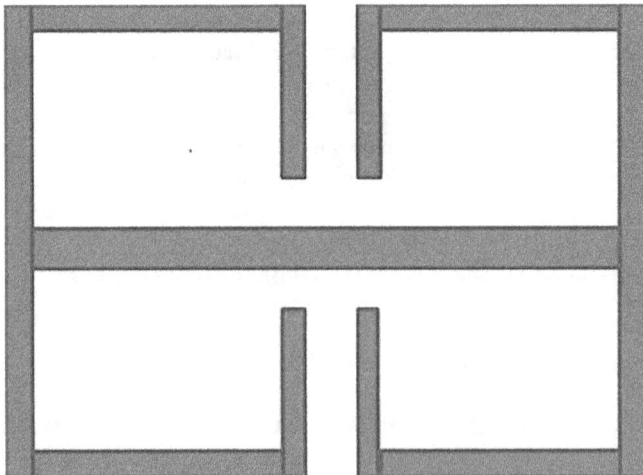

Figure 16.6 Unit cell of a dual-layer FSS in an ISM antenna [41].

incorporated in antennas to improve the antenna radiation efficiency, bandwidth, and gain. The ground plane consists of a periodic structure of slot arrays, and nonlinear optimization has been used to design unit cells.

16.5.2 EM Absorber

Today the world perceives a momentous increase in the use of mobile devices, thanks to the latest innovations in wireless communication systems. But hospitals, clinical laboratories, and other sensitive institutions could be affected by these signals and need to be isolated. Telemedicine involves a number of wireless devices to treat patients remotely, and hospitals do contain devices operating in the ISM band. We need an absorber to eliminate unwanted radiation in hospitals; otherwise, it would interfere with medical devices. The communication system in hospitals requires that critical data like electrocardiograph (ECG) and electroencephalogram (EEG) data be transmitted with accuracy and should be free from interference. Also the safety of the patients is of high concern. The antenna used inside hospital communication systems should not have a high SAR. An FSS-made absorber is the ideal choice, as it can provide the desired shielding. A novel convoluted FSS geometry has been designed with angular stability and polarization to block ISM and Unlicensed National Information Infrastructure (UNII) radio bands [43]. The four-legged FSS is designed to provide shielding from 5 to 6 GHz. The attenuation level is from 10 to 20 dB at the ISM band and 10 to 30 dB at the UNII band. Also the FSS design is compact and lightweight, presenting angular stability and polarization independence.

Telemedicine involves a number of wireless devices to treat patients remotely, and hospitals contain devices operating in the ISM band. The idea behind using the FSS as an absorber is to transform a building wall into FSSs. To achieve a good stop and absorption behavior in a wireless local area network (WLAN), two new frequency selective absorbers have been proposed [44]. An idea to develop absorber for biomedical and sensing applications is proposed [45]. The authors have investigated the performance of three U-shaped FSSs. The quad-shaped FSS has polarization independency, and this makes it a suitable candidate for biomedical sensing applications. When FSS is used for sensing, there is an interface between the biological environment and microwave components. Researchers are gearing up to design lightweight, durable, reliable, and highly compatible biomedical sensors. FSS could be a promising candidate, and a double U-shaped FSS provides 90% of absorbance.

16.5.3 FSS for Biomedical Sensing

Biodegradable medical implants dissolve in biofluids, leaving no trace when placed inside the body. Many radio frequency (RF) circuits are designed using biocompatible and water-soluble materials.

The FSS embedded in biodegradable materials for biological and sensing applications is designed in the shape of flower-shaped slots. Three important conditions are considered in the design of a biodegradable FSS. The structure should be insensitive to polarization and angle of incidence, as the sensors may rotate inside the body. FSS-based polarization-insensitive devices have been designed [46]. They should have sharp resonance such that they are sensitive to environmental changes. When dissolved, the structure should exhibit stable resonance. The proposed FSS consists of 16 unit cells in a 4 × 4 array. The conductor is a combination of zinc and magnesium, and a thin silk film forms the substrate. The resonance frequency is 3.25 GHz for different incident angles and polarization, which is a clear indication of angular and polarization stability. Though in this paper the fabrication is not given, the proposed FSS can potentially serve as an idea for biological tracking and sensing applications. Electromagnetic radiation in the THz frequency range is very much useful for sensing chemical and biochemical compounds [47]. FSS facilitates the identification of biomolecules' composition or chemicals by detecting the resonant absorption of molecular or phonon resonances. Dielectric loading causes the resonance frequency to shift under the inductance and capacitance of the resonating structures. The shift in resonant frequency is observed as sensitivity in high terahertz sensors. Two things are necessary to achieve high sensitivity. The FSS can be employed to detect microorganisms by their change in resonance frequency. First, FSS as sensors should exhibit a sharp edge in the frequency response to enable the detection of small changes in the dielectric emission. Second, there should be a high concentration of an electric field to facilitate sensing even a small amount of the sample. FSS in terahertz sensing has been successfully demonstrated in the form of an asymmetric double split ring (aDSR). The sensing environment consists of an analyte emulating the organic system with a dielectric constant similar to organic systems and a film thickness of 10 nm in biosensing applications. The dual resonance feature (DRF) shows a shift in resonance frequency at 5 GHz when the sample is placed at the spot where the E-field concentration is high and the DRF frequency shifts to a lower frequency by 4 GHz.

16.5.3.1 Artificial Magnetic Conductors (AMC)

Sievenpiper was the first to introduce frequency selective structure (FSS) as a high-impedance surface, which was called an MAC [46–49]. The architecture is seen as a transmission line with a substrate providing hindrance to the propagation of surface waves constituting the forbidden frequency band. The surface waves are trapped in the band gap, thereby reducing undesired waves in a substrate. An AMC is a surface with large surface impedance. Ideally, the AMC forms the ground plane, which is used in the forefront to suppress surface waves inside the substrate [50]. The AMC is a perfect magnetic conductor that cannot support an electric current and prohibits the propagation of surface waves inside the substrate.

For early detection and prevention of disease, wearable medical devices serve as a boon to the medical field. The antenna is the prominent component in wireless modules. The antenna performance should not change in the presence of the human body. The antenna needs to be isolated from the body, and the high impedance surface is used to enhance antenna gain [51]. AMCs are also a class of FSS that reflects incident waves without phase reversal. The incident waves are similar to a perfect magnetic conductor increasing the radiation for the antenna placed parallel and close to the surface.

16.5.4 Dosimeter Tag

In hospitals, radiation dosimeters are worn on the patient's body to monitor or measure the dose uptake of ionizing radiation. In [52] a 2.45 GHz radio frequency identification (RFID) dosimeter tag with inkjet-printed dipole antenna is constructed on an elastic Kapton substrate using AMCs. When a dosimeter tag is used in a blood environment, the radiation of the antenna gets altered and affects the performance, as blood is a lossy environment.

AMC as a ground plane quarantines the patient's body from undesired exposure to electromagnetic radiation and eliminates impedance mismatch due to the proximity of human tissue. The AMC-integrated antenna is flexible and supports broadside radiation with a gain measuring from 4.1 dBiA to 4.8 dBi. A square loop unit cell shown in Figure 16.7 is chosen as an AMC

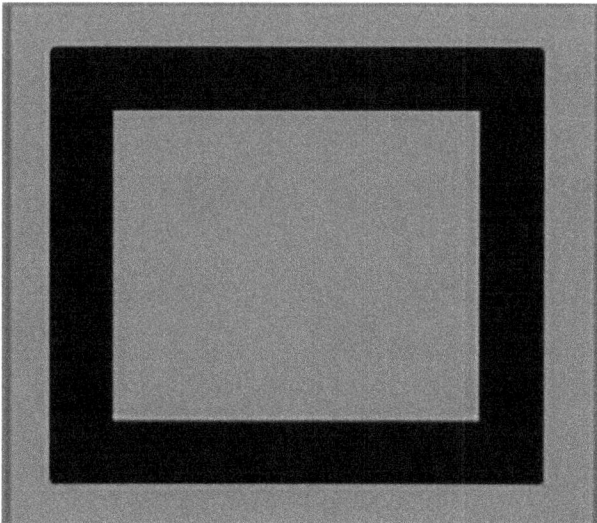

Figure 16.7 Unit cell of AMC in dosimeter [52].

at 2.45 GHz. To validate the use of AMC, the authors have experimentally verified this by studying the radiation characteristics and reflection coefficient of the antenna by placing the antenna on a blood bag. The reflection coefficient does not show variation from 2 GHz to 3 GHz, and also the antenna shows broadside radiation with a variation of 0.7 dBi in negative and positive values. The antenna is connected to a rectifier to be used as a wireless device. The study also proved that the antenna with AMC showed the same reflection coefficient in bending conditions in the blood bag. Thus, AMC proves to be a perfect candidate for enhancing the performance of the antenna in dosimeters when they are to be used in a lossy environment.

Though research on biomedical antennas has increased to an unprecedented level, the researchers face problems like identifying the appropriate fabrication techniques and designing low-profile antennae. Jerusalem cross (JC) is a promising candidate to be employed as an AMC. JC offers high angular stability when illuminated obliquely, and the structure is well suited for biomedical applications. The AMC consists of 3 × 3 unit cells of a slotted JC AMC resonating at the same frequency as the antenna [53]. The AMC is embedded on an elastic vinyl substrate with 1.5 mm thickness and a dielectric of 2.5. The authors have successfully implemented this in their research and reported an improvement of gain of 3.7 dBi in the antenna. The unit cell of the JC is shown in Figure 16 8. The antenna gain measured is 4.8 dBi, while it was only 1.1 dBi without AMC. There is a 64% reduction in SAR, which validates that AMCs are a good candidate for SAR reduction.

Figure 16.8 Unit cell of Jerusalem cross in low-SAR antenna [53].

The AMC-inspired antenna reduces electromagnetic coupling in wireless body networks (WBANs) and hence finds very good application in the biomedical field. The antenna is a low profile integrated with the AMC for the 5.8 GHz ISM band [54]. The monopole antenna dimension is 27×34 mm^2. The antenna is placed on an AMC layer of 4×6 units, and the overall dimension is 102×68 mm^2. These structures are designed on a 1.8-mm-thick Pellon fabric substrate. The AMC structure is studied by reflection phase characteristic and zero reflection phase power at 5–8 GHz. The achieved bandwidth is 390 MHz. When the AMC is placed below the antenna, the resonant frequency is shifted to 4.8 GHz with a 14.47 dB increase in S_{11}. The increase in FBR is achieved by the in-phase reflection property of the AMC. The unit cell of the AMC is shown in Figure 16.9. The SAR value is minimized, making the antenna robust and less sensitive to the loading effects of the human body. The measured gain for the proposed AMC is 6.12 dBi compared to 2.65 for the monopole. The simulated gain was 7.34 dBi for AMC, whereas with the monopole antenna, it was only 3.18 dBi. The deformation of the structure due to bending and crumbling is studied under the E-plane and H-plane. When the AMC antenna is bent in the H-plane direction, it causes a shift in resonant frequency. The antenna performance on the body is investigated under different human body models: an averaged human arm phantom and a multilayer human tissue model consisting of four layers of skin, fat, muscle, and bone tissue. Then the antenna is placed on the arm of the Ella model. The antenna retains its impedance-matching properties in the presence of the AMC. The calculated SAR is an under-regulated value.

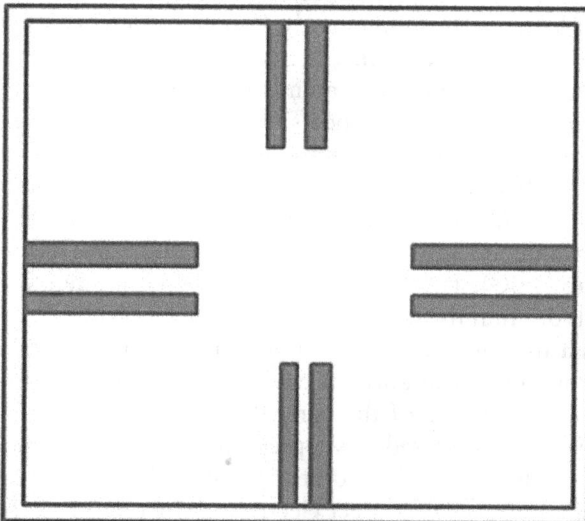

Figure 16.9 Unit cell of the AMC in a WBAN [54].

16.5.5 Wearable Medical Devices

Works on wearable RFID are becoming popular nowadays due to modulated FSSs. A WBAN connects all actuators and sensors located in different parts of the body and on clothes. A wearable antenna fabricated on a denim substrate with enhanced bandwidth is used for WBAN applications [55]. While using the antenna for WBAN applications, there is a reduction in antenna efficiency due to the proximity of the body. A transponder is constructed with an array of dipoles loaded with a varactor diode to modulate the backscatter response [56]. Using the modulated FSS principle, a breathing sensor has been designed at the 2.45 GHz ISM band. A temperature sensor with a negative temperature coefficient of resistance is used. When there is a variation in the temperature of the airflow, the frequency of the oscillator changes and modulates the RCS of the FSS. The FSS consists of an array of dipoles. The transponder consists of several FSSs, and it is integrated with a headband located around the head [57]. The device could be used to cure apnea where a patient experience breathing cessation while asleep.

16.5.6 Mobile Body Area Network

A coplanar waveguide (CPW) antenna incorporated with EBG-FSS at 2.45 GHZ is proposed for mobile body area network (MBAN) applications. MBAN is essential in health sectors for monitoring ECG, EEG, heart rate, and blood pressure. During a pandemic or other event, telemedicine is an alternative way for patients to get medical advice without going to hospitals. The antenna used for MBANs should be lightweight, low profile, miniaturized, and conformal to the human body. Due to capacitive coupling between the human body and the antenna, the frequency of the antenna is altered, and the radiation pattern of the antenna changes due to the absorption by the human body. A high-impedance surface is implemented as it provides high isolation between the antenna and body tissue. EBG is used to successfully nullify the effect of harmful electromagnetic energy from entering the human body. A CPW antenna in a C shape with an additional C slot etched is designed on a denim substrate [58]. A square loop acts as an electromagnetic band gap (EBG). The EBG FSS is constructed on denim material as a substrate with 0.7 mm thickness.

The antenna used for an MBAN should have good impedance matching and good return loss. The authors have validated the superiority of EBG FSS by studying the phase of the current in the antenna and the EBG-FSS acting as ground. EBG is used to suppress unwanted electromagnetic radiation entering the human body and hence reduces SAR. When an antenna is placed over a conventional perfect electric conductor (PEC), there is a 180 degree phase between the current in the antenna and the image current in PEC, and the currents do not add up; hence the reflection coefficient is very high at −10 dB. On the other hand, the current in the antenna and EBG-FSS are in the same phase, and hence the current in the antenna and EBG-FSS add up, and the return loss is less than −10 dB. The authors have also studied the bending condition of the antenna in free space and the human body condition. The EBG-FSS integrated antenna is the best choice to be used in a WBAN for telemedicine. The square loop design is chosen as the EBG-FSS, as it gives a larger inductance value. The EBG-FSS unit cell is designed with the capacitance and inductance as shown in equations 16.2 and 16.3 [58]:

$$c_s = \frac{W \varepsilon_0 \left(1 + \varepsilon_r\right) \cosh^{-1}\left(\dfrac{W + g}{g}\right)}{\pi} \tag{16.2}$$

$$L_s = l_n \frac{\mu_0}{4\pi} \ln\left\{ 1 + \frac{32h^2}{\omega_n^2}\left[1 + \sqrt{1 + \left(\frac{\pi\omega_n^2}{8h^2}\right)^2} \right] \right\} \tag{16.3}$$

ε_0 is free space permittivity
μ_0 is free space permeability
ε_r is dielectric constant

W is the perimeter of the conductive material, and g is the gap size between the unit cells. C_S and L_S form the capacitance and inductance of the equivalent circuit. ln is the length of the strip, w is the width of the strip, h is the thickness of the substrate, and g is the gap. After designing the square loop, it is integrated with the CPW antenna with an optimum separation between the antenna and FSS with 1-mm-thick foam of permittivity 1.05 and loss tangent 0.003. This separation gives the optimum S_{11}. Integrating the antenna with EBG structures shows better reflection coefficients. The antenna gain is increased from 1.74 dBi to 6.55 dBi. The reflection coefficient is found to be −10 dB due to the phase matching between the antenna and PEC giving a constructive result and good impedance matching. Due to the suppression of backward radiation by employing EBG-FSS, the FBR is improved by 13 dB and the gain by 6.55 dBi. The performance of the antenna is carried out in three parts of the human body, and the results agreed well with the simulated results showing a decrease in SAR by more than 95%.

16.5.7 MRI

MRI is used to study the anatomy of the human body. It forms an integral part of the medical equipment in modern hospitals. Identifying deadly diseases like cancer at an early stage can save millions of people. MRI operates in a strong magnetic field. The antenna used in MRI should have a narrow directive beam of 5 dBi in the broadside direction in the desired frequency for mechanical steering principles. This is achieved using an FSS where the gain of the antenna is increased by the concept of in-phase reflection. The author designed an array of 6 × 4 rectangular slots. The unit cell is designed such that the FSS should reflect at all frequencies but transmission should be strictly at 10.6 frequencies only. In [59] the author has come out with a novel cylindrical substrate with Teflon. For the ultra-wideband (UWB) antenna, the authors have designed a triangular-shaped patch antenna with flare. The antenna is slotted for miniaturization and is placed on the cylindrical substrate. The ground is also partially deflected to give good impedance matching and stable gain in the UWB range. Several FSS unit cells are arranged in an array and are placed on the inner surface of the cylinder at a distance of 20 mm. The FSS layer is responsible for increasing the gain of the antenna by in-phase reflection in the frequency range of 7 GHz to 19 GHz. The FSS acts as a reflector for all frequencies and adds constructively with the radiated wave and reduces the side lobe radiation; therefore a high FBR is obtained. The FSS reflector reduces the backward wave and thereby meets the requirements of the directive beam in the MRI. The unit cell of the FSS is shown in Figure 16.10. With the implementation of FSS, the antenna gain is increased from 2.5 dBi to 6.4 dBi at 3.1 GHz. There is also an increase in bandwidth of 82.3% when FSS is incorporated in the UWB antenna.

One of the main problems in MRI is a high mutual coupling that exists between two tuned coils. When coupling occurs, there is a shift in the resonance frequency of the tuned coil. Hence, FSS is used to decouple the dual-tuned coil [60]. FSS is included in the dual-tuned coil to suppress the mutual coupling and thereby reduce the frequency shift. The FSS acts as a band-stop filter.

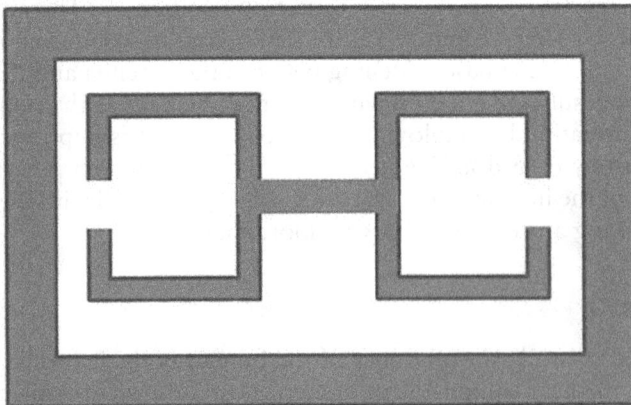

Figure 16.10 Unit cell of FSS in MRI [59].

16.5.8 Concentrating EM Energy into Target Tissue

The need for concentrating the EM field exactly at the target tissue forces researchers to design an FSS that provides the optimal solution to enhance the energy transmission into the human body. For example, in treating prostate cancer, a small microwave antenna inserted inside the body emits a small amount of energy that heats the excess prostate cancer tissue. This type of application requires FSS where energy could be concentrated in the tissue. In [61] the authors have successfully demonstrated how FSS could be used for therapeutic purposes. When FSS is integrated with an antenna, the electromagnetic field is concentrated in human tissue under investigation for the therapeutic modality. The designed unit cell consists of three layers separated by the substrate of TLY 5A polytetrafluoroethylene (PTFE). The first layer consists of capacitive grating supported by meandering slits. The second layer is an inductive strip, and the final layer is a square ring. All the layers are designed on a small size of 5 mm × 4 mm with a total height of 2.54 mm. A perfect condition can be satisfied when the reflection loss around the boundary of the dielectric media is reduced. This facilitates the maximum amount of electromagnetic energy to be focused on the target tissue. The extension of this work is reported in [62] where the unit cell is implanted in a distilled water layer. Though the unit cell is experimentally verified by simulation, the research could pave the way for medical BANs, biomedical imaging systems, and microwave hyperthermia systems.

Table 16.1 shows the summary of the structures used in medical applications.

Table 16.1 Summary of FSS Structures Applied in Medical Applications and Their Performance

FSS Structures	Performance
	The return loss of the ISM band antenna is increased by 2 dB at the resonant frequency by the inclusion of anchor-shaped. The is placed at a distance of 10 mm for optimum results [41].
	An array of the dipole is used in the dosimeter tag to isolate the human body from antenna radiation and also to increase the gain of the antenna used in the tag. Here FSS acts as an artificial magnetic conductor and when implemented with an antenna, broadside radiation gains of 4.1 dBi to 4.8 dBi are achieved [52].

(Continued)

Table 16.1 (Continued)

FSS Structures	Performance
	FSS as an AMC is used to reduce SAR. The Jerusalem cross is designed to work as an AMC. Jerusalem cross is a good candidate as there is a 64% reduction in SAR. The reported SAR values is 0.683 W/kg [53].
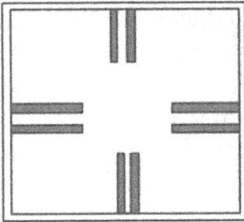	FSS as AMC finds extensive application in WBAN. AMC suppresses side lobe radiation and increases FBR. The measured gain of monopole with AMC is 6.12 dB compared to 2.65 dBi without AMC [54].
	FSS is used as RFID sensor. An array of dipoles is used as the FSS [56].
 square loop	Integrating EBG-FSS gives good FBR around 13 dB. Antenna gain is increased from 1.74 dBi to 6.55 dBi. Good impedance matching with S_{11} <−10 dB. SAR value is reduced by 95% [58].
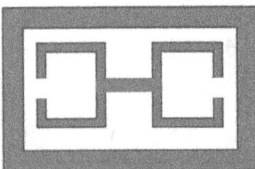	An array of 6 × 4 rectangular slots is used as FSS which provides in-phase reflection to increase the FBR of the UWB antenna in MRI. The antenna accomplishes a directive beam when FSS is included inside the cylindrical structure. Gain is increased from 2.5 dBi to 6.4 dBi [59].

16.6 FUTURE IMPROVEMENTS

For future generations that are going to handle numerous frequencies and protocols, FSSs are seen as a promising candidate to enhance the performance of RF front-end systems. As more and more research is carried out in FSS design, new implementation and new fabrication procedures of FSS will be adopted. The future of terahertz waves is slowly being explored and demands terahertz components. With fabrication of terahertz components becoming easier now, the future wireless communication can see FSS operating in this region. Terahertz FSSs are incorporated in antenna to improve the return loss and radiation characteristics of antenna used in terahertz imaging. Another promising area that has future applications for FSS is reconfigurability.

16.7 CONCLUSION

Today the world is witnessing a pandemic, and there is a need to isolate doctors from patients due to the spread of contagious diseases. Telemedicine is the only option where doctors can remotely monitor patients. As telemedicine requires wireless systems and sensing devices, these performances can be increased by an FSS. Hence, an attempt has been made to review the literature where FSS has played a significant role in increasing the performance of the antenna. In the future, telemedicine is the world. This review starts with the theory, simulation, and measurement setup of FSS in an anechoic chamber. Next, a comprehensive view of FSS in various applications in the medical field is presented. This chapter gives a clear picture of the application of FSS in the clinical environment, as the antenna is one of the key elements in modern wireless medical equipment, starting from the performance enhancement of ISM antenna to the analysis of the characteristics of FSS as an AMC in a flexible compact textile antenna, FSS-EBG in MRI, and terahertz sensing.

REFERENCES

[1] B. A. Munk, 2000. *Frequency Selective Surfaces: Theory and Design*. New York: John Wiley and Sons., Inc. https://www.wiley.com/en-us/Frequency+Selective+S urfaces:+Theory+and+Design-p-9780471370475.

[2] F. Costa, A. Monorchio, and G. Manara, 2014. "An overview of equivalent circuit modeling techniques of frequency selective surfaces and metasurfaces," *ACES Journal*, vol. 29, no. 12.

[3] T. Cwik, R. Mittra, K. Lang, and T. Wu, 1987, April. "Frequency selective screens," *IEEE Antennas and Propagation Society Newsletter*, vol. 29, no. 2, pp. 5–10. doi: 10.1109/MAP.1987.27905.

[4] J. C. Vardaxoglou, 1997. *Frequency-Selective Surfaces: Analysis and Design*. Taunton: Research Studies Press.

[5] R. Mittra, C. H. Chan, and T. Cwik, 1988. "Techniques for analyzing frequency selective surfaces a review," *Proceedings of the IEEE*, vol. 76, no. 12, pp. 1593–1615.

[6] www.maximizemarketresearch.com/market-report/global-metamaterials-market/24404/

[7] V. D. Agrawal and W. A. Imbriale, 1979. "Design of a dichroic Cassegrain subreflector," *IEEE Transactions on Antennas and Propagation*, vol. AP-27, no. 4, pp. 466–473.

[8] L. E. Comtesse, R. J. Langley, E. A. Parkera, and J. C. Vardaxoglou, 1987. "Frequency selective surfaces in dual and triple band reflector antennas," *European Microwave Conference*, 1987, pp. 208–213.

[9] G. H. Schennum, 1973. "Frequency selective surfaces for multiple frequency antennas," *Microwave Journal*, vol. 16, no. 5, pp. 55–57.

[10] T. K. Wu, 1994. "Four-band frequency selective surface with double-square-loop patch," *IEEE Transactions on Antennas and Propagation*, vol. 42, no. 12, pp. 1659–1663.

[11] Wenhua Xu, Yun He, Peng Kong, Jialin Li, Haibing Xu, Ling Miao, Shaowei Bie, and Jianjun Jiang, 2015, November 14. "An ultra-thin broadband active frequency selective surface absorber for ultrahigh-frequency applications," *Journal of Applied Physics*, vol. 118, no. 18, 184903.

[12] K. Zhang, W. Jiang, and S. Gong, 2017. "Design bandpass frequency selective surface absorber using LC resonators," *IEEE Antennas and Wireless Propagation Letters*, vol. 16, pp. 2586–2589,

[13] Pingyou Wang, Pu Tang, Wuqiong Luo, Ziyuan He, Lutong Li, and H. Senhang, 2016. "Design of dual-band frequency selective surface for antenna RCS reduction," *Progress in Electromagnetic Research Symposium (PIERS)*, Shanghai, 8–11.

[14] I. S. Syed, Y. Ranga, L. Matekovits, K. P. Esselle, and S. Hay, 2014. "A single-layer frequency-selective surface for ultrawideband electromagnetic shielding," *IEEE Transactions on Electromagnetic Compatibility*, vol. 56, no. 6, pp. 1404–1411.

[15] Bora Doken and Mesut Kartal, 2017. "Easily optimizable dual-band frequency-selective surface design," *IEEE Antennas and Wireless Propagation Letters*, vol. 16.

[16] X. Ri-Hui and L. Jiu-Sheng, 2018. "Frequency selective surface for terahertz bandpass filter," *Journal of Infrared, Millimeter, and Terahertz Waves*, vol. 39, pp. 1039–1046.

[17] B. Lin, F. Li, Q. Zheng, and Y. Zen, 2009. "Design and simulation of a miniature thick-screen frequency selective surface radome," *IEEE Antennas and Wireless Propagation Letters*, vol. 8, pp. 1065–1068.

[18] K. K. Varikuntla and R. Singaravelu, 2018. "Design of a novel 2.5D frequency selective surface element using fibonacci spiral for radome application," *2018 Asia-Pacific Microwave Conference* (APMC), Kyoto, pp. 1289–1291.

[19] D. J. Kozakoff, 2009. *Analysis of Radome Enclosed Antennas*. Norwood, MA: Artech House.

[20] N. Liu, X. Sheng, C. Zhang, and D. Guo, 2018. "Design of frequency selective surface structure with high angular stability for radome application," *IEEE Antennas & Wireless Propagation Letters*, vol. 17, no. 1, pp. 138–141.

[21] K. K. Varikuntla and R. Singara Velu, 2019. "Design and development of angularly stable and polarisation rotating FSS radome based on substrate-integrated waveguide technology," *IET Microwaves, Antennas & Propagation*, vol. 13, no. 4, pp. 478–484.

[22] Y. Zhu, Y. Chen, and S. Yang, 2019. "Decoupling and low-profile design of dual-band dual-polarized base station antennas using frequency-selective surface," *IEEE Transactions on Antennas and Propagation*, vol. 67, no. 8, pp. 5272–5281.

[23] Y. Li, P. Ren, and Z. Xiang, 2019. "A dual-passband frequency selective surface for 5G communication," *IEEE Antennas and Wireless Propagation Letters*, vol. 18, no. 12, pp. 2597–2601.

[24] R. C. Hadarig, M. E. de Cos Gomez, Y. Alvarez, and F. Las-Heras, 2010. "Novel bow-tie—AMC combination for 5.8-GHz RFID tags usable with metallic objects," *IEEE Antennas and Wireless Propagation Letters*, vol. 9, pp. 1217–1220.

[25] G. Goussetis, A. P. Feresidis, and J. C. Vardaxoglou, 2006. "Tailoring the AMC and EBG characteristics of periodic metallic arrays printed on grounded dielectric substrate," *IEEE Transactions on Antennas and Propagation*, vol. 54, no. 1, pp. 82–89.

[26] A. Y. I. Ashyap, Zuhairiah Zainal Abidin, Samsul Haimi Dahlan, Huda A. Majid, Shaharil Mohd Shah, Muhammad Ramlee Kamarudin, and Akram Alomainy, 2017. "Compact and low-profile textile EBG-based antenna for wearable medical applications," *IEEE Antennas and Wireless Propagation Letters*, vol. 16, pp. 2550–2553.

[27] S. M. Saeed, C. A. Balanis, C. R. Birtcher, A. C. Durgun, and H. N. Shaman, 2017. "Wearable flexible reconfigurable antenna integrated with artificial magnetic conductor," *IEEE Antennas and Wireless Propagation Letters*, vol. 16, pp. 2396–2399.

[28] Mohammed T. Al Haddad, 2016. "Design of frequency selective surface for mobile signal shielding," MS thesis, Elec Engg., IUG. Univ., Gaza., Palestine.

[29] T. K. Wu, 1995. *Frequency Selective Surfaces and Grid Array*. New York: John Wiley and Sons, Inc.

[30] S. Manoharan, P. Ramasamy, and R. Singaravelu, 2021. "A quad-band fractal antenna with metamaterial resonator-backed ground for sub-6 GHz, C and X band applications," *Applied Physics A*, vol. 127, p. 703. https://doi.org/10.1007/s00339-021-04862-6

[31] A. Bhattacharya, B. Roy, R. Caldeirinha, and A. Bhattacharjee, 2019. "Low-profile, extremely wideband, dual-band-notched MIMO antenna for UWB applications," *International Journal of Microwave and Wireless Technologies*, vol. 11, no. 7, pp. 719–728. https://doi.org/10.1017/S1759078719000266

[32] Ankan Bhattacharya, Bappadittya Roy, Santosh K. Chowdhury and Anup K. Bhattacharjee, 2019. "Computational and experimental analysis of a low-profile, isolation-enhanced, band-notch UWB-MIMO antenna," *Journal of Computational Electronics*, vol. 18, pp. 680–688.https://doi.org/10.1007/s10825-019-01309-3

[33] D. Õoglu, 2014. "Chiral frequency selective surfaces comprised of multiple conducting strips per unit cell," *IET Microwaves, Antennas & Propagation*, vol. 8, pp. 621–626.

[34] D. H. Werner and D. Lee, 2000. "Design of dual polarised multiband frequency selective surfaces using fractal elements," *Electronics Letters*, vol. 36, pp. 487–488.

[35] A. P. Feresidis and J. C. Vardaxoglou, 2001. "High gain planar antenna using optimised partially reflective surfaces," *IEE Proceedings – Microwaves, Antennas and Propagation*, vol. 148, no. 6, pp. 345–350.

[36] A. Pirhadi, F. Keshmiri, M. Hakkak, and M. Tayarani, 2007. "Analysis and design of dual band high directive EBG resonator antenna using square loop FSS as superstrate layer," *Progress in Electromagnetics Research*, vol. 70, pp. 1–20.

[37] H. Attia, M. L. Abdelghani, and T. A. Denidni, 2017. "Wideband and high-gain millimeter-wave antenna based on FSS fabry—perot cavity," *IEEE Transactions on Antennas and Propagation*, vol. 65, no. 10, pp. 5589–5594.

[38] Z. Sharipov, F. Günes, A. S. Türk, M. A. Belen, P. Mahouti and S. Demirel, 2016 "Microstrip frequency selective surface for use in horn filtenna," *IEEE Radar Methods and Systems Workshop (RMSW)*, Kiev, 2016, pp. 107–109.

[39] G. Yuehe, K. P. Esselle, and T. S. Bird, 2012. "The use of simple thin partially reflective surfaces with positive reflection phase gradients to design wideband, low-profile EBG resonator antennas," *IEEE Transactions on Antennas and Propagation*, vol. 60, pp. 743–750.

[40] M. L. Abdelghani, H. Attia, and T. A. Denidni, 2017. "Dual- and wideband fabry—pérot resonator antenna for WLAN applications," *IEEE Antennas and Wireless Propagation Letters*, vol. 16, pp. 473–476.

[41] M. A. Belen, 2018. "Performance enhancement of a microstrip patch antenna using dual-layer frequency-selective surface for ISM band applications," *Microwave and Optical Technology Letters*, vol. 60, no. 11, pp. 2730–2734.

[42] Z. Shenzheng and W. Encheng, 2016. "A novel ISM band antenna with frequency selective surface structure," *IEEE International Conference of Online Analysis and Computing Science* (ICOACS), Chongqing, 2016, pp. 277–280.

[43] Vitor Fernandes Barros, Francisco Carlos G. da Silva Segundo, Antonio Luiz P. S. Campos, and Sandro Gonçalves da Silva, 2017. "A novel simple convoluted geometry to design frequency selective surfaces for applications at ISM and UNII bands," *Journal of Microwaves, Optoelectronics and Electromagnetic Applications*, vol. 16, no. 2.

[44] Mesut Kartal Bora Döken, 2016. "A new frequency selective absorber surface at the unlicensed 2.4-GHz ISM band," *Microwave and Optical Technology Letters*, vol. 58, no. 10.

[45] Mehmet Bakir, Kemal Delihacioglu, Muharrem Karaaslan, Furkan Dincer, and Cumali Sabah, 2016. "U-shaped frequency selective surfaces for single- and dual-band applications together with absorber and sensor configurations," *IET Microwaves, Antennas & Propagation*, vol. 10, no. 3, pp. 1–8.

[46] G. S. Paul and K. Mandal, 2019. "Polarization-insensitive and angularly stable compact ultrawide stop-band frequency selective surface," *IEEE Antennas and Wireless Propagation Letters*, vol. 18, no. 9, pp. 1917–1921.

[47] C. Debus and P. H. Bolivar, 2007. "Frequency selective surfaces for high-sensitivity terahertz sensors," *2007 Conference on Lasers and Electro-Optics (CLEO)*, Baltimore, MD, 2007, pp. 1–2.

[48] D. F. Sievenpiper, 1999. "High-impedance electromagnetic surfaces," PhD thesis, Univ. Calif., Los Angeles, CA,

[49] D. Sievenpiper, L. Zhang, R. F. Jimenez Broas, N. G. Alexpolous, and E. Yablonvitch, 1999. "High-impedance electromagnetic surfaces with a forbidden

frequency band," *IEEE Transactions on Microwave Theory and Techniques*, vol. 47, no. 11, pp. 2059–2074.

[50] Farhad Bayatpur, 2001. "Metamaterial inspired frequency selective surface," PhD dissertation, Electrical Engg. Dept., The Univ of Michigan.

[51] S. M. Saeed, C. A. Balanis, C. R. Birtcher, A. C. Durgun and H. N. Shaman, 2017. "Wearable flexible reconfigurable antenna integrated with artificial magnetic conductor," *IEEE Antennas and Wireless Propagation Letters*, vol. 16, pp. 2396–2399.

[52] O. M. Sanusi, F. A. Ghaffar, A. Shamim, M. Vaseem, Y. Wang, and L. Roy, 2019. "Development of a 2.45 GHz antenna for flexible compact radiation dosimeter tags," *IEEE Transactions on Antennas and Propagation*, vol. 67, no. 8, pp. 5063–5072.

[53] H. R. Raad, A. I. Abbosh, H. M. Al-Rizzo, and D. G. Rucker, 2013. "Flexible and compact AMC based antenna for telemedicine applications," *IEEE Transactions on Antennas and Propagation*, vol. 61, no. 2, pp. 524–531.

[54] A. Alemaryeen and S. Noghanian, 2019. "On-body low-profile textile antenna with artificial magnetic conductor," *IEEE Transactions on Antennas and Propagation*, vol. 67, no. 6, pp. 3649–3656.

[55] A. De, B. Roy, A. Bhattacharya, and A. K. Bhattacharjee, 2021. "Bandwidth-enhanced ultra-wide band wearable textile antenna for various WBAN and Internet of Things (IoT) applications," *Radio Science*, vol. 56, p. e2021RS007315. https://doi.org/10.1029/2021RS007315

[56] A. Lorenzo, R. Lázaro, R. Villarino and D. Girbau, 2016. "Modulated frequency selective surfaces for wearable RFID and sensors applications," *IEEE Transaction on Microwave Theory and Techniques*, vol. 64, no. 10, pp. 4447–4456.

[57] S. Milici, J. Lorenzo, A. Lázaro, R. Villarino, and D. Girbau, 2017. "Wireless breathing sensor based on wearable modulated frequency selective surface," *IEEE Sensors Journal*, vol. 17, no. 5, pp. 1285–1292.

[58] A. Y. I. Ashyap, Zuhairiah Zainal Abidin, Samsul Haimi Dahlan, Huda A. Majid, Muhammad Ramlee Kamarudin, Akram Alomainy, Raed A. Abd-Alhameed, Jamal Sulieman Kosha, and James M. Noras, 2018. "Highly efficient wearable CPW antenna enabled by EBG-FSS structure for medical body area network applications," *IEEE Access*, vol. 6, pp. 77529–77541.

[59] Ahmed I. Imran and Taha A. Elwi, 2017. "A cylindrical wideband slotted patch antenna loaded with frequency selective surface for MRI applications," *Engineering Science and Technology*, vol. 20, no. 3, pp. 990–996.

[60] N. Fontana, F. Costa, G. Tiberi, L. Nigro, and A. Monorchio, 2017. "Distributed trap FSS filter for dual tuned RF MRI coil decoupling at 7.0T," *International Conference on Electromagnetics in Advanced Applications (ICEAA)*, Verona, pp. 1229–1231.

[61] S. P. Singh, A. K. Jha and M. J. Akhtar, 2016. "Design of broadband superstrate FSS for terahertz imaging and testing applications," *2016 IEEE MTT-S International Microwave and RF Conference* (IMaRC), New Delhi, pp. 1–4.

[62] Praphat Arnmanee and Chuwong Phongcharoenpanich, 2018. "Improved microstrip antenna with HIS elements and FSS superstrate for 2.4GHz band applications," *International Journal of Antennas and Propagation*, vol. 2018, Article ID 9145373 (11 pages).

Chapter 17

Scattering Matrices and Their Applications in Microwave Engineering

Shipra Bhatia and Manoj Sarkar

17.1 INTRODUCTION

At low frequencies, the transmitted signal wavelength is much higher than the physical length of the network. So, parameters such as impedance (Z), admittance (Y), hybrid (h), and ABCD parameters are used, where voltage and current are used to define the measurable input and output variables. The current and voltage are not well defined at high frequencies because of comparatively large or the same physical length as the wavelength [1]. The issues with the low-frequency parameters are defined as follows:

- The perfect open or short circuit is difficult at high frequencies because it can cause oscillations in the case of sensitive active devices.
- Non-linearity of some passive components, especially resistance at high frequency.
- Switching/measurement of voltage is very difficult, especially in the terahertz (THz) frequency range.
- Voltage and current measurement equipment are not available at high frequencies.
- Visualization of radio frequency wave propagation is difficult with the previously mentioned low-frequency parameters.
- Due to reflection at higher frequencies, analyzing the circuit using the defined parameters becomes difficult.

To overcome these issues at high frequencies, a scattering (S) matrix is introduced which linearly relates the amplitude of reflected waves with that of incident waves at a microwave junction. The first physical meaning of the scattering matrix was given in the year 1965 by Chow and Cassigbol, who found the usefulness of the scattering matrix in many cases. Scattering matrices are used when traveling waves associated with their power scatter through the ports in a microwave junction. The elements of the scattering matrix are known as scattering parameters or scattering coefficients or S-parameters because they relate to the scattered waves from the network

DOI: 10.1201/9781003459880-17

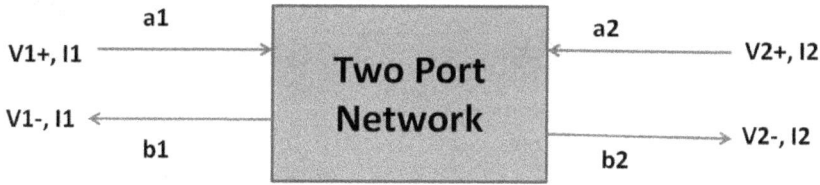

Figure 17.1 Two-port network with port components.

when waves are incident upon them. S-parameters consist of both magnitude and phase and can be explained as the ratio of voltages at a port when a single port is excited, keeping the rest of the ports properly matched. Figure 17.1 displays the two-port network where port 1 defines incident voltage V_1+, reflected voltage V_1-, and current I1 and port 2 defines incident voltage V_2+, reflected voltage V_2-, and current I2.

Here, a1 and a2 are signals entering the ports, whereas b1 and b2 are signals leaving the ports. The incident and reflected voltages are related by S-parameters as shown in equation (17.1):

$$S_{11} = \frac{V_1-}{V_1+} \text{ when } V_2+=0 \qquad S_{12} = \frac{V_1-}{V_2+} \text{ when } V_1+=0$$

$$S_{21} = \frac{V_2-}{V_1+} \text{ when } V_2+=0 \qquad S_{22} = \frac{V_2-}{V_2+} \text{ when } V_1+=0$$

(17.1)

where S_{11} and S_{22} are the reflection coefficients when ports are terminated with matched load, and S_{12} and S_{21} are transmission coefficients or attenuation from one port to another [2]. Similarly, for an N-port network, the earlier relationship is generalized as:

$$[V_N -] = [S] [V_N +]$$

(17.2)

17.2 QUANTIFYING AND ANALYZING THE INPUT SIGNAL AND POWER FLOW

Figure 17.2 portrays an N-port network with the input voltage 'Vg', resistance 'Rg', input signal 'a', and output signal 'b'. The network is analyzed and the input and output signal are determined in terms of voltage 'V' and network impedance 'Z' in equations (17.3)-(17.7) [3].

$$\text{Input signal } a = \frac{1}{2}\left[\frac{V}{\sqrt{Z}} + I\sqrt{Z}\right]$$

(17.3)

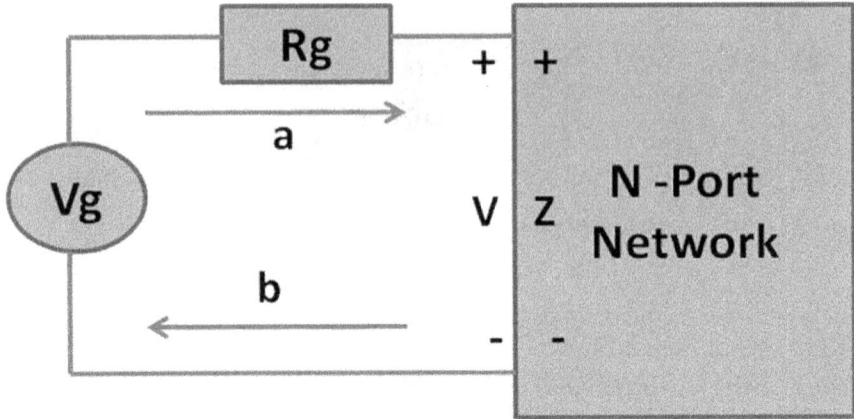

Figure 17.2 N-port network for quantifying the input signal.

$$a = \frac{1}{2\sqrt{Z}}\left[V_g - IR_g + \frac{Z * V_g}{R_g + Z_L}\right] \tag{17.4}$$

$$a = \frac{1}{2\sqrt{Z}}\left[V_g - \left(\frac{V_g}{R_g + Z_L}\right)R_g + \frac{Z}{R_g + Z_L}\right] \tag{17.5}$$

$$a = \frac{V_g}{2\sqrt{Z}}\left[1 + \frac{Z - R_g}{R_g + Z_L}\right] \tag{17.6}$$

Here Z_L is load impedance. If $Rg = Z$ and $\dfrac{Z - R_g}{R_g + Z_L} = 0$, the previous equation reduces to $a = \dfrac{V_g}{2\sqrt{Z}}$

$$\begin{pmatrix} a1 \\ a2 \end{pmatrix} = \begin{pmatrix} \dfrac{V_1 +}{\sqrt{2Z_1}} \\ \dfrac{V_2 +}{\sqrt{2Z_2}} \end{pmatrix} \quad \text{and} \quad \begin{pmatrix} b1 \\ b2 \end{pmatrix} = \begin{pmatrix} \dfrac{V_1 -}{\sqrt{2Z_1}} \\ \dfrac{V_2 -}{\sqrt{2Z_2}} \end{pmatrix} \tag{17.7}$$

The total power flow in the network is defined by voltage 'V' and current 'I' in equation (17.8):

$$P = \frac{1}{2} Re\left[VI^* \right] \tag{17.8}$$

$$P = \frac{1}{2} Re\left[(a_n + b_n)\sqrt{Z} \right]\left[\frac{a_n^* - b_n^*}{\sqrt{Z}} \right] \tag{17.9}$$

$$P = \frac{1}{2}\left(|a_n|^2 - |b_n|^2 \right) \tag{17.10}$$

Input power at the nth port, assuming no reflection $= P_i = \frac{1}{2} |a_n|^2$

$$= \frac{1}{2}\left(\frac{V_g^2}{4Z_o} \right) = \frac{V_g^2}{8Z_o} \tag{17.11}$$

Equation (17.11) denotes the maximum power which can be fed to the network by the source. Also, the reflected power at the nth port is given by $P_r = \frac{1}{2} |b_n|^2$. Here a_n and b_n are normalized amplitudes of the waves at the nth port. The normalization of the signal is required to obtain a symmetrical scattering matrix for the reciprocal network. Due to different impedance levels in the network, the S-matrix will be asymmetrical in the absence of normalization [4].

17.3 ORIGIN OF SCATTERING PARAMETERS FROM THE TRANSMISSION LINE

Voltage 'V' and current 'I' in the transmission line [5] are defined in equations (17.12) and (17.13).

$$V(z) = V_r e^{\gamma z} - V_i e^{-\gamma z} \tag{17.12}$$

$$I(z) = \frac{-1}{Z} \frac{dV}{dZ} \tag{17.13}$$

$$I(z) = \frac{-1}{Z}\left(V_r \gamma e^{\gamma z} - V_i \gamma e^{-\gamma z} \right) = \frac{\gamma}{Z}\left(V_r e^{\gamma z} + V_i \gamma e^{-\gamma z} \right) \tag{17.14}$$

In equation (7.14), V_r and V_i are the voltage of reflected and incident waves, and γ is the propagation constant. The evolution of S-parameters can be explained in equations (17.15)-(17.17):

$$V(z) = V_i e^{-\gamma z} + V_r e^{\gamma z} \tag{17.15}$$

$$I(z) = \frac{1}{Zo}\left(V_i e^{-\gamma z} + V_r e^{\gamma z}\right) \tag{17.16}$$

$$Zo * I(z) = V_i e^{-\gamma z} + V_r e^{\gamma z} \tag{17.17}$$

Here Zo is the characteristic impedance. Adding and subtracting equations (17.15) and (17.17), we obtain:

$$\frac{V + Zo * I}{2} = V_i e^{-\gamma z} \text{ and } \frac{V - Zo * I}{2} = V_r e^{\gamma z} \tag{17.18}$$

Transforming:

$$V_i e^{-\gamma z} = \frac{\sqrt{Zo}}{2}\left(\frac{V}{\sqrt{Zo}} + \sqrt{Zo} * I\right) \tag{17.19}$$

$$V_r e^{\gamma z} = \frac{\sqrt{Zo}}{2}\left(\frac{V}{\sqrt{Zo}} - \sqrt{Zo} * I\right) \tag{17.20}$$

$$a = \frac{1}{2}\left(\frac{V}{\sqrt{Zo}} + \sqrt{Zo} * I\right) \tag{17.21}$$

$$b = \frac{1}{2}\left(\frac{V}{\sqrt{Zo}} - \sqrt{Zo} * I\right) \tag{17.22}$$

Comparing equation (17.19) with (17.21) and equation (17.20) with (17.22), we obtain the amplitude of incident and reflected power waves.

$$a = \frac{V_i e^{-\gamma z}}{\sqrt{Zo}} \text{ and } b = \frac{V_r e^{\gamma z}}{\sqrt{Zo}} \tag{17.23}$$

17.4 TERMS RELATED TO ANY NETWORK

- **Reciprocal network:** A network with only passive components such as resistance, inductance, and capacitance is known as a reciprocal network. In other words, if the voltage of the current at one port is equal to that at another port and vice versa, the network is said to be reciprocal [2].

$$|S_{11}|^2 + |S_{12}|^2 = 1 \qquad (17.24)$$

- **Symmetric network:** A network is said to be symmetric if its input impedance is equal to its output impedance.

$$[S] = [S]^T \text{ or } S_{ij} = S_{ji} \qquad (17.25)$$

- **Lossless network:** A network or circuit that does not contain any heat-dissipative components such as a resistor is defined as a lossless network. This can also be explained as the network that provides an equal amount of total power which is entering and leaving [6].

17.5 S-MATRIX DETERMINED FROM Z- AND Y-MATRIX

Many circuit analysis techniques, methods, and properties are valid for both low-frequency and microwave-frequency ranges. Therefore, sometimes it is required to relate and analyze both low-frequency and high-frequency parameters. The following section gives the relationship of the S-parameter with the Z-parameter and Y-parameter [7].

$$V_N = V_N^+ + V_N^- \text{ and } I_N = I_N^+ - I_N^- \qquad (17.26)$$

Assuming unit impedance, the current term in equation (1726) is rewritten as follows:

$$I_N = V_N^+ - V_N^- \qquad (17.27)$$

$$[V] = [Z][I] \qquad (17.28)$$

$$\left[V_N^+ + V_N^- \right] = [Z]\left[V_N^+ - V_N^- \right] \qquad (17.29)$$

$$V_N^+ \left([Z] - [U] \right) = V_N^- \left([Z] + [U] \right) \qquad (17.30)$$

$$\frac{V_N^-}{V_N^+} = [S] = \frac{[Z] - [U]}{[Z] + [U]} \qquad (17.31)$$

$$[I] = [Y][V] \tag{17.32}$$

$$\left[V_N^+ - V_N^- \right] = [Y]\left[V_N^+ + V_N^- \right] \tag{17.33}$$

$$V_N^+ \left([U] - [Y]\right) = V_N^- \left([U] + [Y]\right) \tag{17.34}$$

$$\frac{V_N^-}{V_N^+} = [S] = \frac{[U] - [Y]}{[U] + [Y]} \tag{17.35}$$

Here, V_N^+ and I_N^+ are the voltage and current of forward traveling waves, whereas V_N^- and I_N^- are for backward traveling waves, [U] is the unitary/identity matrix, [Z] and [Y] are the impedance and admittance matrix of the network [4]. The relationship between impedance, admittance, and scattering parameters offers unique properties for the device at a given frequency:

- The number of elements is equal in Z-, Y-, and S-matrices.
- Z- and S-parameters satisfy the reciprocity property for the reciprocal network.
- The power balance of the lossless network is determined and denoted by the unitary property of the S-matrix. This is not possible in the case of the Z-matrix.
- The phase of S-coefficients changes when the reference plane is changed for S-parameters, but in the case of Z- and Y-parameters, voltage and current are defined as complex impedance. So, any change can lead to variations in both magnitude and phase.

17.6 PROPERTIES OF S-PARAMETERS

The S-parameter offers some unique properties to characterize different types of networks [2–4]. These are explained in detail in this section.

- **Symmetric property for a reciprocal network:** The device which is reciprocal and has the same transmission characteristics can be characterized by a symmetric scattering matrix. From equation (17.31) earlier:

$$[S] = \frac{[Z] - [U]}{[Z] + [U]} = ([Z] - [U])([Z] + [U])^{-1} \tag{17.36}$$

$$[S]^T = \frac{[Z] - [U]}{[Z] + [U]} = ([Z] - [U])^T \{([Z] + [U])^{-1}\}^T \tag{17.37}$$

Since the unitary matrix and its transpose are equal and the network is reciprocal, $[Z] = [Z]^T$. Therefore, $[S] = [S]^T$. Almost all passive components such as a capacitor, resistor, transformer, etc., are reciprocal, whereas, ferrite nanostructures consolidated by spark plasma are non-reciprocal. The amplifier is one kind of active device which is non-reciprocal.

- **Unitary property for a lossless junction:** In the case of a lossless network, the rank of any one column or row of the S-matrix, when multiplied by its complex conjugate, provides a unity S-matrix [8]. For a lossless network with identical and unity characteristic impedance, the average power which can be delivered to the network is defined as:

$$P_{av} = \frac{1}{2} R_e \left\{ [V]^T [I]^* \right\} \tag{17.38}$$

$$= \frac{1}{2} R_e \left\{ \left([V^+]^T + [V^-]^T \right) \left([V^+]^* - [V^-]^* \right) \right\} \tag{17.39}$$

$$= \frac{1}{2} R_e \left\{ \left([V^+]^T [V^+]^* \right) \right\} - \frac{1}{2} R_e \left\{ \left([V^-]^T [V^-]^* \right) \right\} \tag{17.40}$$

For a lossless nth port device, the total incident N-port power must be equal to the total reflection power:

$$\left[V^+ \right]^T \left[V^+ \right]^* = \left[V^- \right]^T \left[V^- \right]^* \tag{17.41}$$

By using the statement $\left[V^- \right] = [S] \left[V^+ \right]$, equation (17.41) can be simplified as:

$$[S]^T [S]^* = [U] \tag{17.42}$$

- **Zero diagonal elements for the perfectly matched network:** For an N-port network, if there is no reflection from any port under perfectly matched conditions, then the diagonal element of the S-matrix will be zero.
- **Phase shift property:** At the lower frequency bands, both magnitude and phase of the signal vary according to different conditions. At higher frequencies, S-parameters are defined for the complex network concerning the position of the port or the reference plane of the network [9]. For a two-port network, S-matrix [S] and new S-matrix [S'] are expressed in equations (17.43) and (17.44). Here $\varnothing_1$, $\varnothing_2$ are the

electrical phase shift denoted by $\beta_1\ell_1$ and $\beta_2\ell_2$, respectively, while others are new variables of the phase shift matrix.

$$[S] = \begin{bmatrix} S_{11} & S_{12} \\ S_{21} & S_{22} \end{bmatrix} \tag{17.43}$$

$$[S'] = \begin{bmatrix} e^{-j\varnothing_1} & 0 \\ 0 & e^{-j\varnothing_2} \end{bmatrix} [S] \begin{bmatrix} e^{-j\varnothing_1} & 0 \\ 0 & e^{-j\varnothing_2} \end{bmatrix} \tag{17.44}$$

- In general, the S-parameter is inherently dependent on complex frequency, load impedance, and source impedance. The S-parameter is varied concerning the certain change in these parameters. However, S-parameters are also determined by the consideration of their symmetries and energy conservation for the lossless network.

17.7 SCATTERING MATRICES OF MICROWAVE COMPONENTS

- **H- and E-plane tee:** These are the three port components where input is provided at one port such that other ports are isolated [10]. Figure 17.3(a) displays the representation of ports for the H-plane tee. Applying a few conditions to obtain the S-matrix for the H-plane tee:
  - As port 3 is matched, $S_{33} = 0$.
  - Applying condition of a reciprocal network, $S_{12} = S_{21}, S_{13} = S_{31}, S_{23} = S_{32}$.
  - As equal power is divided into first and second branches, $S_{13} = S_{23}$.
  - Applying the condition of the unitary matrix, $[S][S]^T = I$.

$$\begin{bmatrix} S_{11} & S_{12} & S_{13} \\ S_{12} & S_{22} & S_{13} \\ S_{13} & S_{13} & 0 \end{bmatrix} \begin{bmatrix} S_{11}^* & S_{12}^* & S_{13}^* \\ S_{12}^* & S_{22}^* & S_{13}^* \\ S_{13}^* & S_{13}^* & 0 \end{bmatrix} = \begin{bmatrix} 1 & 0 & 0 \\ 0 & 1 & 0 \\ 0 & 0 & 1 \end{bmatrix} \tag{17.45}$$

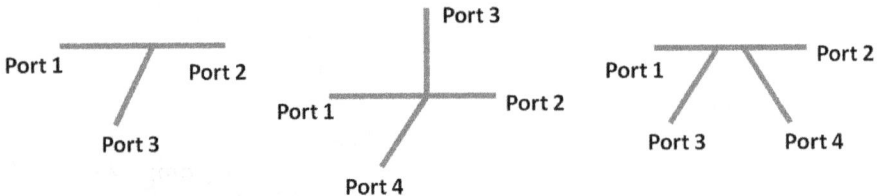

Figure 17.3 Port representation of (a) H-plane tee, (b) magic tee, and (c) directional coupler.

- Applying the reciprocal condition defined earlier in the first row of matrix 1 and multiplying rows with columns, we obtain the final S-matrix for H-plane tee [2].

$$
S = \begin{bmatrix} \dfrac{1}{2} & -\dfrac{1}{2} & \dfrac{1}{\sqrt{2}} \\ -\dfrac{1}{2} & \dfrac{1}{2} & \dfrac{1}{\sqrt{2}} \\ \dfrac{1}{\sqrt{2}} & \dfrac{1}{\sqrt{2}} & 0 \end{bmatrix}
\tag{17.46}
$$

Similarly, the S-matrix of the E-plane tee is:

$$
S = \begin{bmatrix} \dfrac{1}{2} & \dfrac{1}{2} & \dfrac{1}{\sqrt{2}} \\ \dfrac{1}{2} & \dfrac{1}{2} & -\dfrac{1}{\sqrt{2}} \\ \dfrac{1}{\sqrt{2}} & -\dfrac{1}{\sqrt{2}} & 0 \end{bmatrix}
\tag{17.47}
$$

- **Magic/hybrid tee:** It is a four-port device with a combination of E- and H-plane tees. Figure 17.3(b) portrays the port representation of a magic tee [3]. Applying a few conditions to obtain the S-matrix for the magic tee:
  - As ports 3 and 4 are matched, $S_{33} = S_{44} = 0$.
  - Port 1 and 2 are isolated, $S_{12} = S_{21} = 0$; port 3 and 4 are isolated, $S_{34} = S_{43} = 0$.
  - As equal power is divided into both arms, $S_{13} = S_{23}$ and $S_{14} = S_{24}$.
  - Applying the condition of the unitary matrix, $[S]\,[S]^{T} = I$, and proceeding with the calculations the same way as defined earlier:

$$
S = \begin{bmatrix} 0 & 0 & \dfrac{1}{\sqrt{2}} & \dfrac{1}{\sqrt{2}} \\ 0 & 0 & -\dfrac{1}{\sqrt{2}} & \dfrac{1}{\sqrt{2}} \\ \dfrac{1}{\sqrt{2}} & -\dfrac{1}{\sqrt{2}} & 0 & 0 \\ \dfrac{1}{\sqrt{2}} & \dfrac{1}{\sqrt{2}} & 0 & 0 \end{bmatrix}
\tag{17.48}
$$

- **Directional coupler:** It is a four-port waveguide junction with ports 1 and 2 as primary ports and ports 3 and 4 as secondary ports [11]. Input power from port 1 is coupled to port 2 and port 3 but does not affect port 4. Figure 17.3(c) shows the port representation of the directional coupler.
- Applying a few conditions to obtain the four-port scattering matrix for the directional coupler:
  - All ports are matched, $S_{11} = S_{22} = S_{33} = S_{44} = 0$.
  - Ports 1 and 3 are decoupled, $S_{13} = S_{31} = 0$; Ports 2 and 4 are isolated, $S_{24} = S_{42} = 0$.
  - Applying the condition of the unitary matrix, $[S] [S]^T = I$.

$$\begin{bmatrix} 0 & S_{12} & 0 & S_{14} \\ S_{12} & 0 & S_{23} & 0 \\ 0 & S_{23} & 0 & S_{34} \\ S_{41} & 0 & S_{34} & 0 \end{bmatrix} \begin{bmatrix} 0 & S_{12}^* & 0 & S_{14}^* \\ S_{12}^* & 0 & S_{23}^* & 0 \\ 0 & S_{23}^* & 0 & S_{34}^* \\ S_{41}^* & 0 & S_{34}^* & 0 \end{bmatrix} = \begin{bmatrix} 1 & 0 & 0 & 0 \\ 0 & 1 & 0 & 0 \\ 0 & 0 & 1 & 0 \\ 0 & 0 & 0 & 1 \end{bmatrix} \quad (17.49)$$

- Applying the reciprocal condition in every row of matrix 1 to obtain $S_{14} = S_{23}$, $S_{12} = S_{34}$, and $S_{23} = S_{41}$.
- Assuming $S_{12} = S_{34} = p$ and $S_{23} = S_{41} = q$.

$$S = \begin{bmatrix} 0 & p & 0 & q \\ p & 0 & q & 0 \\ 0 & q & 0 & p \\ q & 0 & p & 0 \end{bmatrix} \quad (17.50)$$

- **Isolator:** It is a two-port device that isolates the component from reflections [4, 12]. In the isolator, the transmission of the signal is in one direction from port 1 to port 2. So, the value of S_{21} is 1, while others are kept at 0. The S-parameter matrix for the isolator is defined next which shows that [S] is not symmetric and hence it is a non-reciprocal device.

$$S = \begin{bmatrix} 0 & 0 \\ 1 & 0 \end{bmatrix} \quad (17.51)$$

- **Circulator:** It is a device with multiple ports where the power circulates from one port to another in the forward direction but not in the backward direction [13, 14]. For a device with three ports, power flows from ports 1 to 2 and then to 3. So, the value of $S_{21} = S_{32} = S_{13}$ is 1 and the others are 0. The S-parameter matrix for the circulator is defined as follows:

$$S = \begin{bmatrix} 0 & 0 & 1 \\ 1 & 0 & 0 \\ 0 & 1 & 0 \end{bmatrix} \tag{17.52}$$

- The scattering matrix for a uniform transmission line is defined as:

$$S = \begin{bmatrix} 0 & e^{-j\beta l} \\ e^{-j\beta l} & 0 \end{bmatrix} \tag{17.53}$$

17.8 MEASUREMENT OF SCATTERING PARAMETERS

A vector network analyzer (VNA) is the equipment that measures the scattering parameters for any microwave device. The significance of scattering matrices concerning microwave antennas is depicted in [15–18]. The resonance of the antenna can be defined as the frequency with the lowest value of reflection coefficient Γ defined by |Sii| and the amount of coupling between the two nearby elements using insertion loss |Sij| in terms of transmission coefficient T. Here i and j are two different ports of antenna elements. The reflection coefficient is the ratio of reflected and incident voltage, whereas the transmission coefficient defines the ratio of transmitted and incident voltage [19]. Also, the bandwidth of the antenna can be measured using the −10 dB points of the reflection coefficient curve from the VNA. Therefore, the scattering parameters in the antenna portray the reflection behavior, coupling between the elements, and impedance bandwidth.

$$Return\,loss = -20\log|\Gamma| \;\; dB = -20\log|S_{11}| \;\; dB \tag{17.54}$$

$$Insertion\,loss = -20\log|T| \;\; dB = -20\log|S_{21}| \;\; dB \tag{17.55}$$

As the VNA usually has two ports, it can easily measure a two-port device using a 2×2 S-parameter matrix. For a multiport device such as a multiple-input, multiple-output (MIMO) antennae, two ports of the antenna can be measured using the VNA, keeping the rest of the ports terminated with a 50-ohms load [20]. Various MIMO parameters such as envelope correlation coefficient (ECC), diversity gain (DG), total active reflection coefficient (TARC), mean effective gain (MEG), and channel capacity loss (CCL) can be defined using S-parameters and measured from the VNA itself. The correlation between the antenna elements is defined and characterized using ECC. The DG shows the effect of diversity and can be obtained using ECC. The TARC manipulates the S-parameters of a MIMO network and displays a single curve with all the necessary information. The MEG defines the effect of the environment on antenna characteristics. The CCL provides the limit at which the transmitted message does not suffer any loss.

$$ECC_{ij} = \frac{|S_{ii}^{*}S_{ij} + S_{ji}^{*}S_{jj}|^2}{\left(1 - |S_{ii}|^2 - |S_{ji}|^2\right)\left(1 - |S_{jj}|^2 - |S_{ij}|^2\right)} \tag{17.56}$$

$$DG_{ij} = 10 * \sqrt{(1 - |ECC_{ij}|^2)} \tag{17.57}$$

$$TARC_{ij} = \frac{\sqrt{(|S_{ii} + S_{ij}e^{j\theta}|^2) + (|S_{ji} + S_{jj}e^{j\theta}|^2)}}{\sqrt{2}} \tag{17.58}$$

$$MEG = 0.5 * \left(1 - \sum_{j=1}^{N} |S_{ij}|^2\right) \tag{17.59}$$

$$MEG1 = 0.5 * (1 - |S_{11}|^2 - |S_{12}|^2) \text{ and } MEG2 = 0.5 * (1 - |S_{21}|^2 - |S_{22}|^2) \tag{17.60}$$

$$CCL = -\log_2 \begin{bmatrix} \rho_{11} & \rho_{12} \\ \rho_{21} & \rho_{22} \end{bmatrix} \tag{17.61}$$

where $\rho_{ii} = \left(1 - |S_{ii}|^2 - |S_{ij}|^2\right)$ and $\rho_{ij} = -(S_{ii}^{*}S_{ij} + S_{ji}^{*}S_{jj})$ for i, j = 1, or 2.

17.9 SIGNAL FLOW GRAPH

Samuel Jefferson Mason was the first person who originated the flow graph analyzing theory in the early 1950s. It is a pictorial or graphical representation used for analyzing cascaded components in any multiport microwave network using scattering parameters [2]. The S-parameter not only gives insight into meaningful interception of any network performance measurement but has also created an inherent data set for the simplification of any network [20]. This is comparatively a new technique for high-frequency spectrum applications, where it follows the incident and reflection of a wave throughout the network [21].

A two-port network and its pictorial illustration are shown in Figure 17.4. The figure shows that every network system is represented by two nodes, one is input signal a_i which is entering the network, and the other one is the

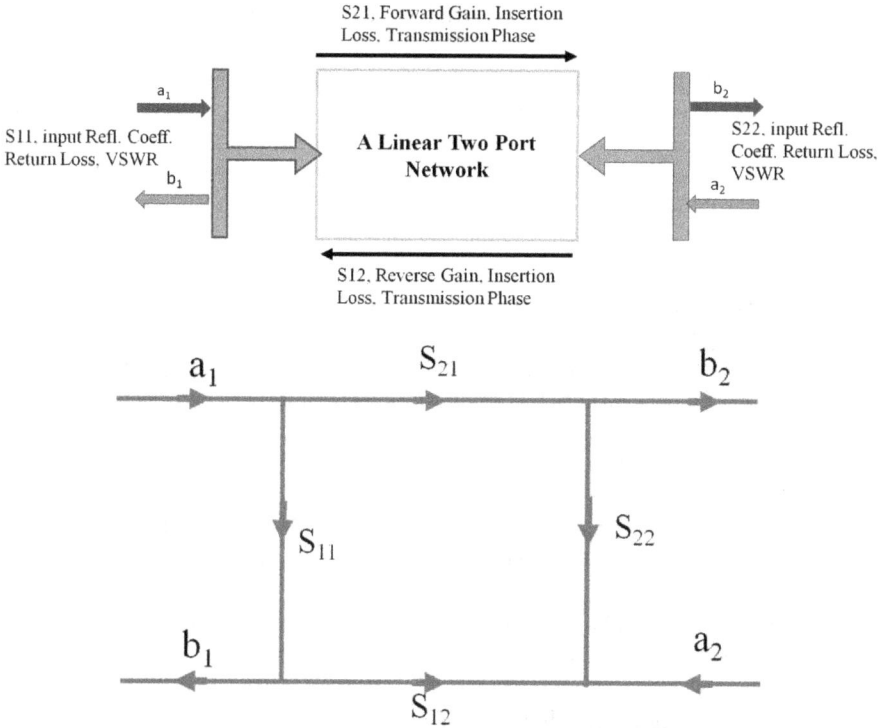

Figure 17.4 (a) Representation of network and (b) pictorial illustration.

output signal b_i, which departs from the network. The S_{ij} term in the figure indicates the iteration of the input and output transfer function at two different ports as well as at the same port. The graphical representation of this network is explained as follows [4]:

1. The signal of amplitude a_1 entering at port 1 splits into two paths. The porting of the wave, which is incident through S_{11}, is an incident power wave, and the rest of the power returns through port 1 with amplitude $b_1 = S_{11}a_1 + S_{12}a_2$.

2. The rest of the power wave is transmitted through S_{21} toward the output node and flows out from node $b_2 = S_{21}a_1 + S_{22}a_2$.

3. However, if at the output port, any non-zero reflection coefficients are connected, then the wave which follows out from node b_2 will be reflected in the two-port network through node a_2 via S_{22}.

4. The rest of the power will be dispatched from port 1 toward port 2 through S_{12}.

There are some useful terms associated with the signal flow graphs:

1. **Node:** The node is defined by the point that represents the signal. The sum of all entering signals into the node is equal to the voltage at the node. A node can be further classified into three categories:
   a. **Input node:** The input node, also called the source node, consists of only the emerged branch.
   b. **Output node:** It is also defined as a sink node, which consists of only an incoming branch.
   c. **Mixed node:** A mixed node consists of both incoming and emerged branches.
2. **Branch:** A branch generally defines the two different segments of node joining. Every branch has an associated S-parameter coefficient. The branch arrow indicates the power flow and gain direction.
3. **Transmittance:** When the signal is traveling between nodes, the obtained gain between the signal is called transmittance.
4. **Open and closed path:** The open path will start at one node and ends at the other node, whereas the closed path will start and stop at the same node.
5. **Forward path:** The forward path should start from the input node and reach out the output node without crossing the node more than once.
6. **Non-touching loop:** These are the loops that do not have any common node.

A signal flow graph is significantly useful to interconnect several devices and to avoid mathematical derivation [22]. The basic rule for constructing the signal flow graph for the characterization of any high-frequency device is mentioned as follows:

1. Every variable is represented by a node and every port is described by a node pair. For example, a_1 and b_1 are the pair of port 1 and similarly b_2 and a_2 are the pair of port 2.
2. Nodes are generally connected with different branches which imply the direction of power flow.

However, while the network has been described in the pictorial form, some rules are defined next to analyze the network. These rules are quite useful for signal flow graphs instead of using algebraic manipulation [3, 23].

1. **Series rule:** For the signal flow graph with one incoming and outgoing wave at a single node, it can be represented as a branch whose coefficient will be the summation of the coefficient of the initial branch. This rule has been depicted in Figure 17.5(a).

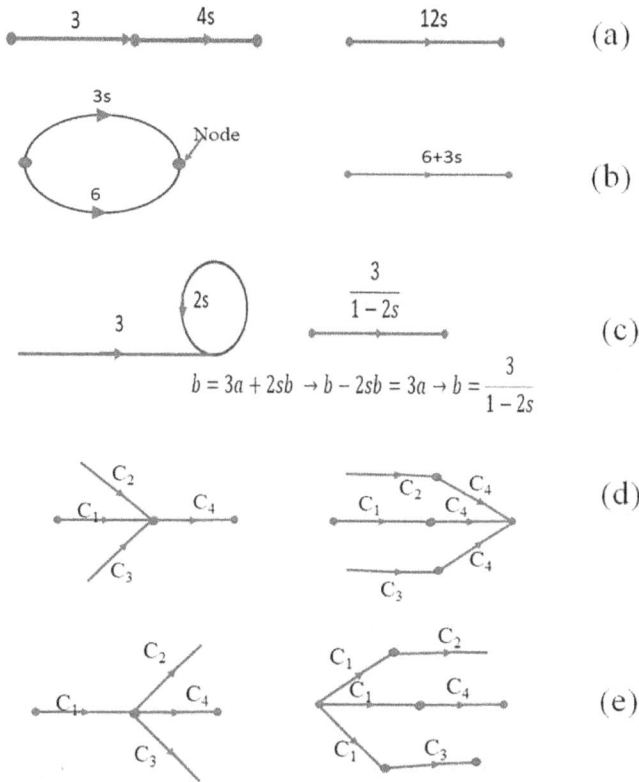

Figure 17.5 Graphical representation of rules of flow graphs.

2. **Parallel rule:** If two branches enter into the same node, they can be treated as a single branch whose coefficient will be the addition of the original branches. This rule has been illustrated in Figure 17.5(b).

3. **Self-loop rule:** When the branch starts and ends at a particular node, it forms a self-loop. If the self-loop coefficient is denoted by S_{xx}, it can be eliminated by multiplication of the branch coefficients feeding the node by $1/(1 - S_{xx})$. This is explained in Figure 17.5(c).

4. **Combining rule:** If two or more incoming branches combine at a node to form a common branch, then the resultant branch is the combination of the incoming and outgoing branches. Figure 17.5(d) depicts this rule.

5. **Splitting rule:** If a node has a single incoming branch and one or more branches that leave the node, then the incoming branch can be split and combined with the outgoing branch. This can be explained using Figure 17.5(e).

Mason's gain formula: This is the most useful and most important rule that allows us to find out the overall transfer function of the signal flow graph by 'inspection' [24]. This is explained using a simple two-port network flow graph depicted in Figure 17.6. The network consists of b_s, which is the independent variable whose power wave emerges from the input source. This network is analyzed in terms of a path loop. S_{21} is the value of the path which flows out from b_s and reaches the b_2 node; S_{11} is the value of the path which is coming from b_s to b_1; and S_{21}, S_{12}, and Γ_L is the value from path b_s to b_1. The first-order loop is defined as lines that are followed in sequence while coming to a closure, and the loop cannot pass more than one node in the same direction. Loop value is generally defined by the multiplication of all coefficients that are encountered while traversing the loop [23, 25].

In the previous representation, three first-order loops are shown, namely Γ_s, S_{11} for a_1-b_1; S_{22} for b_2-a_2; and S_{21}, Γ_s, S_{12}, and Γ_L for a_1-b_2-a_2-b_1. A second-order loop is defined as the multiplication of any two previous-order non-touching loops. These are the second-order loop S_{11}, Γ_s, S_{22}, Γ_L. The third-order loop is not present in this structure. In general, an nth-order loop is the product of any non-touching n first-order loops, which should not touch. The higher-order loop is demonstrated similarly except for the very complex system; they are seldom obstructing in practice. Mason's rule is defined as the ratio of a dependent variable parameter to an independent variable parameter, so the rule has been expressed symbolically as follows [3]:

$$T = \frac{P_1\left[1 - \Sigma L(1)^{(1)} + \Sigma L(2)^{(1)} - \cdots\cdots\right] + P_2\left[1 - \ldots\cdots\right]}{1 - \Sigma L(1) + \Sigma L(2) - \Sigma L(3) + \cdots\cdots} \tag{17.62}$$

Here,

1. P_1, P_2, and P_3 denote the paths from the independent variable parameter to the node whose value is to be calculated. Here $\Sigma L(1)$ and $\Sigma L(2)$ define the addition of all first- and second-order loops.

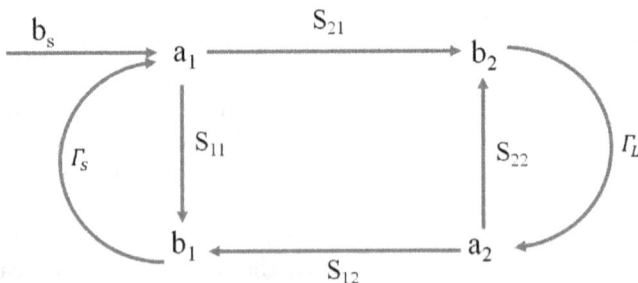

Figure 17.6 Graphical representation of two-port network.

2. $\Sigma L(1)^{(1)}, \Sigma L(2)^{(1)}$ defines the addition of all first-order and second-order loops, which does not overlap the first path between the node representing the two variables.

3. $\Sigma L(1)(2), \Sigma L(2)(2)$ shows the addition of all first-order and second-order loop products, which do not overlap with the second path among two variable notes.

17.9.1 The Flow Graph Representation for Different Conditions

Different types of flow graph representation concerning the different active and passive elements are portrayed as follows [2, 3]:

- **Flow graph for source:** Figure 17.7(a) portrays the flow graph representation for a source. Here, a and b represent incident and reflected waves, and bs denotes the associated power wave whose power is dissipated in load Z_0 connected to the source. Z_s represents the impedance of the source, which converts to Γ_s in the flow graph.

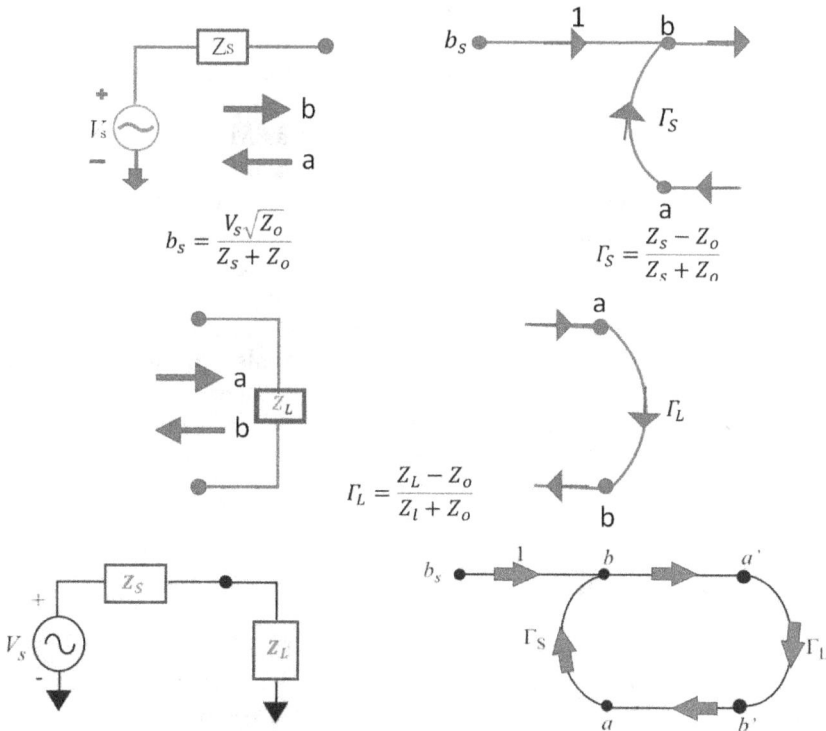

$$b_s = \frac{V_s\sqrt{Z_o}}{Z_s + Z_o}$$

$$\Gamma_s = \frac{Z_s - Z_o}{Z_s + Z_o}$$

$$\Gamma_L = \frac{Z_L - Z_o}{Z_l + Z_o}$$

Figure 17.7 Flow graph representation for the (a) source, (b) load, and (c) composite circuit.

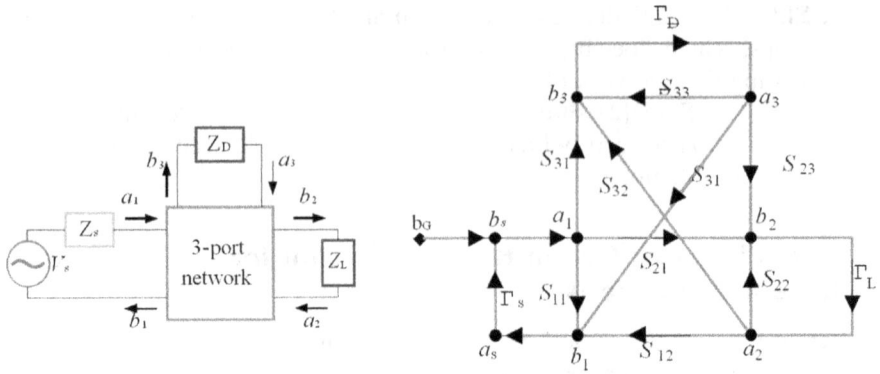

Figure 17.8 Network with three ports and its pictorial graph.

- **Flow graph for load:** Figure 17.7 (b) portrays the flow graph representation for a load Z_L. The load is simply defined as the complex coefficient of the load, which is denoted as Γ_L.
- **Flow graph for a composite circuit:** Figure 17.7(c) portrays the flow graph representation for a composite network with internal voltage source V_s, source impedance Z_s, and load impedance Z_L.

However, in the case of larger networks such as MIMO communication networks, there are multiple ports [26]. So, the scattering matrix of a network with n ports will contain n^2 coefficients. For a better understanding of the concept, an example of a network with its equivalent flow graph is depicted in Figure 17.8. Here, a_1 and b_1 are the incident and reflected waves at port 1. S_{21} and S_{31} are the transmission coefficients that originate from port 1. S_{31} and S_{12} will be emanating from ports 3 and 2, respectively, and combine at port 1. b_3 and b_2 are the reflected signals from port 3 and port 2, respectively, while a_3 and a_2 are the incident wave of the network. In the Figure 7.17, Γ_D, Γ_L are the complex reflection coefficients for the load. S-parameters generally depend on the network type, frequency, and characteristic impedance of the source and load [27].

17.10 CONCLUSIONS

The issues with the low-frequency parameters at high frequencies have led to the introduction of scattering parameters. The scattering parameters can be defined in terms of incident and reflected voltages, where normalization of signals is performed to obtain symmetric properties. The scattering matrix is used to determine the response of the antenna (reflection coefficient, bandwidth, coupling between elements) and MIMO antenna parameters (ECC, DG, TARC, MEG, and CCL). These are measured using a VNA. A scattering matrix is also used to analyze the multiport microwave network using a pictorial representation called a signal flow graph. Future work using S-parameters in microwave circuits needs to determine the impedance of the antenna using the S-parameter method. S-parameters can be used to improve measurement accuracy and integrated circuit (IC) design. S-parameters can also be used in providing in-passivity enforcement, providing back-end optimization, handling electrical delay lines, and fitting algorithms in nm-accurate simulation technology.

REFERENCES

[1] Caspers, F., 2011. RF Engineering Basic Concepts: S-Parameters Cern Yellow Rep., p. 67. https://cds.cern.ch/record/1415639/files/p67.pdf.
[2] Das, S. K., Das, A., 2008. Microwave Engineering. McGraw-Hill Higher Education.
[3] Rizzi, P. A., 1988. Microwave Engineering. Prentice Hall.
[4] Pozar, D. M., 2003. Microwave Engineering. John Wiley & Sons, Limited.
[5] Smith, P. H., 1944. An Improved Transmission Line Calculator. Electronics, vol. 17, p. 130.
[6] Choma, J., Jr., 1985. Electrical Networks: Theory and Analysis. Wiley-Interscience, pp. 509–513.
[7] Collin, R. E., 2001. Foundations of Microwave Engineering, 2nd Edition. Wiley-Interscience.
[8] (Aglient Papers) www.sss-mag.com/pdf/an-95-1.pdf
[9] Choma, J., 2006. Scattering Parameters: Concept, Theory, and Applications. Fall, pp. 213–740.
[10] (Aglient Papers) www.sss-mag.com/pdf/AN154.pdf
[11] Clarke, K., and Hess, D., 1971. Communication Circuits: Analysis and Design. Addison-Wesley Publishing Company. https://www.kemt-old.fei.tuke.sk/STS-Satelitne/comunication_circuits.pdf.
[12] (Studocu) www.studocu.com/in/document/visvesvaraya-technological-university/microwave-and-antennas/eecs242-lect5-sparam-ma/11682549.
[13] (IEEE Long Island Section) www.ieee.li/pdf/viewgraphs/s_parameter_techniques.pdf
[14] Chen, W. K., 1986. Passive and Active Filters: Theory and Implementations. John Wiley & Sons.
[15] De, A., Roy, B., Bhattacharya, A., and Bhattacharjee, A. K., 2021. Bandwidth-Enhanced Ultra-Wideband Wearable Textile Antenna for Various WBAN and Internet of Things (IoT) applications. Radio Science, vol. 56, p. e2021RS007315. https://doi.org/10.1029/2021RS007315
[16] Bhattacharya, A., Roy, B., Caldeirinha, R., and Bhattacharjee, A., 2019. Low-Profile, Extremely Wideband, Dual-Band-Notched MIMO Antenna for UWB Applications. International Journal of Microwave and Wireless Technologies, vol. 11, no. 7, pp. 719–728. https://doi.org/10.1017/S1759078719000266

[17] Bhattacharya, Ankan, Roy, Bappadittya, Chowdhury, Santosh K., and Bhattacharjee, Anup K., 2019. Computational and Experimental Analysis of a Low-Profile, Isolation-Enhanced, Band-Notch UWB-MIMO Antenna. Journal of Computational Electronics, vol. 18, pp. 680–688. https://doi.org/10.1007/s10825-019-01309-3

[18] S. Bhatia and M. V. Deepak Nair, "Frequency Reconfigurable Elliptically Polarized Slotted Diagonally Trimmed Patch Antenna," 2019 TEQIP III Sponsored International Conference on Microwave Integrated Circuits, Photonics and Wireless Networks (IMICPW), Tiruchirappalli, India, 2019, pp. 140–143, doi: 10.1109/IMICPW.2019.8933199.

[19] Forouhar, F., Ali, F., Mahmoud, K., Ali, N., and Mohammad, E., 2022. In the book Introduction to Wireless Communication Circuits (pp. 349–375) Scattering Parameters. doi: 10.1201/9781003338710-11.

[20] Davidovitz, M., 1995. Reconstruction of the S-Matrix for a 3-port using Measurements at Only Two Ports. IEEE Microwave and Guided Wave Letters, vol. 5, no. 10, pp. 349–350. doi: 10.1109/75.465040.

[21] (The University of Vermont) www.uvm.edu/~muse/modules/RFH/MUSE_S-parameters_081003.pdf

[22] (All About Digital Signal Processing) www.dsprelated.com/showarticle/76.php

[23] Poole, C., and Darwazeh, I., 2015. Microwave Active Circuit Analysis and Design. Academic Press.

[24] Mason, S., 1956. Feedback Theory: Further Properties of Signal Flow Graphs. Proceedings of the IRE, vol. 44, pp. 1920–926.

[25] Mason, J., 1953. Feedback Theory—Some Properties of Signal Flow Graphs. Proceedings of the IRE, vol. 41, pp. 1144–1156.

[26] Overfelt, P. L. and White, D. J., 1989. Alternate Forms of the Generalized Composite Scattering Matrix. IEEE Transactions on Microwave Theory and Techniques, vol. 37, no. 8, pp. 1267–1268, doi: 10.1109/22.31092.

[27] (Complex Art of Handling S-Parameters) https://semiengineering.com/the-complex-art-of-handling-s-parameters/

Chapter 18

A Feasibility Study for Biomedical Applications via Microwave Imaging

Sweety Jain

18.1 INTRODUCTION

A brain tumor is the growth of abnormal cells in the human brain. Various types of brain tumors exist such as noncancerous (benign) and cancerous (malignant). However, the growth rate of any tumor may vary. The part of the nervous system affected by the tumor depends upon the location of the tumor in the brain. Earlier detection of it can be lifesaving for an individual. So, tumor detection can be done in two groups: first, it includes detection of a tumor through symptoms, which may or may not be accurate such as headaches, which may be frequent or severe; vomiting; vision problems; tiredness; confusion; personality defects; and various others. Second, this group is more accurate in tumor detection i.e., through microstrip patch antenna (MPA). For instance, an individual feels symptoms of former group but still in confusion whether an individual is suffering from such life-threatening disease or not. Then such an individual can adopt the latter group for accurate detection of a tumor.

To achieve this aim, a microstrip antenna is devised. Before introducing the working principle of the designed antenna, it is necessary to take a look at the research conducted related to it. The antenna is designed on an FR4 substrate having thickness 1.5 mm and size 44 mm × 30 mm while it operates on the band 3.35–12.6 GHz [1]. Another study was conducted in which an antenna operates on the ultra-wideband (UWB) for detection of tumors; this range varied from 3.12 to 10.6 GHz, and this antenna gives low gain [2]. However, a similar study was conducted which provide a maximum return loss using a Vivaldi antenna array [3]. Another design was published in which an antenna is fed by a coplanar waveguide that helps to provide accuracy and efficiency to the antenna [4]. Another antenna structure introduced for detecting brain tumors and breast cancer with the help of different techniques [5–8]. The microstrip antenna is not only boon in the medical field but also in the agricultural field for detecting the moisture content in grains and the detection of salt and sugar [9–18]. Now, to become familiar with the design sensor, the following section is introduced which gives idea related to the size, working principle, and various other parameters.

DOI: 10.1201/9781003459880-18

18.2 DIELECTRIC PROPERTIES OF HUMAN TISSUES

Microwave imaging technology is supposed to be based on the dielectric properties of tumors and other human tissues. According to various studies it is known that tumors and other tissues transmit different microwaves. Therefore, this concept is utilized in this study for detecting cancer cells. For that, an antenna array is conducted to send pulses of microwaves to human tissues which suffer from cancer. As a result of the phenomenon of backscatter, the signals are reflected and later picked up by an array [19]. By this process, it is easy to determine whether tissue contains cancer or not. UW040 is used to maintain the fidelity of the waveform [20]. A planner structure printed circuit board is used to design the antenna [21], also for high resolution and accurate images. A compact antenna needs to transmit on a broad range of frequencies. The dielectric properties of numerous tissues are considered by permittivity [22].

$$\varepsilon \left(\varepsilon = \varepsilon_r + i\sigma / \omega\varepsilon_0 \right) \tag{18.1}$$

Where,
ε_r = dielectric constant
σ = conductivity of the tissue against frequency
ε_0 = dielectric permittivity of vacuum
ω = angular frequency

Because of radio frequency waves, there is a probability occurrence of various hazards which needs to be evaluated. Such hazards can be evaluated from the specific absorption rate (SAR). SAR measures the energy absorption by tissue when exposed to radio frequency. Also SAR is proportional to the squared electric field strength value occurring in the human body. The higher the SAR, the greater the radiation absorption by tissue, which results in a severe impact on the body. SAR can be calculated as [23]:

$$SAR_i = \frac{P_i}{\rho_i} \bullet i \frac{\left| E^2 \right|}{2\rho_i} \tag{18.2}$$

Where,
SAR is watts per kilogram (W/kg)
P is the power loss density (W/m³)
ρ is the density of human tissues (Kg/m³)
E is the electric field strength (V/m)
σ is the conductivity (Siemens/m)
i is the ith tissue

Table 18.1 SAR Exposure Limit Guidelines[25]

Standard	SAR Limit [W/kg]	Average Mass for SAR
ICNIRP	2.0 (f < 10 GHz)	10 g of tissues
FCC/ANSI	1.6 (f < 6 GHz)	1 g of tissues

Another important parameter related to SAR is power loss density. It is defined to get knowledge of the electromagnetic field distribution inside a tissue structure. The field distribution on human tissues can be determined when the antenna reaches human tissues. Hence, there are two methods to determining the SAR i.e., (a) point SAR and (b) averaging SAR. The former is defined as the value that determines the SAR for all grid cells. The latter is defined as a cube with a predefined mass, and power loss density is geometrically interpreted on this cube [23, 24].

In Table 18.1, the International Commission on Non-ionizing Radiation Protection (ICNIRP) guidelines for SAR frequencies above 10 GHz and the Federal Communication Commission (FCC) guidelines for above 6 GHz are given. Also according to FCC guidelines, the application of power density rises when the exposure is above 6 GHz. It is required to be noted down. Here, power density is defined only as power traveling toward tissue and does not define power absorption and field distribution in tissues. Hence, power density has limited application in near-field study [25].

Biological tissues consist layers of skin, fat, muscle, and bone. The structure of tissue contains non-homogeneous dielectric properties. Skin layers have a thickness varying from 50 microns to 1000 micron, and the thickness of fat and muscles are 300 microns to 20,000 microns [26].

The study of dielectric properties of tissues is important because it controls the reflection, propagation, and attenuation of electromagnetic fields in the body. Also, permittivity is inversely proportional to frequency with a rise in permittivity, which brings down the frequency, while conductivity is directly proportional to frequency [27]. The microstrip antenna is also beneficial for use in wireless body are networks (WBANs) and Internet of Things, along with UWB fields [28–30].

18.3 DESIGN AND ANALYSIS OF ANTENNA PERFORMANCE

18.3.1 Antenna Design

General properties of the antenna are shown in Table 18.2. The antenna consists of a T-shaped slot in a patch, and CST software is used for simulation.

Table 18.2 Parameters of Antenna

Parameters	Size
Length of the ground	60 mm
Width of the ground	60 mm
Length of the substrate	60 mm
Width of the substrate	60 mm
Length of the patch	60 mm
Width of the patch	60 mm
Brain (center) radius	30 mm
Brain (top) radius	20 mm
Brain (bottom) radius	20 mm
Tumor (center) radius	5 mm
Tumor (top) radius	4 mm
Tumor (bottom) radius	2 mm

18.3.2 Performance of Antenna

Analysis of the antenna is done using CST software in two ways, which include performance of the antenna with healthy brains and unhealthy brains. On the basis of this, the reflection coefficient, current density, and SAR are determined. Calculation of these parameters indicates the effectiveness of designed antenna, which will be discussed in a later section.

18.3.2.1 Antenna in Contact with Healthy Brain

Figure 18.1 shows what happens when the antenna comes in contact with a healthy brain.

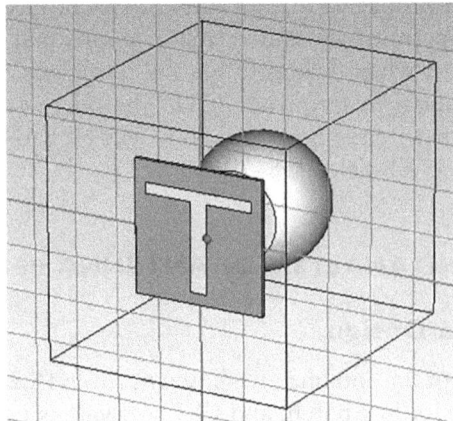

Figure 18.1 Microstrip patch antenna with human head (healthy brain).

Table 18.3 Healthy Brain

Parameters	Value on 2.4 GHz
Reflection coefficient	−18.68 dB
Current density	7.76 A/m²
SAR (10 g)	0.00412 W/kg

After simulation, the performance of the antenna is found as shown in Table 18.3.

18.3.2.2 Antenna in Contact with Unhealthy Brain

Figure 18.2 shows what happens when the antenna comes in contact with a unhealthy brain.

After simulation, the performance of the antenna is found as shown in Table 18.4.

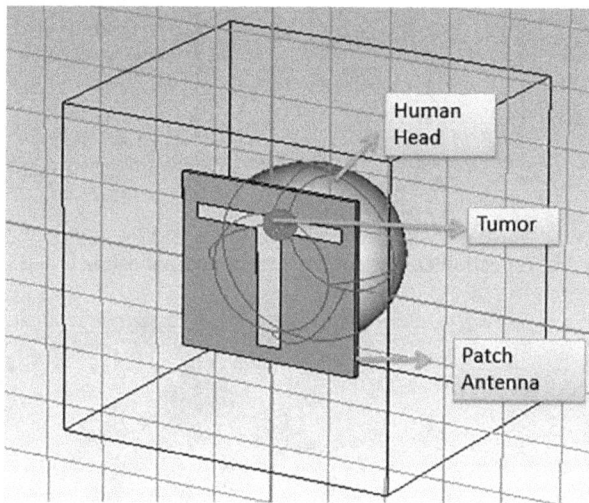

Figure 18.2 Microstrip patch antenna with human head (unhealthy brain).

Table 18.4 Unhealthy Brain

Parameters	Value on 2.4 GHz
Reflection coefficient	−1.66 dB
Current density	22.8 A/m²
SAR (10 g)	0.0363 W/kg

18.3.2.3 Comparison and Effectiveness of Antenna

When the performance of the antenna with a healthy and unhealthy brain is compared, then it is found that the antenna is effective in fulfilling the aim for which it was designed.

(i) **Reflection coefficient:** The reflection coefficient in a healthy brain is found to be lower than in an unhealthy brain, as shown in Figure 18.3. The study shows that whenever the reflection coefficient is lower for a healthy brain compared with unhealthy, it signifies that the antenna is accurate for this parameter.

(ii) **Current density:** It is also found to be lower in a healthy brain as compared to an unhealthy brain, as shown in Figure 18.4(a) and (b). It signifies that the antenna is effective in terms of performance.

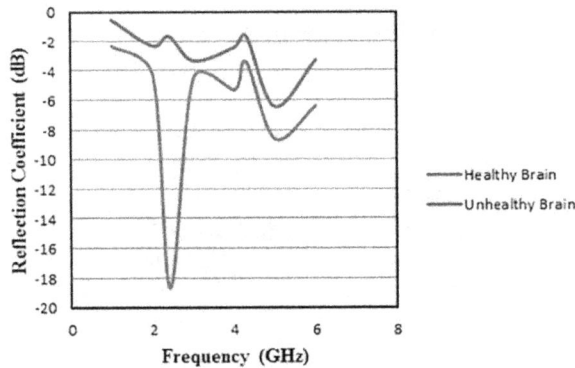

Figure 18.3 Effect of reflection coefficient on healthy and unhealthy brains.

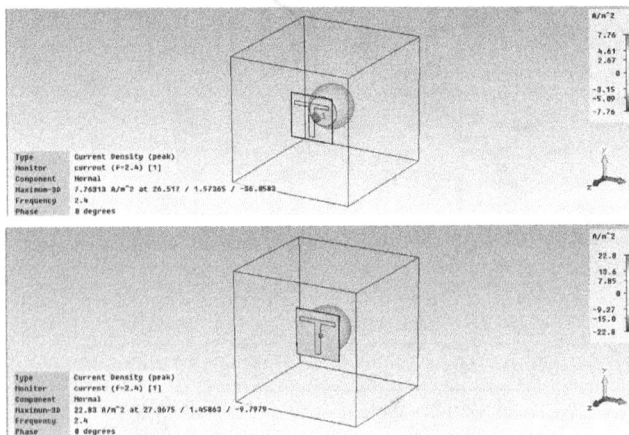

Figure 18.4 Effect of current density on (a) healthy and (b) unhealthy brain.

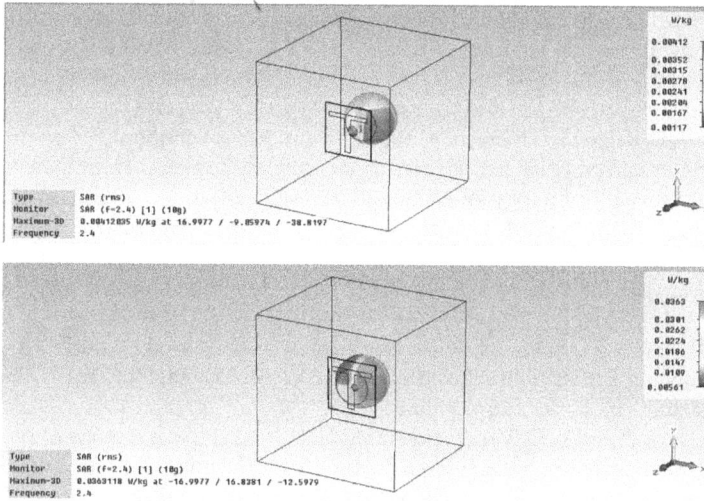

Figure 18.5 Effect of SAR on (a) healthy and (b) unhealthy brain.

(iii) **SAR:** It is found that in a healthy brain, SAR is decreased, while it is increased in an unhealthy brain. Hence, the antenna satisfies this condition also, as shown in Figure 18.5(a) and (b).

By analyzing these parameters, it can be concluded that the designed antenna is fully effective in detecting tumors.

18.4 CONCLUSION

This antenna can help in the earlier detection of tumors. So, the inference can be made that the antenna is satisfying all of the criteria which are necessary for its performance. Also, it is low cost, easy to maintain, and easy to fabricate on a low-cost FR4 substrate. Hence, it is recommended to adopt this MPA due to its high accuracy and better results. With a few modifications, the design can be used for the early detection of stones in different internal organs or to detect fractures in bone.

REFERENCES

1. Adhitya Satria Pratama, Basari, Muhammad Firdaus S. Lubis, Fitri Yuli Zulkifli and Eko Tjipto Rahardjo, "A UWB Antenna for Microwave Brain Imaging," Proceedings of IEEE 4th Asia-Pacific Conference on Antennas and Propagation, pp. 202–208, 2015.

2. Haoyu Zhang, Brian Flynn, Ahmet T. Erdogan and Tughrul Arslan, "Microwave Imaging for Brain Tumour Detection using an UWB Vivaldi Antenna Array," Proceedings of Loughborough Antennas and Propagation Conference, pp. 1–4, 2012.

3. A. Afyf, L. Bellarbi, F. Riouch, A. Errachid and M. A. Sennouni, "Flexible antenna array for early breast cancer detection using radiometric technique," International Journal of Biology and Biomedical Engineering, vol. 10, pp. 10–17, 2016.

4. Suriya, R. Nandhini, M. Leepika, and S. R. Praveen, "Microstrip patch antenna for brain tumour detection," International Journal of Scientific Research and Review, vol. 7, no. 3, pp. 63–66, 2018. http://dynamicpublisher.org/gallery/ijsrr-d194.pdf.

5. E. C. Fear, S. C. Hagness, P. M. Meaney, M. Okoniewski, and M. A. Stuchly, "Enhancing breast tumor detection with near-field imaging," IEEE Microwave Magazine, vol. 3, no. 1, pp. 48–56, 2002.

6. X. Li, E. J. Bond, B. D. Van Veen, and S. C. Hagness, "An overview of ultra-wideband microwave imaging via space-time beamforming for early-stage breast-cancer detection," IEEE Antennas and Propagation Magazine, vol. 47, no. 1, pp. 19–34, 2005.

7. M. A. Hemandez-Lopez, M. Quintillan-Gonzalez, S. Gonzalez Garcıa, A. Rubio Bretones, and R. Gomez Martın, "A rotating array of antennas for confocal microwave breast imaging," Microwave and Optical Technology Letters, vol. 39, no. 4, pp. 307–311, 2003.

8. Y. Huo, R. Bansal, and Q. Zhu, "Modeling of noninvasive microwave characterization of breast tumors," IEEE Transactions on Biomedical Engineering, vol. 51, no. 7, pp. 1089–1094, 2004.

9. Sweety Jain, "Early detection of salt and sugar by microstrip moisture sensor based on direct transmission method," Wireless Personal Communication, vol. 122, no. 1, pp. 593–601, 2022. DOI: 10.1007/s11277-021-08914-1.

10. Sweety Jain, "Determination of moisture content from microstrip moisture sensor with minimum mean relative error," Smart Antennas, pp. 345–357, 2022. https://doi.org/10.1007/978-3-030-76636-8_26.

11. Sweety Jain, Pankaj Kumar Mishra, and Vandana Vikas Thakare, "Design and analysis of dual-frequency microwave moisture sensor based on rectangular microstrip antenna" Materials Today Proceeding Elsevier, vol. 47, part 18, pp. 6441–6448, 2021. https://doi.org/10.1016/j.matpr.2021.08.179.

12. Sweety Jain, Pankaj Kumar Mishra, and Vandana Vikas Thakare, "The analysis and design of circular microstrip moisture sensor for rice grain," Materials Today Proceeding Elsevier, vol. 47, part 18, pp. 6449–6456, 2021. https://doi.org/10.1016/j.matpr.2021.08.180

13. Vivek Singh Kushwah and Sweety Jain, "Enhancement of reflection coefficient for circular ring microstrip sensor," Materials Today Proceeding Elsevier, 2021. https://doi.org/10.1016/j.matpr.2021.05.106, ISSN: 2214-7853

14. Sweety Jain, Pankaj Kumar Mishra, Vandana Vikas Thakare, and Jyoti Mishra, "Design and Analysis of moisture content of hevea latex rubber using microstrip patch antenna with DGS," Materials Today Proceeding Elsevier, vol. 29, part 2, pp. 581–586, 2020. https://doi.org/10.1016/j.matpr.2020.07.312,

15. Sweety Jain, Pankaj Kumar Mishra, Jyoti Mishra, and Vandana Vikas Thakare, "Design and analysis of H-Shape patch sensor for rice quality detection," Materials Today Proceeding Elsevier, vol. 22, part 2, pp. 556–560, 2020. https://doi.org/10.1016/j.matpr.2020.07.317

16. Sweety Jain, Pankaj Kumar Mishra, Vandana Vikas Thakare, and Jyoti Mishra, "Design of microstrip moisture sensor for determination of moisture content in rice with improved mean relative error," Microwave and Optical Technology Letters, vol. 61, no. 7, pp. 1764–1768, 2019. https://doi.org/10.1002/mop.31763.

17. Sweety Jain, Pankaj Kumar Mishra, Vandana Vikas Thakare, and Jyoti Mishra, "Microstrip moisture sensor based on microstrip patch antenna", Progress in Electromagnetic Research-M, vol. 76, pp. 175–185, 2018. https://doi.org/10.2528/PIERM18092602.

18. Sweety Jain, Pankaj Kumar Mishra, and Vandana Vikas Thakare, "Analysis and optimal design of moisture sensor for rice grain moisture measurement", American Institute of Physics, pp. 1–3, 2018. https://doi.org/10.1063/1.5028775, 060005.

19. A. Abu Bakar, D. Ireland, A. M. Abbosh and Y. Wang, "Experimental assessment of microwave diagnostic tool for ultra-wideband breast cancer detection," Progress in Electromagnetics Research M, vol. 23, pp. 109–121, 2012.

20. E. C. Fear, P. M. Meaney, and M. Stuchly, "Microwaves for breast cancer detection? Potentials," IEEE, vol. 22, pp. 12–18, 2003.

21. J.-H. Lu and F.-C. Tsai, "Planar internal LTE/WWAN monopole antenna for tablet computer application," IEEE Transactions on Antennas and Propagation, vol. 61, pp. 4358–4363, 2013.

22. M. S. Islam, S. Kibria, and M. T. Islam, "Experimental breast phantoms for estimation of breast tumor using microwave imaging systems," IEEE Access, vol. 6, pp. 78587–78597, 2018. https://doi.org/10.1109/ACCESS.2018.2885087.

23. A. A. Aly and M. Piket May, "FDTD computation for SAR induced in human head due to exposure to EMF from mobile phone," Advanced Computing an International Journal, vol. 5, 2014.

24. F. Gustrau and A. Bahr, "W-band investigation of material parameters, SAR distribution, and thermal response in human tissue," IEEE Transactions on Microwave Theory and Techniques, vol. 50, pp. 2393–2400, 2002.

25. T. Wu, T. S. Rappaport, and C. M. Collins, "The human body and millimeter wave wireless communication systems: Interactions and implications," IEEE International Conference on Communications, London, 2015.

26. D. Miklavčič, N. Pavšelj, and F. X. Hart. "Electric properties of tissues," In Wiley Encyclopedia of Biomedical Engineering, M. Akay (Ed.), 2006. https://doi.org/10.1002/9780471740360.ebs0403.

27. Wang, J. and Q. Wang, Body Area Communications: Channel Modeling, Communication Systems, and EMC, 1st Edition, John Wiley & Sons, Singapore, 2013.

28. A. De, B. Roy, A. Bhattacharya, and A. K. Bhattacharjee, "Bandwidth-enhanced ultra-wide band wearable textile antenna for various WBAN and Internet of Things (IoT) applications," Radio Science, vol. 56, p. e2021RS007315, 2021. https://doi.org/10.1029/2021RS007315

29. A. Bhattacharya, B. Roy, R. Caldeirinha, and A. Bhattacharjee, "Low-profile, extremely wideband, dual-band-notched MIMO antenna for UWB applications," International Journal of Microwave and Wireless Technologies, vol. 11, no. 7, pp. 719–728. doi:10.1017/S1759078719000266

30. Ankan Bhattacharya, Bappadittya Roy, Santosh K. Chowdhury and Anup K. Bhattacharjee, "Computational and experimental analysis of a low-profile, isolation-enhanced, band-notch UWB-MIMO antenna," Journal of Computational Electronics, vol. 18, pp. 680–688, 2019. https://doi.org/10.1007/s10825-019-01309-3

Chapter 19

Review of Current Advancements in Microwave UWB Filter

Partha Protim Kalita, Akash Buragohain,
Yatish Beria, and Gouree Shankar Das

19.1 INTRODUCTION

Ultra-wideband (UWB) technology has been gaining significant attention in recent years due to its ability to support high-speed data transfer and its potential for a wide range of applications such as wireless communications, radio location, and high-speed data transfer systems. In 2002 the Federal Communication Commission (FCC) permitted the unlicensed use of the UWB frequency strictly within the range from 3.1 to 10.6 GHz [1] along with an power restriction of −41.3 dBm/MHz. After that, the development of UWB system increased rapidly. A UWB communication system consists of different components. One key component of UWB systems is the UWB filter, which is designed to pass the desired UWB signals while rejecting unwanted narrowband interference. UWB filters play a critical role in ensuring the reliable and efficient operation of UWB systems. They are designed to have a wide bandwidth and a flat passband response, which allows them to pass the full bandwidth of the UWB signal. Additionally, UWB filters are designed to reject narrowband interference, which is a common problem in UWB systems due to the presence of other communication systems that operate in close proximity. In this chapter, a comprehensive overview of UWB filters, including their history and evolution, types of filters, filter design, performance evaluation, and applications, is presented. The chapter starts by discussing the background of UWB technology and the development of UWB filters, including the advantages, disadvantages, applications of UWB filters, and how they are used in wireless communications, radio location, and high-speed data transfer systems. The performance of the UWB filter is evaluated based on different parameters such as insertion loss, return loss, wide passband, group delay, and significant out-of-band rejection. Microstrip filters are widely used to design UWB filters due to their advantages, such as low cost, compact size, and ease of integration. Also, coplanar waveguide (CPW) [2] and substrate-integrated waveguide (SIW) [3] are used to design UWB filters. The CPW filters offer excellent performance with low insertion loss and high rejection levels. SIW filters provide superior performance with

DOI: 10.1201/9781003459880-19

low loss, wide bandwidth, and compact size. This chapter discusses different techniques for designing UWB filters using microstrip lines. Some techniques include cascading low- and high-pass filters, multimode resonators, ring resonators, etc., and papers from various researchers are also reviewed.

By the end of this chapter, the reader will have a good understanding of the importance of UWB filters and how they contribute to the success of UWB technology. Whether a researcher, engineer, or student, this chapter will provide valuable information and insights on the topic of UWB filters.

19.2 UWB TECHNOLOGY

UWB is a generic term to describe any radio system with a large amount of bandwidth. UWB technology is a wireless communication technology that operates by transmitting high-frequency, low-power radio signals over a large frequency range. UWB systems use pulses of short durations (nanoseconds to picoseconds) to transmit low-power radio signals. UWB technology is not a new technology and has been around for over 100 years, but its development and use have rapidly increased in the past few decades due to the advancement of the technology. In 1901 Italian innovator Guglielmo Marconi first introduced UWB technology for the long-distance transmission of radio signals [4]. With the development of the linear time-invariant system in 1960, the growth of UWB technology started [5]. In the 1960s, the US military began experimenting with UWB technology for radar applications. UWB pulses were found to have a very low probability of being intercepted, making them difficult to detect by enemy radars. The US military continued to develop and use UWB technology for various applications, including ground-penetrating radar, mine detection, and target identification. After that UWB was used strictly for military applications [6] as a very secure connection until 2002. In 2002 the FCC approved the unlicensed commercial use of UWB communication systems for indoor and handheld applications strictly between the range of 3.1 and 10.6 GHz with a power restriction of -41.3 dBm/MHz [1]. UWB systems use short pulses for transmitting a signal, which can operate without a radio frequency (RF) mixing stage [7]. Today, UWB technology is an area of active research and development, with ongoing efforts to improve its performance, reliability, and cost-effectiveness. UWB technology is also being explored for its potential in emerging applications, such as Internet of Things (IoT), 5G wireless communication, and augmented reality.

19.3 ADVANTAGES

UWB technology has several advantages over other wireless communication technologies. A large channel capacity is one of the major advantages of UWB, which is defined as the maximum amount of data that can be transmitted per second over a communication channel. UWB technology is capable of achieving very high data transfer rates, up to several gigabits per second, with distances of 1 to 10 meters, making it ideal for short-range applications that require fast data transfer. Also, the UWB system is a low-cost and less complex system [7]. UWB technology requires very little power to operate, which can significantly extend the battery life of devices using this technology. This makes it an excellent option for applications that require long battery life, such as handheld devices, wearable devices, sensors, and IoT devices. The power level of UWB signals is very low, so existing radio services cannot recognize it, reducing interference with existing systems. So, existing radio communications can work along with UWB communication. UWB technology is capable of providing accurate distance and location tracking, with accuracy up to a few centimeters. UWB pulses are time modulated with codes unique to each transmitter/receiver pair. Therefore, UWB systems hold significant promise of achieving a highly secure communication system with a low probability of intercept and detection.

19.4 DISADVANTAGES

With numerous advantages of UWB technology, there are also some disadvantages, which are a major challenge in developing UWB systems. The major disadvantage of the UWB system is its limited range. Hence, we can use this system to communicate only within a few meters. Since UWB pulses have a very limited time domain, high-speed analog to digital converter (ADC) and high-speed digital signal processing (DSP) are necessary for UWB systems to digitize and analyze UWB data. Wideband antennas are required for UWB systems. Wideband antennas are larger and more expensive than narrowband antennas; therefore, creating a compact, low-cost antenna is essential for the widespread adoption of UWB technology. UWB technology can be more expensive than other wireless technologies, which can be a barrier to adoption for some applications. This is due to the need for specialized hardware and software to implement UWB technology. Due to the vast RF bandwidth used by UWB signals, interference with existing narrowbands becomes a severe issue. For instance, because global positioning system (GPS) signals often have low power densities, they are vulnerable to interference from UWB.

19.5 APPLICATIONS

UWB technology has several applications for short-range and indoor communication due to its various advantages. For short-range transfer of a large amount of data, UWB technology can be widely used. High-resolution video streaming can be smoothly performed using the UWB system. Using UWB systems, we can also connect all computer peripherals wirelessly. UWB technology can be used to detect the exact location of an object with high accuracy in short ranges. UWB technology can operate in very complex environments like populated places such as malls, stations, and hospitals for faster and more effective communication between people. It is also applicable in radar applications, which include automotive sensors, collision avoidance sensors, etc. Using UWB technology, cars can locate and track the motion of things close to them, enabling functions like increased air bag activation and better near-collision avoidance, object detection, and keyless entry systems [8]. Since UWB can detect objects behind walls, it can be used for rescue operations, security, and medical operations; it also can be used in IoT applications, such as smart homes, industrial automation, and asset tracking.

19.6 UWB FILTER

To construct a UWB system, different components such as UWB antenna [9, 10], UWB filter [2, 3, 11–36], low noise amplifier, UWB pulse generator, etc., are used. In this chapter, the UWB filter is discussed. A UWB filter is an electronic component designed to selectively pass or block a specific range of frequencies within the UWB frequency spectrum. UWB filters are designed to operate within this wide frequency band and eliminate unwanted noise and interference that can degrade the performance of UWB devices. There are different types of UWB filters, such as low pass, high pass, band pass, and notch filter. A low-pass filter restricts a higher frequency and permits a low frequency below a certain limit; a high-pass filter restricts a low-frequency signal and allows only high frequency; a band pass filter operates only in the certain region of frequency; and a notch filter rejects a certain frequency. The design of UWB filters can be challenging due to the complexity of the UWB spectrum. UWB filters must be able to operate over a wide frequency range and provide high attenuation of unwanted frequencies. The performance of UWB filters is typically characterized by their insertion loss, which measures the amount of attenuation they provide, and their return loss, which measures how well they match the impedance of the connected components.

UWB filters are an important component of UWB technology in modern communication systems because they help to reject unwanted frequencies which may interfere with the signal and degrade the signal quality. By reducing the amount of noise and interference in a UWB signal, UWB filters can help to improve the overall signal quality, reducing errors and improving the reliability of wireless communication systems. In many countries there are strict regulations on the use of frequencies for wireless communication system, and UWB filters can help to ensure compliance with these regulations by limiting the signal bandwidth to specific frequencies. UWB technology operates over a wide range of frequencies at very high speed and can transmit a large scale of data. UWB filters manage to achieve this wide bandwidth required for UWB communication by providing precise filtering at specific frequencies.

19.7 PERFORMANCE ANALYSIS OF UWB FILTERS

Insertion loss: The amount of energy lost by a signal as it moves over a medium or a cable link is known as insertion loss. In S, the parameter matrix, the magnitude of S_{12} is called the insertion loss, having input at port 1 and output at port 2. It is expressed in dB (decibels). If V_t is the transmitted voltage and V_r is the received voltage, then insertion loss can be defined by equation (19.1) [37]:

$$IL = -20log_{10}\left|\frac{V_t}{V_r}\right| = -20log_{10}|S_{12}| \qquad (19.1)$$

Return loss: Due to the discontinuities in the transmission line, some signal power is always reflected or returned to the source when a signal is transmitted through it. Return loss can be defined as the ratio of reflected power to incident power in decibels (dB) [38]. If P_{out} is the reflected output power and P_{in} is the incident input power, then return loss is given by equation (19.2):

$$\text{Return Loss}(dB) = 10log_{10}\frac{P_{out}}{P_{in}} \qquad (19.2)$$

Fractional bandwidth: The fractional bandwidth of a filter is a measure of how wide the bandwidth of the filter is. If the filter operates at a center frequency f_c between the lower frequency f_1 and upper frequency f_2, then the fractional bandwidth (FBW) [39] is given by equation (19.3):

$$FBW = \frac{f_2 - f_1}{f_c} \qquad (19.3)$$

Where f_c is given by equation (19.4):

$$f_c = \frac{f_1 + f_2}{2} \qquad (19.4)$$

The FBW can be expressed in terms of percentages between 0% and 200%. The higher percentage indicates a wider bandwidth of the filter. Wideband antennas/filters typically have a fractional bandwidth of 20% or more. Filters with FBW of greater than 50% are referred to as UWB filters [39].

Roll-off rate (ROR): The rate of change in gain or steepness of the curve in a filter's stop band is measured by the ROR. It is usually expressed in decibels per octave or decibels per decade. For good filter performance, the ROR must be high. The ROR of a filter will eventually reach 6 dB per octave per pole, or 20 dB per decade per pole, as a general observation [40].

Group delay: All frequency components of a signal are delayed when they pass through a filter. Group delay in a filter is the time delay of the signal through the device under test as a function of frequency. Group delay is measured in seconds. For an ideal filter, the phase will be linear and the group delay will be constant. It is the negative derivative (or slope) of phase response versus frequency [41]. It provides the transmission characteristics of UWB pulses and indicates to what degree it may be distorted or dispersed.

Out-of-band rejection level: The out-of-band rejection of a UWB filter refers to its ability to attenuate or suppress undesired frequencies outside its intended passband. The specific out-of-band rejection performance of a UWB filter can depend on various factors, such as the filter type, order, and design parameters. The typical out-of-band rejection of a UWB filter is in the range of 20 dB or higher.

Transmission zeroes: A transmission zero of a filter refers to a frequency at which the filter's transfer function has a value of zero. At this frequency, the filter has no effect on the input signal, meaning it does not attenuate or amplify that particular frequency. As a result, the signal is transmitted through the filter without any distortion at the transmission zero frequency. The presence or absence of transmission zeros can affect the filter's phase response, group delay, and other performance characteristics.

19.8 DIFFERENT TECHNIQUES TO DEVELOP UWB FILTERS

In general, filters can be designed and constructed using lumped elements like inductors, capacitors, resistors, etc., or distributed elements like microstrip lines and waveguides. It is challenging to construct filters using lumped elements for UWB communication to operate at the microwave region. This is due to the parasitic effect caused by each element (inductor or capacitor) on other elements [42]. Therefore, UWB filters are designed using distributed elements with microstrip line, strip line, and waveguide techniques. In this chapter, the filters constructed using only microstrip structures are discussed.

Microstrip structure: The microstrip structure consists of a ground plane made of copper with a substrate on top of the ground plane. The thickness of the substrate is h with dielectric constant ε_r. Above the substrate the microstrip lines are designed, which are also made of copper having a width w. The thickness of both the ground plane and microstrip line is t. Figure 19.1 shows a microstrip structure consisting of a single microstrip line.

Defected ground structure: A defected ground structure (DGS) is created by introducing an imperfection on the ground plane of a microstrip structure. The defected ground plane structure refers to the etching of cladding material (usually copper) from the ground plane of a microstrip line in a certain form. Due to the defects, the current distribution of the ground plane changes, or it changes the effective inductance and effective capacitance. By introducing DGS, the circuit size becomes compact; it is easy to design and fabricate and also easy to realize its equivalent circuit. DGSs are designed in various shapes such as dumbbell, rectangular, 'U', 'V', 'H', etc. Nowadays, DGSs are widely used in active and passive devices. DGS is used to design filters, microwave amplifiers, CPWs, and antennas [43].

The development of UWB filters started in 2002 after the FCC approved the unlicensed use of the ultra-wide frequency strictly in the range of 3.1 to 10.6 GHz. Different design structures of UWB filters have been proposed over the years by many researchers. In 2003 an UWB filter [12] was reported which was designed in the range of 3.1 to 10.6 GHz and fabricated on a lossy composite substrate. In 2004 a design method was proposed for a tunable UWB filter using a ring resonator structure with a stub which achieved 86.6% of the relative bandwidth [13]. In [14] a bandpass filter was designed by cascading a low-pass filter (LPF) and high-pass filter (HPF) based on a

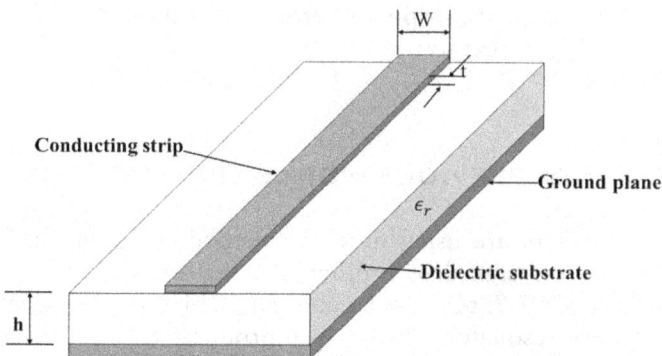

Figure 19.1 Microstrip structure [11].

CPW. To create a bandpass filter by cascading an LPF and HPF [14–17], the output of the LPF is connected to the input of the HPF. The LPF passes low frequencies and attenuates high frequencies, while the HPF passes high frequencies and attenuates low frequencies. The design is simple, but the problem with this type of filter is the circuit size is larger. The size of the filter is also large with this method. Hence to design the UWB filter, some other filters have been proposed such as the multimode resonator (MMR) [18–22]. MMRs have multiple resonant modes, which can be utilized to create a filter response with multiple passbands and stopbands. The stepped impedance resonator (SIR) is one of the examples of MMR, which consists of transmission lines of different widths. MMRs offer a compact size and high selectivity, making them suitable for UWB applications where space and selectivity are important. This [22] filter is constructed with an MMR using novel dumbbell-shaped stubs and one-arm-folded interdigital coupled lines in the input and output sides. The MMR consists of three pairs of shunt dumbbell stubs and a high impedance microstrip line. Another ring-shaped resonator [23–27] is also widely used by researchers to design a UWB filter. It consists of a square or circular-shaped ring made of conducting material. They are compact and can be fabricated using printed circuit board technology.

In terms of UWB communication, many pre-existing narrowband radio signals (like Wi-MAX, wireless area network [WLAN], Wi-Fi6, satellite communication link signal, etc.) can cause interference with the UWB signals. This can degrade the signal quality and reduce the system performance. To remove this interference, notches at certain frequencies are required in a UWB filter. Many research works have been done previously to design a single- or multinotch UWB bandpass filters. Notches are mostly generated by loading horizontal or vertical free stubs or SIRs or other resonators at a proper position in the original structure. In 2006 a paper [28] was presented on a UWB bandpass filter in the frequency range of 2.8 to 10.2 GHz introducing a notched band to avoid the interference between the UWB communication system and the existing radio system. The filter consists of a broadside-coupled microstrip-CPW structure and exhibited an excellent UWB property. In the next section, different filters are analyzed on the basis of performance.

19.9 REVIEW OF PREVIOUS WORKS IN UWB FILTERS

Different researchers are using different methods to design UWB filters. Some UWB filters designed with different structures are discussed in this section. A basic UWB filter was designed in 2014 using an open stub and short stub loaded resonator (SLR), which produces two transmission zeros at 3.01 and 10.68 GHz. A slot of a half-wavelength spiral shape is used to implement the frequency notch at 5.1 GHz, and an inward folded resonator is placed near the open SLR to provide notch rejection at 8.0 GHz [29]. In 2016, a new stub loaded MMR was used to produce a UWB bandpass filter (BPF) with a notched band filter at 8.0 GHz. The MMR's parameters may be changed to control the five resonant modes, which consist of two odd modes and three even modes [30]. In 2018, a modified elliptical-ring and multimode stub-loaded resonator UWB BPF in the frequency band 3.77 to 10.42 GHz was realized, and by adding two bends in the middle resonator, a notch band centered at 6.86 GHz was created [31]. In 2019, a novel resonator having T-shaped microstrip patches in the top and bottom layers and a circular coupling slot in the middle layer was used to design a wideband UWB filter a with notch band at 5.8 GHz with insertion and return loss better than –2.0 dB and –15.0 dB [32]. Yinchuan Xiao et al. proposed a UWB filter in 2022 with a wide upper stopband from 10.6 to 30.0 GHz [33]. The structure of the filter consisted of two rectangular stub resonators separated by a quarter-wavelength parallel coupled line. A notch is generated at 8 GHz to reject interference from shortwave satellite communications. Also, an insertion loss of less than 0.3 dB and return loss of more than 17 dB was achieved in the passband. In 2023 Basit et al. designed a compact UWB BPF with two stopband notches generated at frequencies of 3.5 GHz and 7.5 GHz [34]. The structure of the filter consists of microstrip parallel coupled lines with a rectangular stub resonator and spur line structure. The filter attains an insertion loss of –1.1 dB and return loss lower than –18 dB throughout the passband with a fractional bandwidth of about 119.4%.

19.10 UWB FILTER CASCADING A LOW-PASS AND HIGH-PASS FILTER

In [15] a UWB filter is constructed by cascading the LPF and HPF into each other. A stepped impedance structure is used by varying the width of the microstrip line to achieve the LPF structure and HPF to get a lower stopband, which is achieved by short-circuited stubs connected at a high-impedance section. Here two filters are designed in a substrate of thickness $h = 0.508$ mm and dielectric constant 2.2. The geometry of the first filter consists of three low-impedance patches and four short-circuited stubs, as shown in Figure 19.2(a). A passband of 6 to 10 GHz is achieved covering the upper range of the UWB region.

The second filter is designed on the same substrate consisting of four low-impedance patches and two short-circuited stubs, as shown in Figure 19.2(b). The passband is designed for the range of 3 to 10 GHz. The return loss was found to be greater than 15 dB from 5 GHz to 9.5 GHz with good upper stopband rejection.

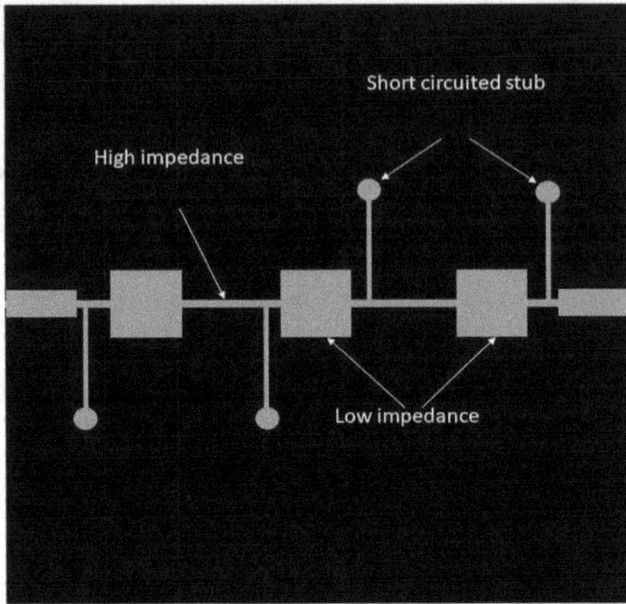

Figure 19.2(a) UWB filter with four short-circuited stubs **[15]**.

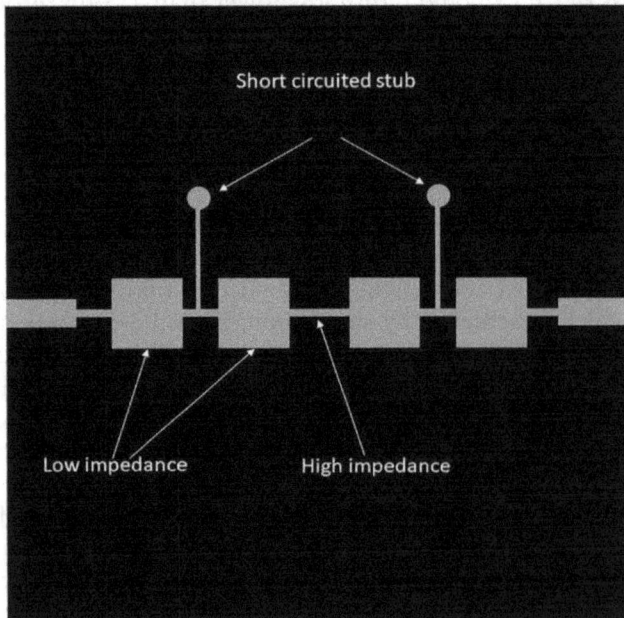

Figure 19.2(b) UWB filter with two short-circuited stubs [15].

19.11 MMR FOR DESIGNING A UWB FILTER WITH MULTIPLE NOTCHES

By cascading the LPR and HPR, the size of the filter increases. Hence, to minimize the size MMRs are used. Here [21], an MMR is designed with a high-and low-impedance structure and two vertical open stubs are used on both sides of the MMR. The MMR comprises an SIR and a uniform impedance resonator (UIR). The MMR is excited through two parallel coupled lines, and on the ground plane two rectangular etched apertures are used underneath the interdigital coupled lines to improve the coupling. The filter is designed in a 0.8-mm-thick Arlon substrate with a dielectric constant of 2.5. This filter is operated in the frequency range of 2.86 GHz to 11.2 GHz with an insertion loss variation of 0.25 to 0.6 dB in the passband. Three notches are generated to restrict the interference caused by Wi-Fi6, extended C-band, and X-band satellite communication for TV networks. Two methods are used to generate notches: a comb-shaped resonator is place below the UIR of the MMR, and one arm is connected to the end of two interdigital coupled lines. The design of the final structure is shown in Figure 19.3. This results in notches generated at 6 GHz, 6.53 GHz, and 8.35 GHz with attenuation of –16.16 dB, –26.47 dB, and –15.22 dB, respectively. This UWB filter also achieved a wide upper stopband of 5.37 GHz in the range of 10.97 GHz to 16.34 GHz.

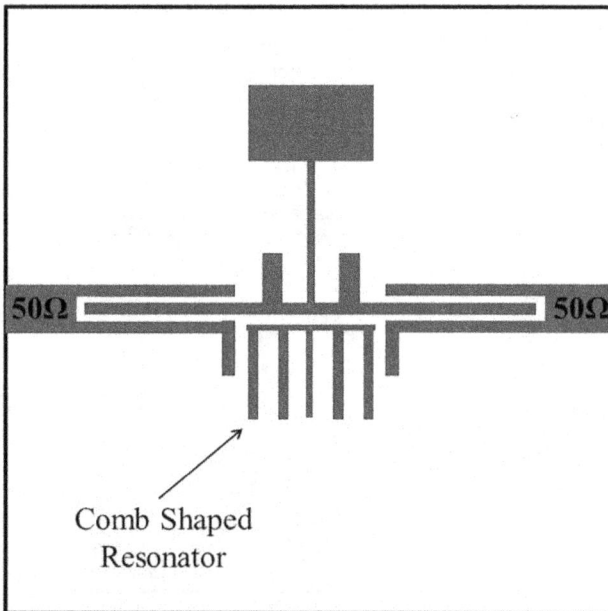

Figure 19.3 MMR with two open stubs and comb-shaped resonator [21].

19.12 RING RESONATOR WITH QUAD T STUB-LOADED STRUCTURE

In this paper [23], a planar UWB bandpass filter is designed. A modified ring resonator is implemented by introducing four T-shaped stubs to the inner surface of the ring, as shown in Figure 19.4. The ring is connected by parallel coupled lines to the feed line. The proposed filter is fabricated on a Taconic RF-35 substrate with a relative dielectric constant of 3.5, and the thickness is 0.508 mm. The passband of the filter spreads from 3.6 to 10.9 GHz. The simulated and fabricated results are in good accordance with one another. The insertion loss is less than 1.3 dB, and the return loss is more than 12 dB over the passband. Also, a very flat group delay throughout the passband is observed, which is less than 0.34 ns within the passband.

19.13 SQUARE RING RESONATOR WITH EXTENDED STOP BAND

Different researchers are trying to implement square rings to design UWB filters. In [24], researchers designed a square ring of very small width, and two parallel coupled lines were placed at two sides of the square ring connected to the 50 Ω transmission line. A pair of quarter-wavelength short-circuited stubs corresponding to the center frequency of UWB at 6.85 GHz were connected to the upper and lower sides of the ring. This improves the rejection level of the filter. To further improve the out-of-band rejection level, four open-circuited stubs were introduced at the four corners of the square ring, as shown in Figure 19.5. This resulted in extension of the upper

Figure 19.4 Circular ring resonator filter with four T-shaped stubs [23].

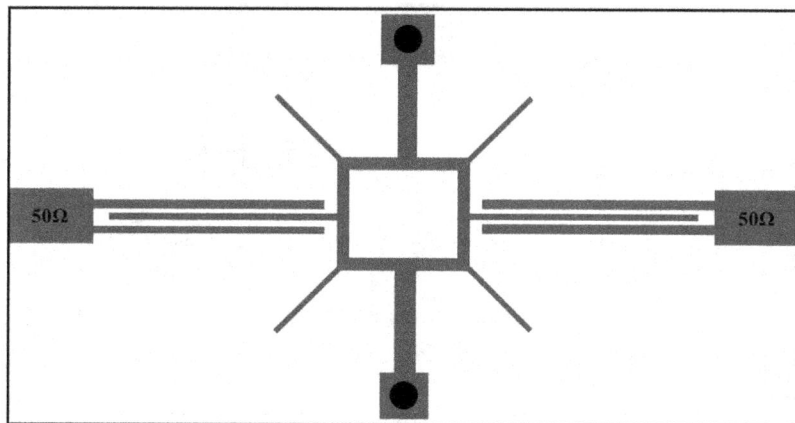

Figure 19.5 Square ring resonator filter with four open stubs [24].

stopband up to 20 GHz with an insertion loss of more than 18 dB. The –3 dB passband is observed in the frequency range of 3.1 to 10.6 GHz with a fractional bandwidth of 118%. The variation of group delay of the filter is less, between 0.15 and 0.65 ns.

19.14 UWB FILTER IMPLEMENTING A DEFECTED GROUND STRUCTURE

To design a bandpass filter, a DGS is also used by many researchers. The DGS is an etching of a small portion of the ground plane in various shapes. In this paper [35], a square loop–shaped DGS is used to design a compact and high-performance filter. The filter is composed of two metal faces with two microstrip lines, and to enhance the out-of-band rejection, a short-circuited stub is used. Also, a notch is generated in the passband to reject unwanted frequency by using open stubs in the microstrip. The DGS is implemented in between the two ports, which is in the shape of a square ring (Figure 19.6). The 3 dB passband is found to be in the frequency range of the UWB signal from 3.1 GHz to 10.6 GHz, with a return loss of higher than 12 dB and insertion loss of less than 1.2 dB. Also, the maximum experimental group delay is found to be 0.2 ns. Due to the open stubs, a stop band is created at 5.8 GHz. Here a new approach to design a UWB filter using DGS is used, and the simulated results are in accordance with the fabricated results. DGS-based UWB filters are always easier to construct and are smaller in size than standard UWB filters that use MMRs.

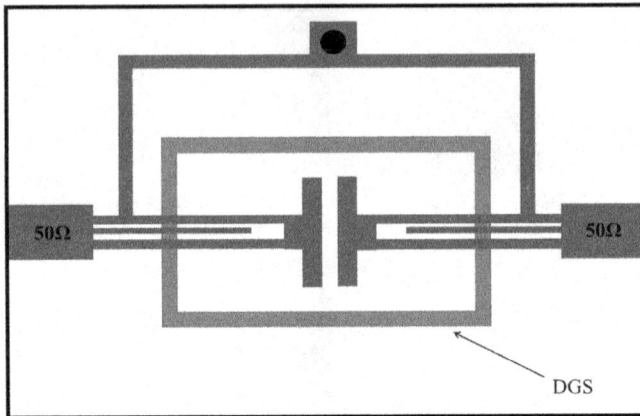

Figure 19.6 UWB filter with square ring DGS [35].

19.15 A TRIPLE-NOTCHED BAND UWB FILTER

A triple-notched band compact UWB filter is designed [36] with good selectivity. To generate a triple-notch bandpass filter, microstrip parallel coupled lines with two open-circuited stubs are used. A wide passband with an increased coupling degree has been found to appear. By implementing an open-circuited stub to the parallel coupled line structure, a notch is generated. A low-impedance ring is introduced at the center, and a symmetrical parallel coupled line is placed on both sides. The passband of the filter can be controlled by changing the radius of the ring. To generate a triple notch in the passband, asymmetric dual line coupling is used for the feed line. In Figure 19.7 the final structure of the proposed filter is shown. The fabrication of the proposed filter was done using an RT/Duroid 5880 substrate having a

Figure 19.7 A circular ring-shaped UWB filter [36].

relative dielectric constant of 2.2 with a thickness of 0.78 mm. The effective size of the UWB filter is found to be 21.6 mm × 6 mm. From the fabricated filter, the passband is found to be in the range from 4.3 to 10.2 GHz, which has very low insertion loss below 1.0 dB. Three notches are generated at 5.9 GHz, 8.0 GHz, and 9.0 GHz with an insertion loss of more than 15 dB. Also, a flat group delay with a maximum of 0.3 ns variation is achieved in the passband except in the three notches.

19.16 CONCLUSION AND FUTURE SCOPE

In conclusion, UWB technology is a type of wireless communication technology that operates in a large frequency range of 3.1 GHz to 10.6 GHZ. It is widely used due to its advantages such as high data rates, ability to track and detect objects, and low power consumption. The UWB filter is a major component of UWB technology and can help improve the performance of UWB systems by providing efficient signal transmission and reception. These filters are designed to pass desired frequencies or suppress unwanted frequencies and improve signal quality by reducing interference and noise. They come in various designs such as microstrip, CPW, and substrate-integrated waveguide. In this chapter various microstrip line filters were discussed. The choice of filter design depends on various factors such as size, power consumption, frequency range, and insertion loss. UWB filters have several applications in wireless communication systems, radar systems, and medical devices.

UWB technology is mostly being used in smaller devices; hence a more compact UWB filter is needed. Researchers are trying to reduce the size of the filter using various techniques so that it can be integrated into small devices. In order to achieve more efficient and cost-effective communication system, UWB filters need to be better integrated with other components such as antennas and amplifiers. Researchers are exploring ways to develop UWB filters that can be integrated directly onto the same chip as other wireless components, resulting in a more compact and efficient system. Bhattacharya et al. [44] developed a dual-notched band antenna integrated with a UWB filter called a filtenna operating in the frequency range of 3.1 to 10.6 GHz. Using such filtenna, a more compact and low-cost design can be developed. To improve the performance of the filter, researchers are using different new materials such as metamaterials to design filters. It can be another area of research in the field of UWB filters. The future of UWB technology is very bright. However, the design and implementation of UWB filters remain a challenging task due to the wide frequency range and high demands on performance. Researchers are continually developing new techniques and technologies to overcome these challenges and improve the performance of UWB filters.

REFERENCES

[1] "Revision of Part 15 of the Commission's Rules Regarding Ultra Wideband Transmission Systems," *Federal Communications Commission*, Dec. 27, 2015. www.fcc.gov/document/revision-part-15-commissions-rules-regarding-ultra-wideband-7 (accessed Feb. 15, 2023).

[2] T.-N. Kuo, S.-C. Lin, and C. H. Chen, "Compact Ultra-Wideband Bandpass Filters using Composite Microstrip—Coplanar-Waveguide Structure," *IEEE Trans. Microw. Theory Tech.*, vol. 54, no. 10, pp. 3772–3778, Oct. 2006, doi: 10.1109/TMTT.2006.881624.

[3] Q. Li and T. Yang, "Compact UWB Half-Mode SIW Bandpass Filter with Fully Reconfigurable Single and Dual Notched Bands," *IEEE Trans. Microw. Theory Tech.*, vol. 69, no. 1, pp. 65–74, Jan. 2021, doi: 10.1109/TMTT.2020.3033830.

[4] C. Isak, "What Is Ultra-Wideband and How Does It Work?," *TechAcute*, Jan. 9, 2022. https://techacute.com/what-is-ultra-wideband-how-does-uwb-work/ (accessed Mar. 9, 2023).

[5] H. Nikookar and R. Prasad, "Introduction," in *Introduction to Ultra Wideband for Wireless Communications*, Dordrecht: Springer, 2009, pp. 1–10, doi: 10.1007/978-1-4020-6633-7_1.

[6] D.-S. Kim and H. Tran-Dang, "Ultra-Wideband Technology for Military Applications," in *Industrial Sensors and Controls in Communication Networks: From Wired Technologies to Cloud Computing and the Internet of Things*, D.-S. Kim and H. Tran-Dang, Eds, Cham: Springer International Publishing, 2019, pp. 197–204, doi: 10.1007/978-3-030-04927-0_15.

[7] I. Oppermann, Ed., *UWB Theory and Applications*, Reprinted, Chichester: Wiley, 2006.

[8] K. Siwiak and D. McKeown, *Ultra-Wideband Radio Technology*, Chichester; Hoboken, NJ: John Wiley & Sons, 2004.

[9] A. Bhattacharya, B. Roy, S. K. Chowdhury, and A. K. Bhattacharjee, "Computational and Experimental Analysis of a Low-Profile, Isolation-Enhanced, Band-Notch UWB-MIMO Antenna," *J. Comput. Electron.*, vol. 18, no. 2, pp. 680–688, Jun. 2019, doi: 10.1007/s10825-019-01309-3.

[10] A. Bhattacharya, B. Roy, R. F. S. Caldeirinha, and A. K. Bhattacharjee, "Low-profile, Extremely Wideband, Dual-Band-Notched MIMO Antenna for UWB Applications," *Int. J. Microw. Wirel. Technol.*, vol. 11, no. 7, pp. 719–728, Sep. 2019, doi: 10.1017/S1759078719000266.

[11] T. Hoi and B. Duong, "Designing Wideband Microstrip Bandpass Filter for Satellite Receiver Systems," *Proc. Natl. Conf. Electron. Commun.*, pp. 140–143, Jan. 2013.

[12] A. Saito, H. Harada, and A. Nishikata, "Development of Band Pass Filter for Ultra Wideband (UWB) Communication Systems," in *IEEE Conference on Ultra Wideband Systems and Technologies*, Nov. 2003, pp. 76–80, doi: 10.1109/UWBST.2003.1267806.

[13] H. Ishida and K. Araki, "A Design of Tunable UWB Filters," in *2004 International Workshop on Ultra Wideband Systems Joint with Conference on Ultra Wideband Systems and Technologies. Joint UWBST & IWUWBS 2004 (IEEE Cat. No. 04EX812)*, May 2004, pp. 424–428, doi: 10.1109/UWBST.2004.1321009.

[14] Y.-S. Lin, W.-C. Ku, C.-H. Wang, and C. H. Chen, "Wideband Coplanar-Waveguide Bandpass Filters with Good Stopband Rejection," *IEEE Microw. Wirel. Compon. Lett.*, vol. 14, no. 9, pp. 422–424, Sep. 2004, doi: 10.1109/LMWC.2004.832069.

[15] Ching-Luh Hsu, Fu-Chieh Hsu, and J.-K. Kuo, "Microstrip Bandpass Filters for Ultra-Wideband (UWB) Wireless Communications," in *IEEE MTT-S International Microwave Symposium Digest*, Long Beach, CA, 2005, pp. 679–682, doi: 10.1109/MWSYM.2005.1516698.

[16] G.-M. Yang, G. Xiao, R. Jin, J. Geng, W. He, and M. Ding, "Design of Ultra-Wide Band (UWB) Bandpass Filter Based on Defected Ground Structure," *Microw. Opt. Technol. Lett.*, vol. 49, no. 6, pp. 1374–1377, 2007, doi: 10.1002/mop.22471.

[17] K. C. Lee, H. T. Su, and W. S. H. Wong, "Realization of a Wideband Bandpass Filter Using Cascaded Lowpass to Highpass Filter," in *2008 International Conference on Microwave and Millimeter Wave Technology*, Apr. 2008, vol. 1, pp. 14–17, doi: 10.1109/ICMMT.2008.4540288.

[18] Q.-X. Chu, X.-H. Wu, and X.-K. Tian, "Novel UWB Bandpass Filter using Stub-Loaded Multiple-Mode Resonator," *IEEE Microw. Wirel. Compon. Lett.*, vol. 21, no. 8, pp. 403–405, Aug. 2011, doi: 10.1109/LMWC.2011.2160526.

[19] Yi-Chyun Chiou, Jen-Tsai Kuo, and Eisenhower Cheng, "Broadband Quasi-Chebyshev Bandpass Filters with Multimode Stepped-Impedance Resonators (SIRs)," *IEEE Trans. Microw. Theory Tech.*, vol. 54, no. 8, pp. 3352–3358, Aug. 2006, doi: 10.1109/TMTT.2006.879131.

[20] R. Li and L. Zhu, "Compact UWB Bandpass Filter using Stub-Loaded Multiple-Mode Resonator," *IEEE Microw. Wirel. Compon. Lett.*, vol. 17, no. 1, pp. 40–42, Jan. 2007, doi: 10.1109/LMWC.2006.887251.

[21] P. Chakraborty, P. P. Shome, J. R. Panda, and A. Deb, "Highly Selective UWB Bandpass Filter with Multi-Notch Characteristics using Comb Shaped Resonator," *Prog. Electromagn. Res. M*, vol. 108, pp. 89–101, 2022, doi: 10.2528/PIERM21112601.

[22] L. Chen, F. Wei, X.-W. Shi, and C.-J. Gao, "An Ultra-Wideband Bandpass Filter with a Notch-Band and Wide Stopband using Dumbbell Stubs," *Prog. Electromagn. Res. Lett.*, vol. 17, pp. 47–53, 2010, doi: 10.2528/PIERL10070103.

[23] K.-D. Xu, Y.-H. Zhang, J. L.-W. Li, W. T. Joines, and Q. H. Liu, "Compact Ultra-Wideband Bandpass Filter using Quad-T-Stub-Loaded Ring Structure," *Microw. Opt. Technol. Lett.*, vol. 56, no. 9, pp. 1988–1991, Sep. 2014, doi: 10.1002/mop.28508.

[24] S. W. Wong, L. Zhu, L. C. Quek, and Z. N. Chen, "A Stopband-Enhanced UWB Bandpass Filter using Short-/Open-Stubs Embedded Ring Resonator," in *2009 Asia Pacific Microwave Conference*, Singapore, Dec. 2009, pp. 913–916, doi: 10.1109/APMC.2009.5384314.

[25] Sheng Sun and Lei Zhu, "Wideband Microstrip Ring Resonator Bandpass Filters under Multiple Resonances," *IEEE Trans. Microw. Theory Tech.*, vol. 55, no. 10, pp. 2176–2182, Oct. 2007, doi: 10.1109/TMTT.2007.906510.

[26] L. Lin, S. Yang, S. Sun, B. Wu, and C. Liang, "Ultra-Wideband Bandpass Filter using Multi-Stub-Loaded Ring Resonator," *Electron. Lett.*, vol. 50, no. 17, pp. 1218–1220, Aug. 2014, doi: 10.1049/el.2014.1256.

[27] Chan Ho Kim and Kai Chang, "Ultra-Wideband (UWB) Ring Resonator Bandpass Filter With a Notched Band," *IEEE Microw. Wirel. Compon. Lett.*, vol. 21, no. 4, pp. 206–208, Apr. 2011, doi: 10.1109/LMWC.2011.2109942.

[28] K. Li, D. Kurita, and T. Matsui, "UWB Bandpass Filters with Multi Notched Bands," in *2006 European Microwave Conference*, Sep. 2006, pp. 591–594, doi: 10.1109/EUMC.2006.281461.

[29] P. Sarkar, R. Ghatak, M. Pal, and D. R. Poddar, "High-Selective Compact UWB Bandpass Filter with Dual Notch Bands," *IEEE Microw. Wirel. Compon. Lett.*, vol. 24, no. 7, pp. 448–450, Jul. 2014, doi: 10.1109/LMWC.2014.2316214.

[30] T. Yan, D. Lu, X.-H. Tang, and J. Xiang, "High-Selectivity UWB Bandpass Filter with a Notched Band using stub-Loaded Multi-Mode Resonator," *AEU—Int. J. Electron. Commun.*, vol. 70, no. 12, pp. 1617–1621, Dec. 2016, doi: 10.1016/j.aeue.2016.09.016.

[31] M. Gandamalla, D. Marathe, and K. Kulat, "Design and Analysis of Compact Single and Dual Notch Ultra Wideband Bandpass Filter," *Prog. Electromagn. Res. M*, vol. 75, pp. 91–102, 2018, doi: 10.2528/PIERM18080705.

[32] X.-C. Ji, W.-S. Ji, L.-Y. Feng, Y.-Y. Tong, and Z.-Y. Zhang, "Design of a Novel Multi-Layer Wideband Bandpass Filter with a Notched Band," *Prog. Electromagn. Res. Lett.*, vol. 82, pp. 9–16, 2019, doi: 10.2528/PIERL18121101.

[33] Y. Xiao, G. Li, B. Chen, Q. Jiang, and Y. Song, "Band-Notched Ultra-Wideband Bandpass Filter with Broad Stopband," in *2022 Asia-Pacific International Symposium on Electromagnetic Compatibility (APEMC)*, Sep. 2022, pp. 548–550, doi: 10.1109/APEMC53576.2022.9888653.

[34] A. Basit, A. Daraz, M. I. Khan, N. Saqib, and G. Zhang, "Design, Modeling, and Implementation of Dual Notched UWB Bandpass Filter Employing Rectangular Stubs and Embedded L-Shaped Structure," *Fractal Fract.*, vol. 7, no. 2, Art. no. 2, Feb. 2023, doi: 10.3390/fractalfract7020112.

[35] J. Liu, W. Ding, J. Chen, and A. Zhang, "New Ultra-Wideband Filter with Sharp Notched Band using Defected Ground Structure," *Prog. Electromagn. Res. Lett.*, vol. 83, pp. 99–105, 2019, doi: 10.2528/PIERL18111302.

[36] X.-M. Shi, X.-L. Xi, Y.-C. Zhao, and H.-L. Yang, "A novel compact ultra-wideband (UWB) bandpass filter with triple-notched bands," *J. Electromagn. Waves Appl.*, vol. 29, no. 9, pp. 1174–1180, Jun. 2015, doi: 10.1080/09205071.2015.1034811.

[37] "Insertion Loss—An Overview | ScienceDirect Topics." www.sciencedirect.com/topics/computer-science/insertion-loss (accessed Feb. 15, 2023).

[38] "What Is Return Loss?—Everything RF." www.everythingrf.com/community/what-is-return-loss (accessed Feb. 15, 2023).

[39] "Antenna-Theory.com—Fractional Bandwidth." www.antenna-theory.com/definitions/fractionalBW.php (accessed Feb. 15, 2023).

[40] "11.4: Filter Order and Poles," *Engineering LibreTexts*, May 2, 2018. https://eng.libretexts.org/Bookshelves/Electrical_Engineering/Electronics/Operational_Amplifiers_and_Linear_Integrated_Circuits_-_Theory_and_Application_(Fiore)/11%3A_Active_Filters/11.04%3A_Section_4- (accessed Feb. 15, 2023).

[41] "What Is Group Delay?—Everything RF." www.everythingrf.com/community/what-is-group-delay (accessed Feb. 15, 2023).

[42] "Microwave Filter," *The Free Dictionary.com*. https://encyclopedia2.thefreedictionary.com/microwave+filter (accessed Feb. 15, 2023).

[43] M. K. Khandelwal, B. K. Kanaujia, and S. Kumar, "Defected Ground Structure: Fundamentals, Analysis, and Applications in Modern Wireless Trends," *Int. J. Antennas Propag.*, vol. 2017, pp. 1–22, 2017, doi: 10.1155/2017/2018527.

[44] A. Bhattacharya, A. De, B. Roy, and A. K. Bhattacharjee, "Investigations on a Low-Profile, Filter Backed, Printed Monopole Antenna for UWB Communication," *Indian J. Pure Appl. Phys. IJPAP*, vol. 58, no. 2, Art. no. 2, Feb. 2020, doi: 10.56042/ijpap.v58i2.24985.

Chapter 20

Design of Broadband Planar Couplers Using an Existing Filter Design Approach

Pratik Mondal, Anumoy Ghosh,
and Tapan Mandal

20.1 INTRODUCTION

Coupling occurs among symmetrical invariable coupled lines, while the odd and even mode (O&E mode) phase velocities of any two coupled-lines are not equal [1, 2]. Such design aspects of asymmetric and symmetric wideband couplers are well accepted. The broadband quadrature hybrids are extensively used in microwave baluns, signal sampling circuits, and phase shifters. The quantity of coupling is primarily determined by the dissimilarity between the phase constants (β) of two asymmetric coupled lines [3]. The band of any two non-symmetrical (forward-wave) coupled lines is always higher than any symmetrical coupled section. In a transverse electromagnetic (TEM) system, any such coupled lines attain the highest level of coupling while they are $\lambda/4$ long. A broad bandwidth is also attained by simply cascading a multiple number of coupled sections [4]. A non-uniform line coupler achieved a higher performance over a multisection coupler because of its flat tapering of spacing. A computational method for implementing any tapered couplers has been reported in [5]. The coupler is implemented by the distribution of a coupling factor through its length. The flat tapering may be estimated by distributing it into multiple hundreds of sections, where each section is approximated to have constant coupling. Dual-layer couplers presented in 2007 [6] were designed by a new rectangular multiaperture structure with a small width. The reported coupler operates for a much wider operating bandwidth (14 to 29 GHz) and at higher frequency compared with the others working in that time. A broadband left-handed coupled-line coupler (backward wave) with flexible coupling is elaborated in [7]. The reported coupler provides a bandwidth of 35% despite the relatively wide line gap of coupled lines around 0.3 mm. The bandwidth was further broadened by wiggling technique as reported in [8]. The wiggled coupled section of the inner edges balances the O&E mode velocities, and the exterior boundary increases even-mode inductance in such a way that the central section can satisfy a high coupling requirement. The reported coupler provides 2 to 18 GHz of the operating band, 15 dB of minimum isolation and 90 ± 100 output

DOI: 10.1201/9781003459880-20

phase differences. Compared with existing unwiggled and non-uniform coupler, the improvement of the proposed coupler in terms of bandwidth was depicted as 6 GHz. Y. Wu et al. in 2005 proposed a new method of stepped-impedance configuration of a coupler with enhanced isolation [9]. Interdigital shaped capacitance was incorporated to equalize O&E phase velocities, which in turns enhances isolation between outputs with a 30% reduction in size compared to the conventional parallel one. In the succeeding year, a broadband monolithic microwave integrated circuits (MMIC) differential coupler was introduced [10]. It depicts a frequency range from 15 to 45 GHz with S_{11} <–20 dB and isolation >15 dB. Further modifications were done to reduce the size by incorporating a slow-wave structure. For that reason, an overall size reduction of around 38% was achieved. Slawomir Gruszczynski et al. presented a broadband coupler with strip line technology [11]. To attain wideband frequency, a three asymmetrical coupled-line section is used. A novel method of broadband tight-coupled coupler design on a printed circuit board (PCB) was also presented [12]. Implementation of coupled-line sections with a floating-plate overlay has been proposed; 6-dB wideband coupler properties were also realized on the same process. In [13], a periodical H-shaped codirectional coupler was reported. Broadband coupling is achieved by accumulating the phase difference as a constant. The presented coupler achieves almost 75% bandwidth with a minimum size. A broadband co-coupler installed with a Y-shaped periodic grounded via was reported in 2013 [14] by J.C. Yen et al. The reported coupler generated an equal phase variation among the E&O modes over a broad frequency regime; thus broad bandwidth was achieved. The design also illustrated the procedure to generate a coupler with an arbitrary coupling level. The proposed coupler provides an operating band (79.2%) with a high coupling level. It is worth mentioning that the reported coupler is one of the smallest of its kind with such a wide bandwidth. Kamil Staszek in that same year [15] investigated a class of wideband symmetrical directional couplers and implemented a coupler consisting of three symmetric coupled sections. In 2015 [16], directional couplers having light coupling and giant directivity were introduced by M. Tran et al. These types of structures with good directivity along with a very broad bandwidth are useful for weak coupling applications. The first coupler having interdigital capacitors obtained a maximum insertion loss of 0.49 dB with an operating band of 44% and directivity of 25 dB in comparison to the second structure having floating stubs in between the coupled sections of the operational bandwidth up to 67% and having the same directivity. The advantage of the second coupler over the first one is less susceptibility to fabrication tolerances.

Recently, a broadband phase-shifter using a multi-mode resonator-transmission line MMR-TL section has been implemented, and its advantages have also been reported [17], which include the broadband property of realizing that a bandpass filter (BPF) can be useful for other types of planar circuitry. In this work, a lightly coupled broadband multiport coupler has been realized by multimode–step impedance resonator (MM-SIR). The ratio of impedance in the MM-SIR monitors the bandwidth of the coupler. Also, the coupling coefficient (k) mostly varies on the spacing present between the coupled sections. Such a type of coupler may be tunable in both the aspects of coupling factors and in bandwidth. A four-port coupler with 10 dB loose coupling with fewer phase and amplitude variations over the entire band and also having adequate isolation among consecutive ports has been designed and demonstrated. It has two coupled line sections, including one 50 ohm TL section and one MMR line section. The design was further improved by placing the same MMR sections on the opposite side of the reference 50 ohm transmission line in order to design a six-port coupler. The reference line works as mirror symmetry, as both the MMR sections are placed at an equal distance. The design is simple, without any bond wire, has a single layer, and is compact in size. This type of circuit might be useful for feeding modern UWB antenna array systems, as mentioned in [18–21].

20.2 OPERATION PRINCIPLE OF FORWARD-WAVE DIRECTIONAL COUPLER

Forward-wave coupling has an objective reality between uniform symmetrical coupled lines introducing a conditional clause that the O&E-mode phase velocities of any coupled transmission lines are unequal. In addition, the backward-wave coupling among the coupled lines can be lowered to a very small value by maintaining a relatively large distance between lines in such a way that the Z_{0e} and Z_{0o} of the coupled lines tends to be equal. These kinds of circuits are commonly known as 'forward-wave (FW) directional couplers (DCs)'.

A directional coupler has two distinct transmission line (TL), the main (primary) arm and the auxiliary (secondary) arm, electromagnetically coupled to each other as portrayed in Figure 20.1. The power entering to the main arm (Port 1; input port) gets divided between Port 2 (direct port) and Port 3 (coupled port), and almost no power comes out in port 4 (isolated port). The power entering port 2 (then input port) is divided between port 1 (then output port) and port 4 (then coupled port) and not in port 3 (then isolated port). These types of couplers are called FW or co-directional couplers.

The coupler is made of symmetrical coupled lines:

$$S_{11} = S_{22} = S_{33} = S_{44} = 0 \tag{20.1}$$

$$S_{12} = S_{21} = S_{34} = S_{43} = \frac{S_{21e} + S_{21o}}{2} = \frac{e^{-j\beta_e l} + e^{-j\beta_o l}}{2} = -je^{-j\frac{(\beta_e + \beta_o)l}{2}} \cos\left[\frac{(\beta_e - \beta_o)l}{2}\right] \tag{20.2}$$

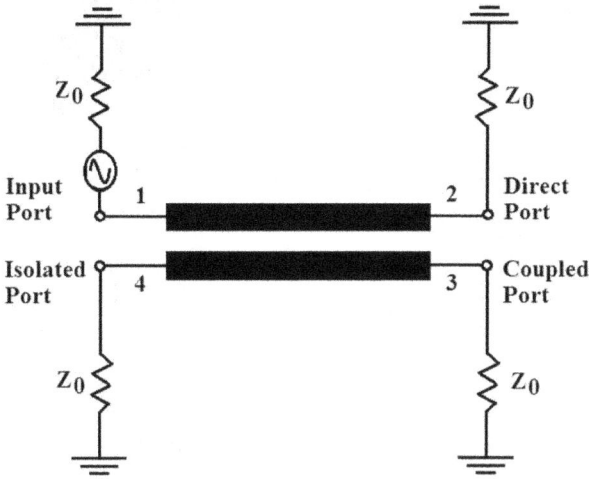

Figure 20.1 Forward-wave coupler using symmetrical coupled lines.

$$S_{14} = S_{41} = S_{23} = S_{32} = \frac{S_{21e} - S_{21o}}{2} = \frac{e^{-j\beta_e l} - e^{-j\beta_o l}}{2} = -je^{-j\frac{(\beta_e + \beta_o)l}{2}} \mathrm{Sin}\left[\frac{(\beta_e - \beta_o)l}{2}\right] \quad (20.3)$$

Here, 'β_e' and 'β_0' denote phase constants of E&O-mode signals, respectively, and 'l' is the length of the coupled section. Thus, fractional power going to port 4 from port 1 is given by:

$$\frac{P_4}{P_1} = |S_{41}^2| = \mathrm{Sin}^2\left[\frac{(\beta_e - \beta_o)l}{2}\right] \quad (20.4)$$

whereas the fractional power coupled to port 2 from port 1 is given by:

$$\frac{P_2}{P_1} = |S_{21}^2| = \mathrm{Cos}^2\left[\frac{(\beta_e - \beta_o)l}{2}\right] \quad (20.5)$$

Hence, for all incident power, from equation (20.4) and (20.5), $|S_{21}|^2 + |S_{41}|^2 = 1$. It is understood that co-couplers can never be obtained using pure TEM lines like coaxial lines. This is because for the pure TEM condition, the O&E-mode propagation constants are identical, and henceforth there is no coupling between port 1 and 4 or between port 2 and 3. FW coupling exists only in non-TEM transmission lines such as dielectric waveguides, metallic waveguides and fine lines and may also subsist in quasi-TEM transmission lines like microstrip and coplanar waveguide (CPW) lines.

20.3 BASICS OF ASYMMETRICAL DIRECTIONAL COUPLERS

Two coupled lines may generate two basic modes, which are termed c and π modes, for asymmetrical coupled lines. In true sense, a structure composed of two coupled lines may support four independent modes: two traveling in the forward direction and two traveling in the backward direction. The c-mode is an even-like mode, while the π-mode is an odd-like mode. The length of the coupler is denoted by:

$$l_g = \frac{\pi}{\beta_c - \beta_\pi} \tag{20.6}$$

where β_c and β_π denote the phase constants of the c and π modes, respectively. Moreover, the R_c and R_π ratio must be chosen as:

$$\frac{R_c}{R_\pi} = \text{Coupling} \pm \sqrt{8} \tag{20.7}$$

The coupler is conventionally implemented by determining the spaces between the two lines in such a way that the ratio of R_c and R_π satisfies the condition. For a predefined width of transmission lines and spaces between them, the inductance and capacitance parameters may be obtained using the coupled microstrip data and a well-established technique [1].

It is worth mentioning that the phase variation among outputs of an asymmetrical directional coupler is not 90 degrees, like symmetrical DC. For a specific frequency, when port 1 is the driven port, the phase deviation between ports 2 and port 4 is of 0 degrees, and the phase difference between outputs is 180 degrees when port 3 is the driven port. Thus, an asymmetrical coupler with end-to-end symmetry satisfies the following relationship:

$$(\angle S41 - \angle S21) + (\angle S23 - \angle S43) = 180 \text{ degrees} \tag{20.8}$$

It is also possible to achieve the desired phase difference between outputs over a wide bandwidth in an asymmetrical DC by using an extra line length as a phase-compensating element.

20.4 DESIGN OF A FORWARD-WAVE DIRECTIONAL COUPLER USING MMR

An FW multimode resonator (MMR)–based four-port coupler is realized using asymmetric configurations. The structure is made of coupled-line sections with one part having an MMR line and the other part is a TL section of 50 ohm impedance. Figure 20.2(a) shows the layout as well as the circuit model of the design.

20.4.1 Design Equations of Asymmetric Coupled Sections

The MMR is basically non-uniform in character, and the mentioned coupler has three asymmetric sections. Two of them are higher-impedance coupled-line parts of the MMR with a 50 Ω TL section (asymmetric coupled section-1) and the remaining one is a low-impedance section of MMR with a 50 ohm microstrip line section (asymmetric coupled section-2). Figure 20.2(a) also reveals the lumped circuit representation of coupled lines of MMR with a 50 ohm TL section of the described coupler for quasi-TEM mode.

The 'c' mode is further analyzed by the parameters: γ_c (the propagation constant of the mode), Z_{c1} (characteristic impedances of MMR line), Z_{c2} (characteristic impedances of 50 ohm transmission line) and R_c (the ratio of voltages on two lines of the c mode, respectively; a similar condition is applied for the π mode).

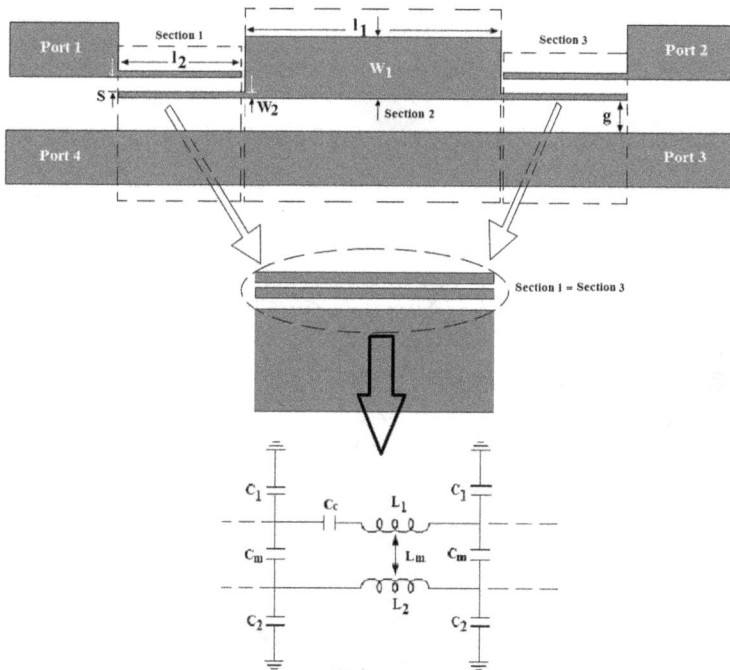

Figure 20.2 (a) Schematic of proposed four-port directional coupler and its lumped circuit model of coupled section-1; (b) simulated S-parameter response of direct and coupled port; (c) isolation and reflection co-efficient of asymmetric coupled section-1.

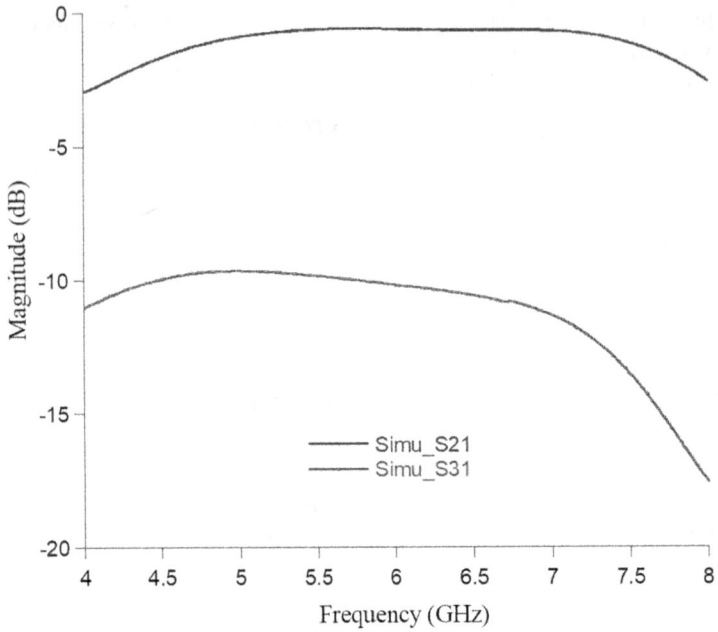

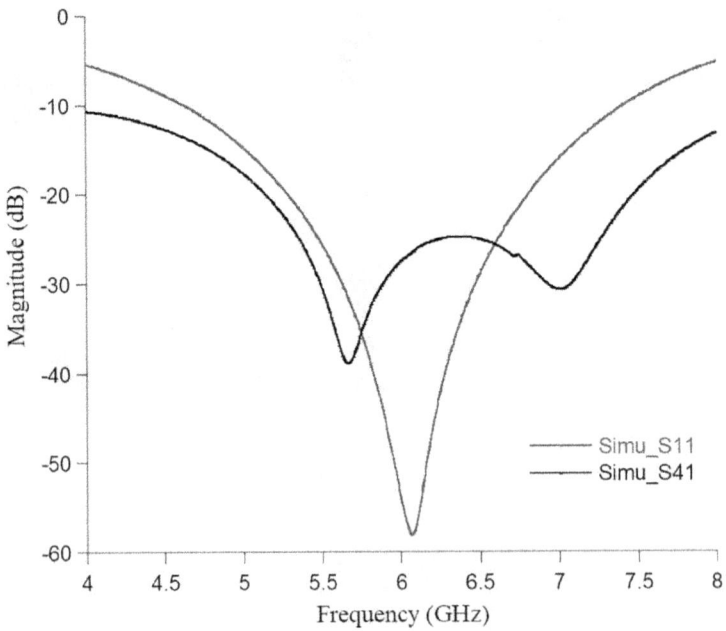

Figure 20.2 (Continued)

The relationship among Z_{c1}, Z_{c2}, $Z_{\pi1}$ and $Z_{\pi2}$ with R_c and R_π, are as follows:

$$\frac{Z_{C2}}{Z_{C1}} = \frac{Z_{\pi2}}{Z_{\pi1}} = -R_C R_\pi \tag{20.9}$$

Seven parameters, i.e., γ_c, Z_{c2} or Z_{c1}, $Z_{\pi2}$ or $Z_{\pi1}$, R_c and R_π, are mandatory to describe the asymmetrical section. For a lossless TEM mode, γ is same for c and π modes:

$$\gamma_C = \gamma_\pi = j\beta \tag{20.10}$$

Considering the quasi-TEM mode of propagation for the proposed structure:

$$R_C = R_\pi = \sqrt{\frac{Z_2}{Z_1}} \tag{20.11}$$

$$Z_1 = \sqrt{\frac{L_1}{C_1} - \frac{1}{\omega^2 C_c C_1}} \quad \text{and} \quad Z_2 = \sqrt{\frac{L_2}{C_2}} \tag{20.12}$$

where, $\quad C_c = \frac{\varepsilon_{eff} + 1}{w'} \times 1_c [A_2 - A_1] \tag{20.13}$

Z_1 and Z_2 are the characteristic impedances of an uncoupled line.

The equation (20.11) is found from equations of interdigital coupled transmission line circuits [17]:

$$A_1 = 4.409 \tanh \left[0.55 \left(\frac{h}{w} \right)^{0.45} \right] \times 10^{-6} \, pF / \mu m \tag{20.14}$$

$$A_2 = 9.92 \tanh \left[0.52 \left(\frac{h}{w} \right)^{0.5} \right] \times 10^{-6} \, pF / \mu m \tag{20.15}$$

where Z_1 and Z_2 are the characteristic impedances of an uncoupled lines 1 (high-impedance coupled section of MMR) and uncoupled line 2 (general microstrip TL), respectively. C_c is the capacitance due to asymmetric coupled section-1. ε_{eff} is the effective dielectric constant of a transmission line of width W and Z_0 is the characteristic impedance with width W':

$$W' = 3S + 4W \tag{20.16}$$

The capacitance matrix may be denoted as:

$$[C] = \begin{bmatrix} C_1 & C_{12} \\ C_{21} & C_2 \end{bmatrix} = \begin{bmatrix} C_1 + C_m & -C_m \\ -C_m C_2 & C_m \end{bmatrix} \tag{20.17}$$

C_m represents the mutual capacitance in the middle of MMR and the strip of microstrip. C_1 and C_2 are the capacitances among coupled lines of the MMR section and TL section and ground. 'c' and 'π' mode are the characteristic impedances of the coupled section of the MMR line (Z_{0C}^{L1}, $Z_{0\pi}^{L1}$), and the general microstrip TL (Z_{0C}^{L2}, $Z_{0\pi}^{L2}$) is obtained as [1]:

$$Z_{0c}^{L1} = \sqrt{\frac{L_1}{C_1} - \frac{1}{\omega^2 C_c C_1}} \tag{20.18}$$

$$Z_{0\pi}^{L1} = \sqrt{\frac{L_1}{(C_1 + 2C_m)} - \frac{1}{\omega^2 C_c (C_1 + 2C_m)}} \tag{20.19}$$

$$Z_{0c}^{L2} = \sqrt{\frac{L_2}{C_2}} \tag{20.20}$$

$$Z_{0c}^{L2} = \sqrt{\frac{L_2}{(C_2 + 2C_m)}} \tag{20.21}$$

In the design mechanism of a predefined coupling-factor level (k), impedance Z_{L1} and Z_{L2} are of lines 1 and 2, respectively. All the impedances may be calculated from equations (20.18) to (20.21):

$$Z_{0c}^{L1} = \sqrt{\frac{Z_a Z_b \sqrt{1-k^2}}{Z_b - k\sqrt{Z_a Z_b}}} \tag{20.22}$$

$$Z_{0\pi}^{L1} = \sqrt{\frac{Z_a Z_b \sqrt{1-k^2}}{Z_b + k\sqrt{Z_a Z_b}}} \tag{20.23}$$

$$Z_{0c}^{L2} = \sqrt{\frac{Z_a Z_b \sqrt{1-k^2}}{Z_a - k\sqrt{Z_a Z_b}}} \tag{20.24}$$

$$Z_{0\pi}^{L2} = \sqrt{\frac{Z_a Z_b \sqrt{1-k^2}}{Z_a + k\sqrt{Z_a Z_b}}} \tag{20.25}$$

Compared with the existing equations of the earlier parameters, different dimensions of asymmetric coupled section-1 are calculated. Similarly, for asymmetric coupled section-2, all the design equations are same except 'C_c', which will be replaced by 'C_g', which is the capacitance due to asymmetric coupled section-2. The capacitive coupling coefficient between the asymmetric lines are denoted by:

$$k_{C_g} = \frac{C_m}{\sqrt{C_1 C_2}} \tag{20.26}$$

The structure of asymmetric coupled section-1 is further simulated using Zealand's IE3D software, and the simulated S-parameters are sketched in Figure 20.1(b) and 1(c). The simulated result shows an operating band from 4.5 GHz up to 7 GHz. The bandwidth is quite high due to its asymmetric nature, though the coupled signal decreases at a higher frequency, as it has only one attenuation zero in the operating band. If the transmission zero is increased, the operating band will be much more stable and wider, which is the prime concern of the proposed design.

The proposed model consists of asymmetrical coupled sections with one section having the MMR unit and another a 50 ohm microstrip transmission line section. Figure 20.3 portrays the circuit model of the MMR and

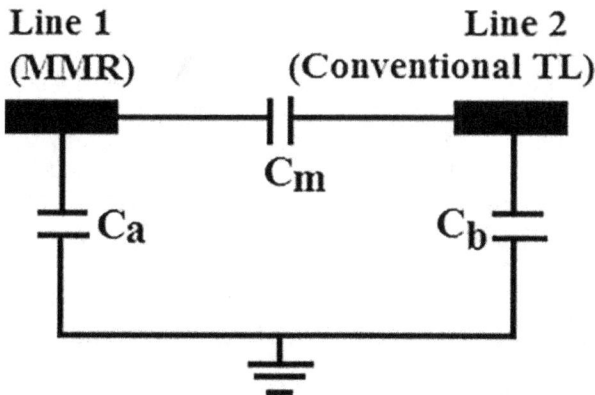

Figure 20.3 Circuit diagram of asymmetric non-uniform coupler.

microstrip conventional TL, which are close to one another. C_m denotes the mutual capacitance between MMR and strip, while there is no ground conductor. C_a and C_b are the capacitances between MMR and microstrip TL and conductive ground, respectively.

As mentioned earlier, it resonates in four discrete frequencies and the design is implemented by a distinct polynomial distribution (like Chebyshev or Butterworth). The attenuation nulls and isolation nulls can also be attained at the desired value as per the band required. The prime advantage of such a mechanism is that the proposed design utilizes the broadbanding facility of an asymmetric line coupler and enables also the tunable broadband mechanism of MM-SIR. To validate the concept, two loose couplers (four-port and six-port) have been designed and illustrated in the later sections.

20.4.2 10 dB Forward-Wave Directional Coupler Using MMR

An MMR-based FW loose DC has been developed at a center frequency of 6.5 GHz with a fractional bandwidth (FBW) of 74% [20]. Normalized poles are computed using Chebyshev polynomial equations as depicted in Table 20.1.

$$f_a = \frac{f_p(1) + f_p(2)}{2} = 1.3847$$

$$f_b = \frac{f_p(4) + f_p(5)}{2} = 0.6153$$

$$\Delta f_{13} = f_a - f_b = 0.7694$$

As for the conventional MMR, f_a resembles f_3/f_2, f_b equivalent to f_1/f_2 and the impedance ratio $(R) = 2.49$ for the design. Considering a glass epoxy FR4 substrate of height 1.58 mm, choosing $W_2 = 0.4$ mm results in $Z_2 = 119.47\Omega$ and $l_2 = 6$ mm for $\theta_2 = 84.28^0$. For R = 2.49, $Z_1 = Z_2/R = 57.03\Omega$. This gives $W_1 = 2.4$ mm and $l_1 = 11.4$ mm for $2\theta_1 = 168.56°$ at 6.8 GHz. As,n = 5 and $L_R = 15$ dB, $g_0 = 1.0$ and $g_1 = 1.2328$, considering FBW = 0.67, $Z_{0e} = 130.02$ and $Z_{0o} = 57.21\Omega$, gaps between adjacent lines s = 0.19 mm.

Table 20.1 Normalized Transmission Poles of MMR-Based Coupler

K	1	2	3	4	5
$F_P(k)$	1.4755	1.2939	1	0.7061	0.5245

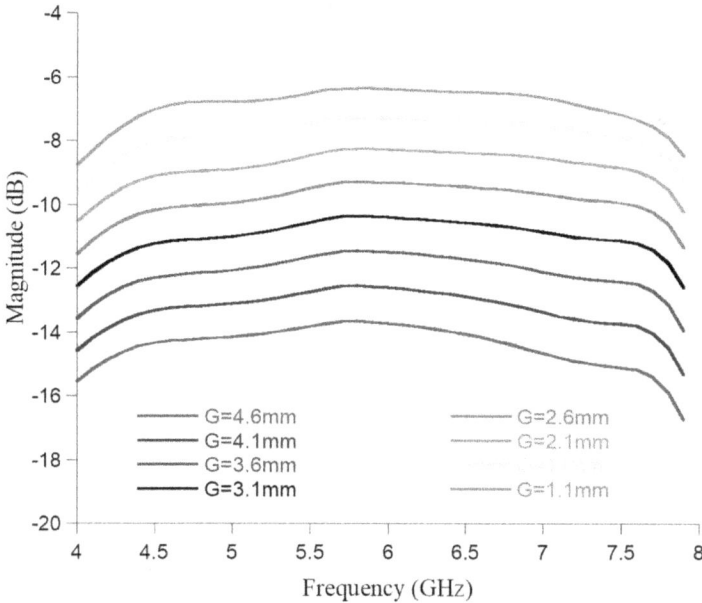

Figure 20.4 Coupling with respect to gap between parallel transmission lines.

The equation (20.26) reveals that 'k' is proportional with the mutual capacitance. Thus, the coupling coefficient depends on the gap (g) between two lines of the DC (same as with the conventional coupler) as illustrated in Figure 20.4. As the gaps decrease, the coupling increases as the external quality factor decreases. Thus, for a 10 dB coupler, the consecutive line gap is considered as 2.6 mm.

The asymmetric and nonuniform-shaped coupler is portrayed in Figure 20.2(a). The software simulated S-parameter responses are portrayed in Figure 20.5(a) and (b). There are four attenuation nulls and three isolation nulls within the specified band of 4 GHz to 8 GHz, and the minimum isolation obtained among ports is 15.2 dB. Though theoretically five attenuation zeros are calculated, to choose the best possible operating band, four zeros are considered.

The photograph of a fabricated prototype on the FR4 substrate of the mentioned coupler is depicted in Figure 20.6(a). It is further measured by Agilent's vector network analyzer (VNA) (model N5091A). The measured result portrays the operating band of 67.2% as in Figure 20.6(b). The measured band is found to be comparable with the simulated results portrayed in Figure 20.6(c). The phase plot also reveals 180° ± 10° phase shift between through a coupled port as for the case of the asymmetric DC. The minor changes found in the measurement are due to the manual fabrication tolerances.

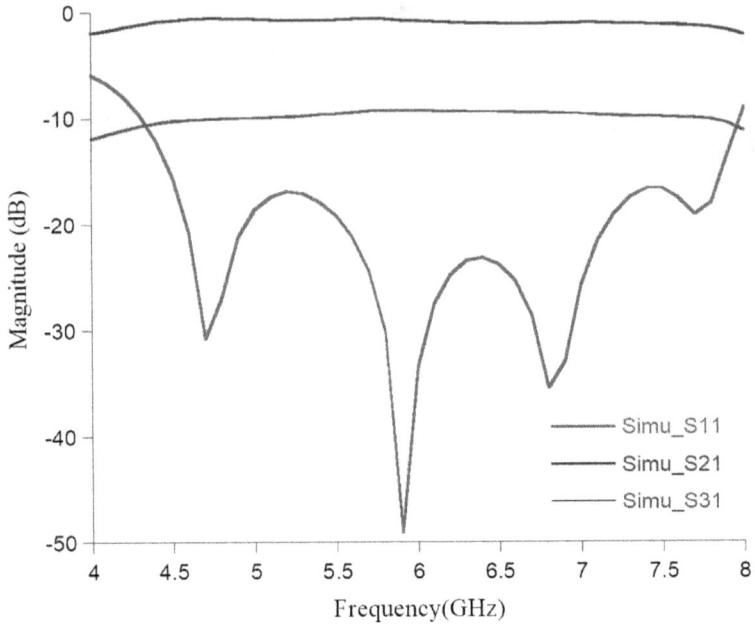

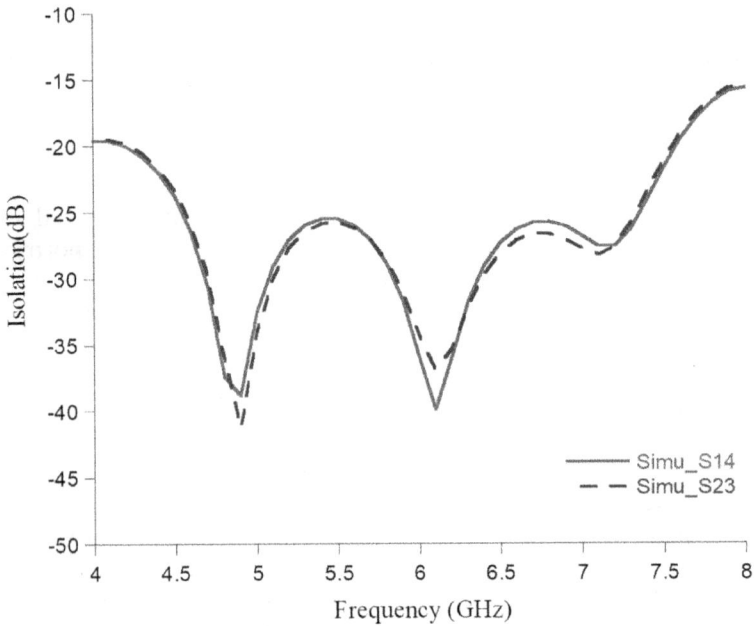

Figure 20.5 (a) Simulated response of S-parameters; (b) isolation between consecutive ports of a designed loose coupler.

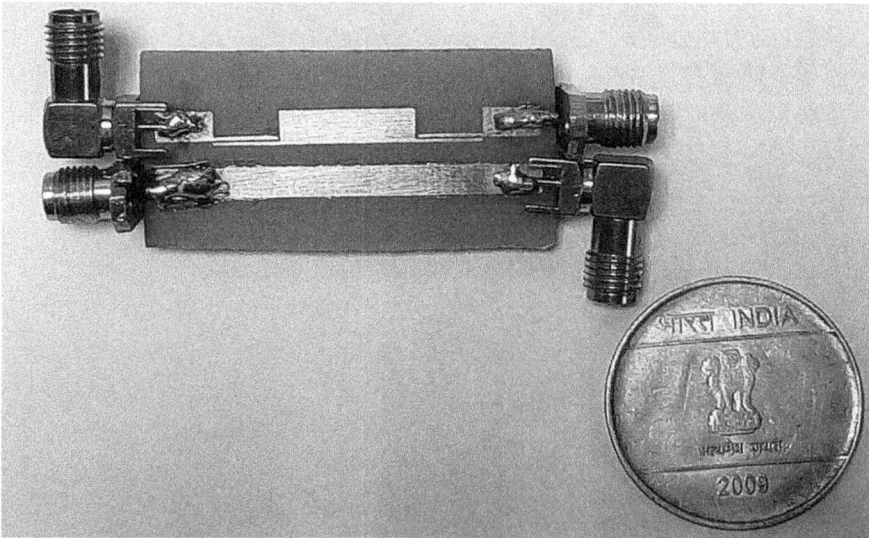

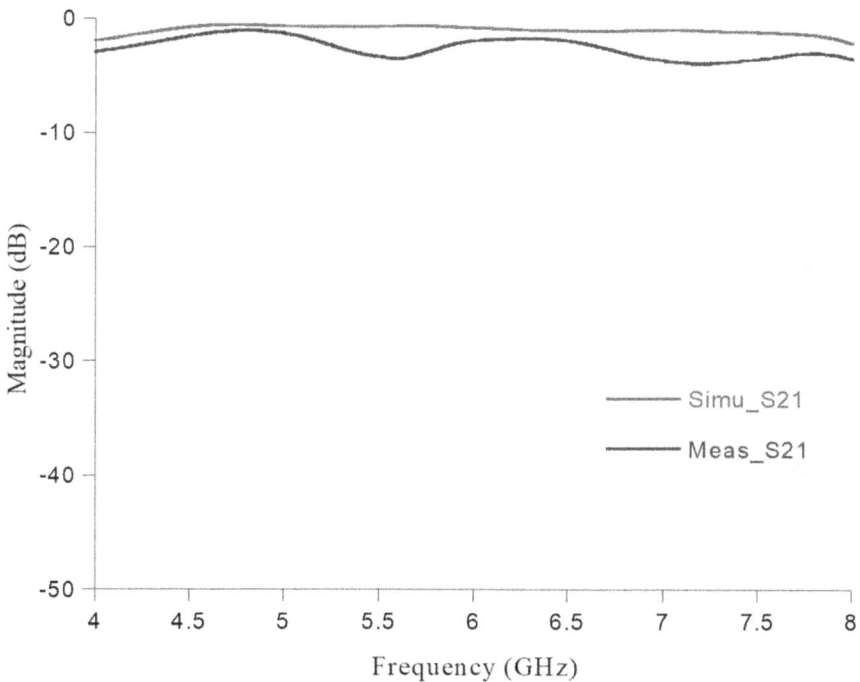

Figure 20.6 (a) Fabricated structure of a four-port wideband co-directional coupler; (b) comparison between simulated and measured scattering parameters at the through port; (c) S-parameters at the coupled port; (d) simulated and measured phase differences.

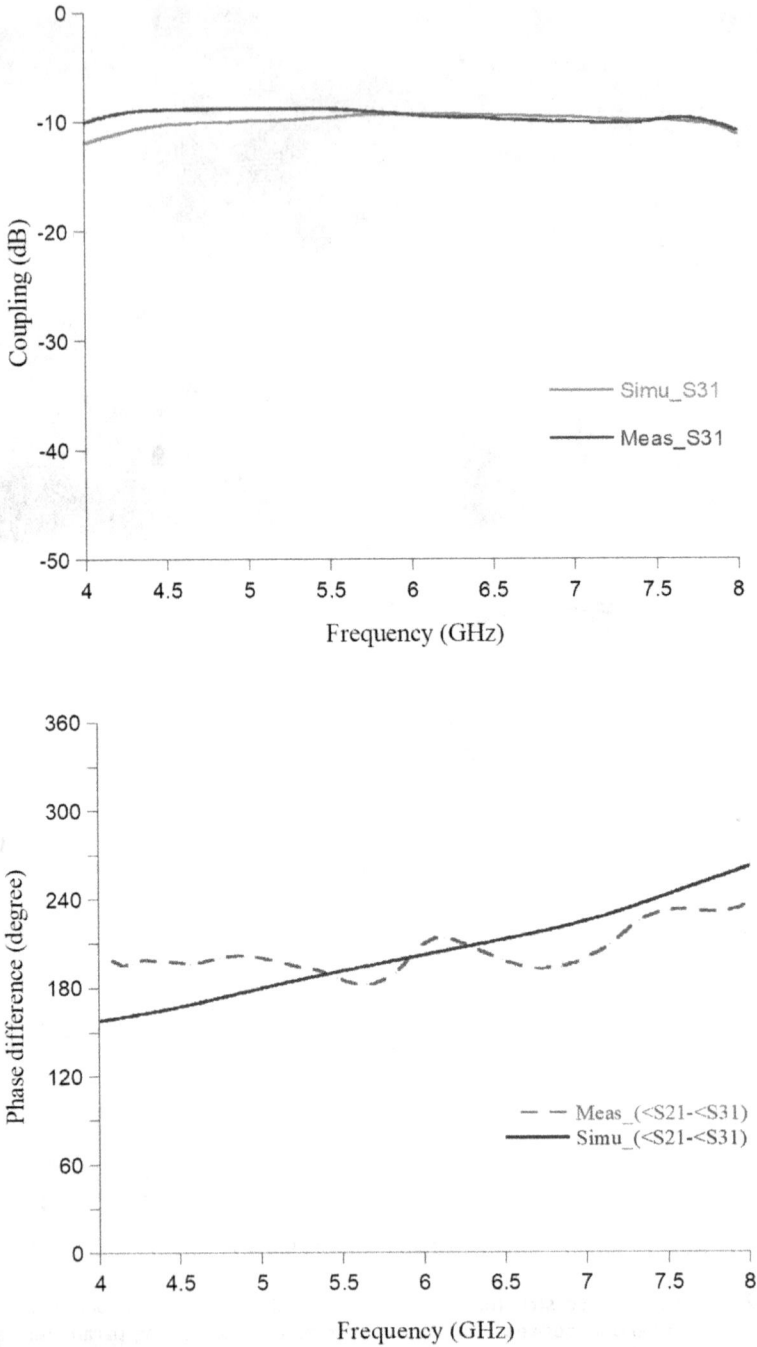

Figure 20.6 (Continued)

20.5 MMR-BASED BROADBAND SIX-PORT COUPLER

A two-line coupler is implementable in different ways and gives a broad range of coupling levels and also phase deviations among ports. Three-conductor coupled-line couplers have already been reported and detailed analysis has also been performed [17]. Utilization in a real-time environment and edge-coupled opposite directional systems is also restricted because of the low coupling level and less than one octave of bandwidth. Applications for a six-port device in reflectometry have also been discussed [18]. A three-line wideband co-directional coupler using comb and herringbone structures has been demonstrated in [19]. Broadbanding (65% FBW) is the widest so far reported with a phase deviation of π radians among outputs.

A six-port coupler has been realized by placing two MMR line sections on both sides of a microstrip 50 ohm reference transmission line as portrayed in Figure 20.7(a). The gap between the MMR line and microstrip line are made equal at both sides, and the overall structure is symmetric with respect to the microstrip signal line.

If a signal is applied at port 1, port 4 will act as the first coupled port having a high coupling level, port 6 will be the second coupled port with a comparatively lower coupling level, port 3 and port 5 will behave as isolated ports and port 2 will be the through port for the proposed design. A similar condition is applied for port 2, port 5 or port 6 as the structure is symmetric in both X and Y directions. If any signal is fed to port 3, it will equally divide into port 2 and port 6, port 4 will behave as through port 3 and port 1 and port 5 will be isolated ports. Thus, the circuit will act as a equal power divider for the previous condition. As both port 2 and port 6 are at equal distance from port 3, it will give am in-phase response at both the arms. A similar condition is applicable for port 4.

From the property of symmetricity, S-parameters: $S_{11} = S_{22} = S_{55} = S_{66}$ and $S_{33} = S_{44}$
Similarly, $S_{12} = S_{21} = S_{56} = S_{65}$

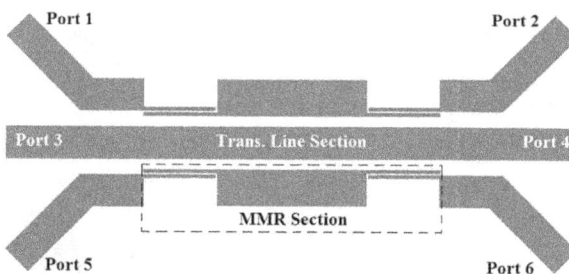

Figure 20.7 (a) Schematic diagram of broadband six-port coupler; (b) simulated S-parameter responses.

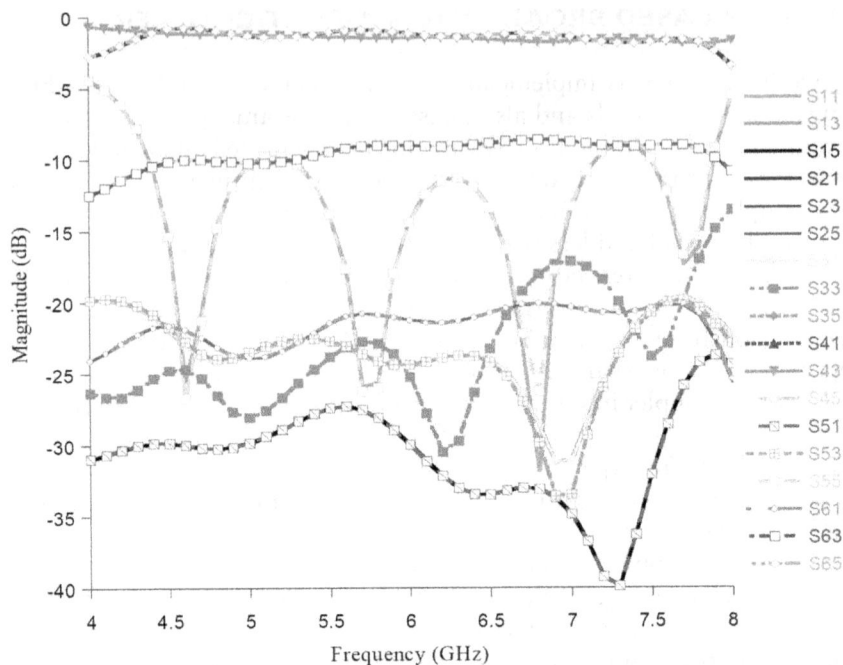

Figure 20.7 (Continued)

Transmission coefficients in first coupled port: $S_{14} = S_{41} = S_{23} = S_{32} = S_{63} = S_{36} = S_{54} = S_{45}$

Transmission coefficients in second coupled port: $S_{16} = S_{61} = S_{25} = S_{52}$

Isolated ports: $S_{13} = S_{31} = S_{53} = S_{35} = S_{42} = S_{24} = S_{46} = S_{64}$ and $S_{15} = S_{26} = S_{51} = S_{62}$

A microstrip six-port coupler using 10 dB coupled resonator sections has been designed using IE3D EM-Simulator software, and the outputs are tabulated in Table 20.2. The length of the coupler is $3\lambda_g/4$ corresponding

Table 20.2 Simulated Output at Different Ports of a Six-Port Coupler (10 dB)

S-matx	1	2	3	4	5	6
1	×	−3 dB	×	−10 dB	×	−22 dB
2	−3 dB	×	−10 dB	×	−22 dB	×
3	×	−10 dB	×	−4 dB	×	−10 dB
4	−10 dB	×	−4 dB	×	−10 dB	×
5	×	−22 dB	×	−10 dB	×	−3 dB
6	−22 dB	×	−10 dB	×	−3 dB	×

to center frequency of 6.7 GHz. The EM-Simulated response is shown in Figure 20.7(b) and depicted in Table 20.2, which also proves the symmetricity condition. Simulated outputs at each port are near similar to the theoretical one.

A prototype of such MMR-based six-port coupler has been fabricated as depicted in Figure 20.8(a), and simulated as well as measured S-parameters are plotted in Figure 20.8(b) to (d). The simulated result shows an almost stable response over the whole operating bandwidth. The isolation between consecutive ports are quite high (>20 dB). The slight deviation in measurement is due to manual fabrication tolerances.

Numerous alternative methods for designing broadband couplers are reported in the literature and compared with the proposed MMR-based six-port coupler as shown in in Table 20.3. The comparison data clarifies that the proposed MMR-based co-directional couplers provide a quite acceptable wide tunable bandwidth with a compact size and low design complexity. In addition, the designed couplers have minimum phase deviation, adequate isolation between output ports, a single layer and via-less design.

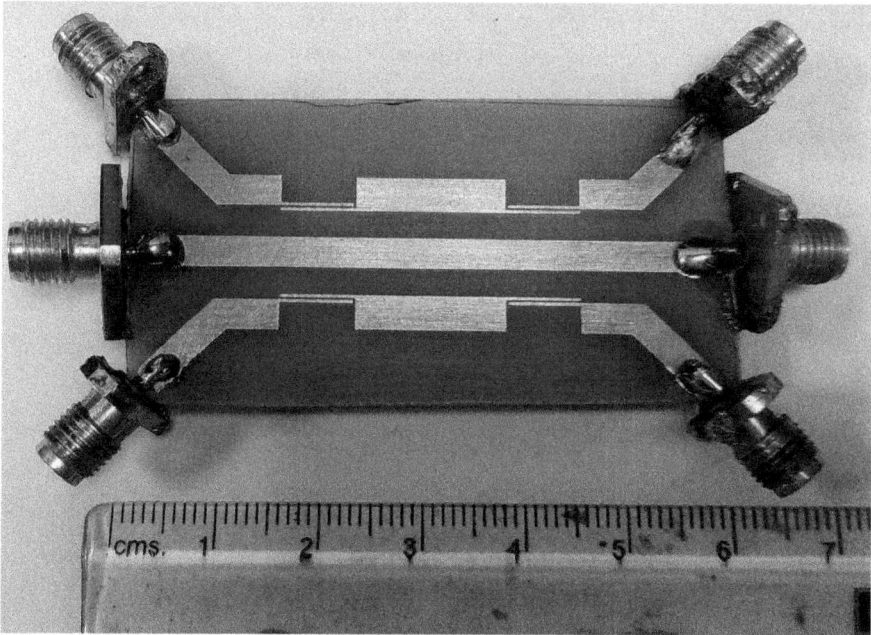

Figure 20.8 (a) Fabricated structure of a six-port coupler; (b) EM-simulated and measured transmission coefficients when the signal is given at port 1; (c) simulated and measured transmission parameters when the signal is given at port 3; (d) simulated and measured isolation parameters.

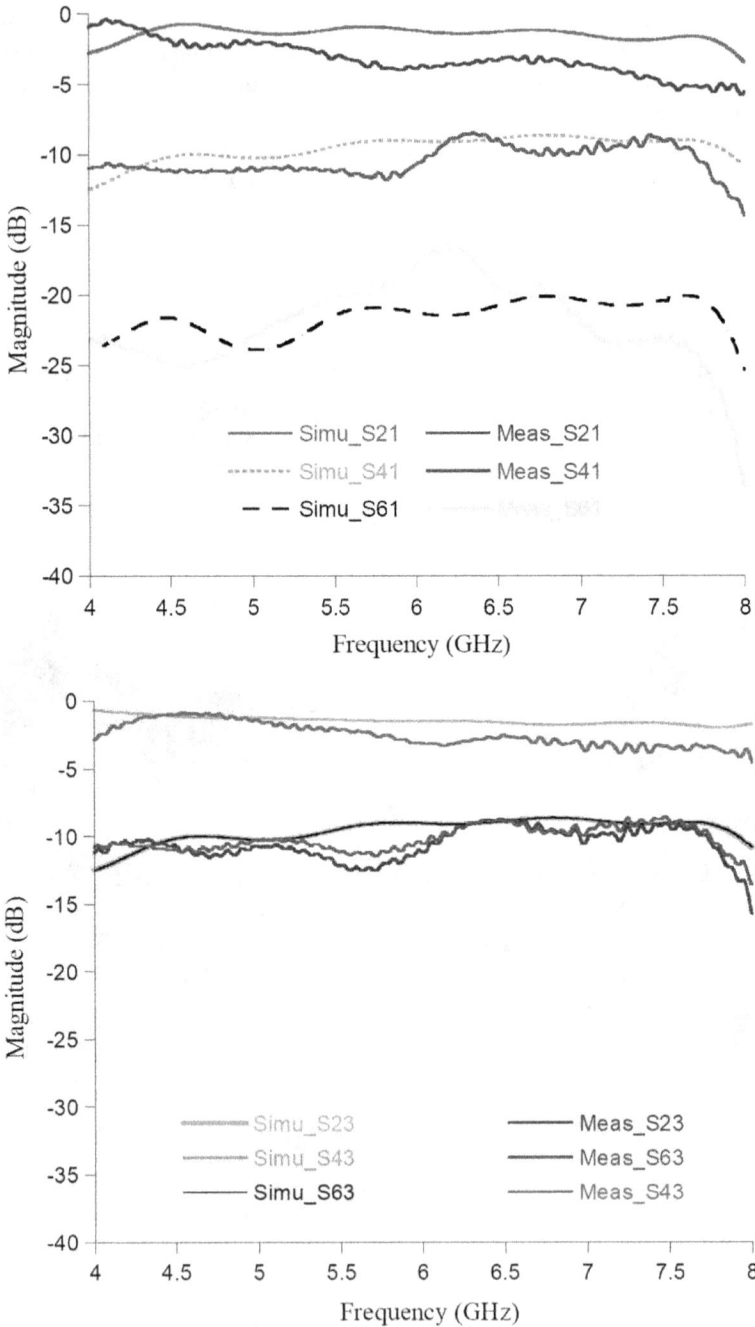

Figure 20.8 (Continued)

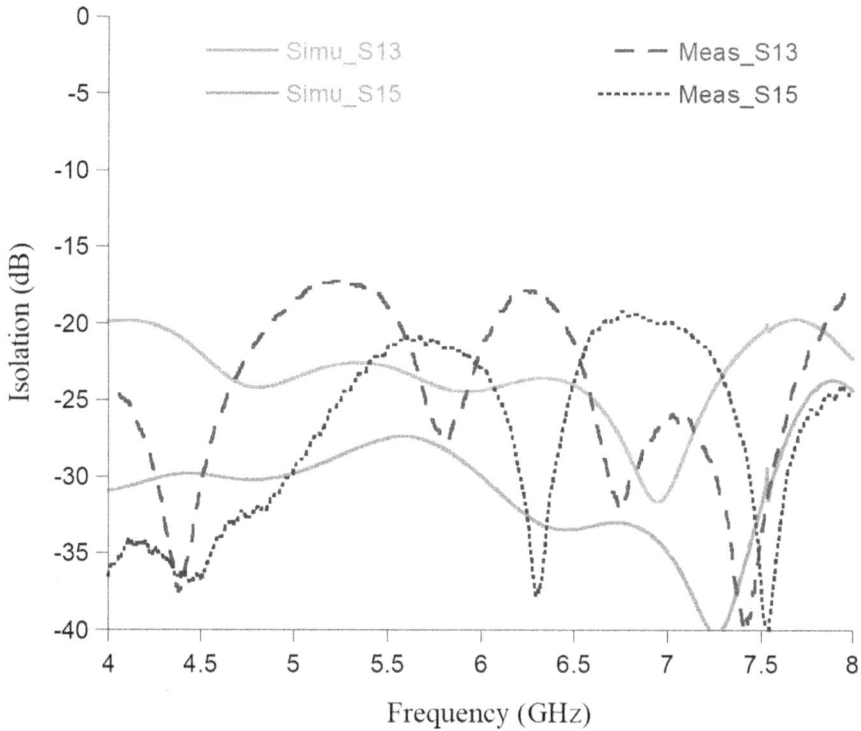

Figure 20.8 (Continued)

Table 20.3 Performance Comparison among Different Broadband Couplers

Ref. No.	Coupling (dB)	FBW	Size	Comments
[3]	0.5	12%	$0.5\lambda_g$ length	Low bandwidth Tight coupling Simple design procedure Need medium real estate
[4]	10±0.5	155%	$3.35\lambda_g \times 0.94\lambda_g$	Very high operating bandwidth Good isolation Loose coupling Large size Single-layer design
[9]	11±1	160%	$2.55\lambda_g \times 0.125\lambda_g$	Very high operating band Moderate isolation Loose coupling Moderately compact in size Single-layer design Moderate design complexity

(Continued)

Table 20.3 (Continued)

Ref. No.	Coupling (dB)	FBW	Size	Comments
[11]	4	100%	0.11 λ_g × 0.27λ_g	High operating bandwidth Extreme compact structure Moderate isolation Tight coupling Multilayer structure High design and fabrication complexity
[13]	6±0.5	93.8%	0.39λ_g × 0.09λ_g	High operating bandwidth Extreme compact structure 6 dB coupler High fabrication difficulty
[15]	0	79.2%	2.02λ_g length	High operating bandwidth Moderate isolation Extreme tight coupling Low loss Average size Average design complexity
Proposed work	10	67.2%	0.7λ_g × 0.32λ_g	High operating bandwidth (tunable) Six-port coupler Moderate isolation Loose coupling Compact in size Monolayer design Design complexity is less

20.6 CONCLUSION

In the present chapter, a broadband loose coupler (four-port) is demonstrated by the MMR technique. Here, the bandwidth is found to be reasonably higher, and it is also tunable in nature, which varies with the impedance ratio. Isolation between outputs is adequate but comparatively lower and can be extended further by using proper chip-resistor circuits. A six-port broadband coupler has also been demonstrated by placing two such MMR sections in an opposite oriented fashion on both sides of the transmission line section. A design example of such multiport couplers using 10 dB coupled sections is physically and experimentally validated. It has benefits like a tunable frequency band, via-less, single layer, no need of bondwire and it doesn't require much real estate. This type of coupler is suitable as a feed network for different wideband and ultra-wideband (UWB) antennas as well as in other MMICs for power distribution. As a future scope of coupler design, reconfigurable couplers can be designed wherein reconfigurability can be realized by varying the level of coupled power to various output ports using some circuitry involving varactor or positive-intrinsic-negative (PIN) diodes.

REFERENCES

[1] Mongia, Rajesh K., J. Hong, Prakash Bhartia, and Inder Jit Bahl. *RF and microwave coupled-line circuits*. Artech House, 2007.

[2] Hsu, Sen-Kuei, Jui-Chih Yen, and Tzong-Lin Wu. "A novel compact forward-wave directional coupler design using periodical patterned ground structure." *IEEE Transactions on Microwave Theory and Techniques* 59, no. 5 (2011): 1249–1257.

[3] Toulios, P. P., and A. C. Todd. "Synthesis of symmetrical TEM-mode directional couplers." *IEEE Transactions on Microwave Theory and Techniques* 13, no. 5 (1965): 536–544.

[4] Tresselt, C. P. "The design and construction of broadband, high-directivity, 90-degree couplers using non-uniform line techniques." *IEEE Transactions on Microwave Theory and Techniques* 14, no. 12 (1966): 647–656.

[5] Willems, David A. "A broadband MMIC quadrature coupler using a braided microstrip structure." In *IEEE MTT-S International Microwave Symposium Digest*, IEEE, 1994, pp. 899–902.

[6] Caloz, Christophe, Atsushi Sanada, Lei Liu, and Tatsuo Itoh. "A broadband left-handed (LH) coupled-line backward coupler with arbitrary coupling level." In *IEEE MTT-S International Microwave Symposium Digest*, IEEE, 2003, vol. 1, pp. 317–320.

[7] Chen, Jia-Liang, Sheng-Fuh Chang, Yng-Huey Jeng, and Chun-Yo Lin. "Wiggly technique for broadband non-uniform line couplers." *Electronics Letters* 39, no. 20 (2003): 1451–1453.

[8] Wu, Y. D., M. L. Her, Y. Z. Wang, and M. W. Hsu. "Stepped-impedance directional coupler with enhanced isolation using interdigital capacitance compensation." *Electronics Letters* 41, no. 10 (2005): 598–599.

[9] Hamed, Karim W., Alois P. Freundorfer, and Yahia M. M. Antar. "A new broadband monolithic passive differential coupler for K/Ka-band applications." *IEEE Transactions on Microwave Theory and Techniques* 54, no. 6 (2006): 2527–2533.

[10] Gruszczynski, Slawomir, and Krzysztof Wincza. "Broadband multisection asymmetric 8.34-dB directional coupler with improved directivity." In *Asia-Pacific Microwave Conference (APMC)*, IEEE, 2007, pp. 1–4.

[11] Chin, Kuo-Sheng, Ming-Chuan Ma, Yi-Ping Chen, and Yi-Chyun Chiang. "Closed-form equations of conventional microstrip couplers applied to design couplers and filters constructed with floating-plate overlay." *IEEE Transactions on Microwave Theory and Techniques* 56, no. 5 (2008): 1172–1179.

[12] Lin, Tong-Hong, Sen-Kuei Hsu, and Tzong-Lin Wu. "A novel broadband forward-wave directional coupler design with periodical H-shaped structure." In *Asia-Pacific Microwave Conference (APMC)*, IEEE, 2011, pp. 1770–1773.

[13] Yen, Jui-Chih, Sen-Kuei Hsu, Tong-Hong Lin, and Tzong-Lin Wu. "A broadband forward-wave directional coupler using periodic Y-shaped ground via structures with arbitrary coupling levels." *IEEE Transactions on Microwave Theory and Techniques* 61, no. 1 (2013): 38–47.

[14] Staszek, Kamil, Krzysztof Wincza, and Slawomir Gruszczynski. "Broadband three-section symmetrical directional couplers with reduced coupling coefficient requirements." *Microwave and Optical Technology Letters* 55, no. 3 (2013): 639–645.

[15] Tran, Minh, Jared Hulme, Sudharsanan Srinivasan, Jonathan Peters, and John Bowers. "Demonstration of a tunable broadband coupler." In *International Photonics Conference (IPC)*, IEEE, 2015, pp. 488–489.

[16] Lyu, Yun-Peng, Lei Zhu, Qiong-Sen Wu, and Chong-Hu Cheng. "Proposal and synthesis design of wideband phase shifters on multimode resonator." *IEEE Transactions on Microwave Theory and Techniques* 64, no. 12 (2016): 4211–4221.

[17] Keshavarz, Rasool, Masoud Movahhedi, and Abdolali Abdipour. "A broadband and compact asymmetrical backward coupled-line coupler with high coupling level." *AEU-International Journal of Electronics and Communications* 66, no. 7 (2012): 569–574.

[18] De, A., B. Roy, A. Bhattacharya, and A. K. Bhattacharjee. "Bandwidth-enhanced ultra-wide band wearable textile antenna for various WBAN and Internet of Things (IoT) applications." *Radio Science* 56 (2021): e2021RS007315.

[19] Bhattacharya, A., B. Roy, R. Caldeirinha, and A. Bhattacharjee. "Low-profile, extremely wideband, dual-band-notched MIMO antenna for UWB applications." *International Journal of Microwave and Wireless Technologies* 11, no. 7 (2019): 719–728.

[20] Bhattacharya, Ankan, Bappadittya Roy, Santosh K. Chowdhury, and Anup K. Bhattacharjee. "Computational and experimental analysis of a low-profile, isolation-enhanced, band-notch UWB-MIMO antenna." *Journal of Computational Electronics* 18 (2019): 680–688.

[21] Bhattacharya, A., A. De, B. Roy, and A. K. Bhattacharjee. "Investigations on a low-profile, filter backed, printed monopole antenna for UWB communication." *Indian Journal of Pure & Applied Physics, CSIR-NISCAIR* 58 (2020): 106–112.

Chapter 21

Sensing of Trapped Survivors Using IR-UWB Radar

Amit Sarkar and Debalina Ghosh

21.1 INTRODUCTION

Ultra-wideband (UWB) radar is frequently used for a variety of purposes, including civil, medical, military, and transportation. Detecting both small and large movements of the human body is a more recent use. The non-contact principle, excellent penetration capabilities, long range, and high spatial resolution of UWB radar have made it more popular recently than continuous-wave (CW) radars, which was one of the earliest technologies employed to observe breathing and cardiac activity. As a result, over the years, a commendable amount of research and development has been put into the fields of fall detection [1], emotion recognition [2], micro-gesture recognition [3], human activity recognition [4], and more employing UWB radar. The estimation of human chest displacement and heartbeat rate is another common application of this radar system. UWB radar has become a valuable tool in instances when real-time human position tracking outside a wall is required. A human-made structure collapsing and trapping numerous people behind the debris is another instance of non-line-of-sight (NLOS). The search and rescue teams' main goal in such a situation is to find and rescue the buried victims. UWB radar has become one of the most popular pieces of equipment in the domain of search and rescue operations during the past few decades. In particular [5, 6] suggested using impulse radio ultra-wideband (IR-UWB) radar to discover people trapped under rubble because of its low power consumption, decreased risk of multipath interference, etc.

After the structures have collapsed, the distribution and orientation of different people trapped inside the rubble are almost random. There may be one person in a particular range [7] from the radar or multiple people at the same location [8]. The postures of each individual will also vary. Also, as a result of the nature of debris, complications arise in detecting minute movements from a living being due to multiple reflections. Hence, a detailed and robust approach is needed to obtain the location and vital signs information of each person. Thus, preprocessing the received signal is very essential as the beginner step. Refs. [9–12] have shown the positive effects

DOI: 10.1201/9781003459880-21

of different pre-processing algorithms, which included removing static and non-static clutters, detrending the received echo, filtering in both slow and fast time dimensions, and finally enhancing the signal amplitude with respect to the clutter. Once the preprocessing is finished, the received signal is passed through a variety of time and/or frequency domain algorithms to extract the vital signs and location of each individual.

21.2 MOTIVATION

There have been a lot of cases where structures such as residential and commercial buildings, bridges, etc., have collapsed all over the world. After the collapse, people get trapped under the rubble, and a majority of them die without getting proper medical treatment in time. Even with the exponential technological advancements, these accidents keep happening every year, and a lot of lives are lost because of it. Based on the reports from Accidental Deaths & Suicides in India published by the National Crime Records Bureau (NCRB) [13], it is seen that a lot of people have died due to collapse of artificial structures in India in a span of just 10 years from 2012 to 2021. The detailed statistics are elaborated in Figure 21.1.

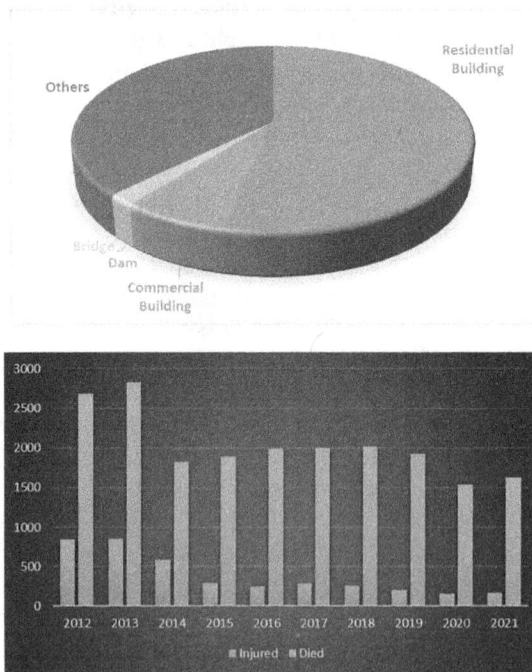

Figure 21.1 Number of deaths and incidents in India from collapse of structures. (a) Types of structures. (b) Number of people injured and dead for a span of 10 years from 2012 to 2021.

21.3 CURRENT TECHNOLOGIES USED

It is crucial to search under the rubble for any trapped living beings after the structures have collapsed. As soon as the search and rescue crew arrives on the scene, they should begin freeing trapped individuals. They use a variety of methods that are at their disposal. In developed countries such as the United States, China, Israel, UK, etc., there are costly UWB radars being used during these scenarios. However, in India the most typical techniques for search and rescue operation are canine search, technical search with acoustic sensors, optic sensors, vibration sensors, and call out. The benefits and disadvantages of these methods for search and rescue purposes are explained in detail in Table 21.1.

21.4 IMPORTANCE OF RADAR

In addition to the techniques mentioned earlier, one more technology has shown to be effective for search and rescue operations. Radio detection and ranging (RADAR) is the name of this technology. Radars have been employed for many purposes for many years, e.g., Doppler radars are able to pick up on even the smallest changes in the surroundings. The ability to detect human breathing is one use for this property. The people in front of them can have their chest motions picked up by the radar. This feature helps with search and rescue activities. The following are some advantages of utilizing a radar device to locate people buried behind debris:

- It can detect persons who are unconscious and unable to speak or make noises to signal for help.
- The rescue workers must manually set vibration, acoustic, and visual sensors as close to the victims as feasible in order for them to function. Unbalanced structures might result from drilling. Additionally, the sensors frequently produce inaccurate results. Radar equipment is usually non-intrusive.

Table 21.1 Advantages and Disadvantages of Current Search and Rescue Techniques

Current Technology	Advantages	Disadvantages
Call out	Trapped person may respond to calls made repeatedly.	Unconscious person will be unable to respond.
Canine search Acoustic sensor Optic sensor	Large areas can be covered in less time.	Cannot work for long stretches of time.
	Simple, robust, reasonably priced, and capable of operation in the dark.	Does not identify the source of the sound and demands complete silence when in use.
	Able to visualize people and surroundings.	Expensive, delicate, and needs instruction for correct use.

When rescue personnel arrive on the scene, they can immediately deploy the radar units to gain a rough picture of where the trapped persons are. Compared to other approaches that may require more time to verify the victims' locations, this is quicker. After that, they can pool their resources to successfully find and rescue victims hidden behind the wreckage.

Despite the aforementioned advantages of radar, there are several situations in which they may not be effective. The following cases are only a few:

- The ability of the electromagnetic (EM) waves to penetrate the debris is compromised if it is covered in water or other conductive materials.
- The environment should be nearly immobile in order to pick up the slight chest motions hidden behind all the debris. But it is frequently noticed that there are lots of individuals strolling around after a structure collapses. Their powerful movements can conceal the target's modest movements. By blocking off the entire collapsed region, the problem may be controlled.

The aforementioned issues are situational. However, using radar devices to find trapped human beings is preferable in most of the circumstances.

21.5 RADAR-RUBBLE-TARGET MODEL

21.5.1 Different Types of UWB Radar

UWB is a radio technology which enables a small range and large bandwidth in communications over a vast area of the radio frequency (RF) spectrum at a very low energy level. Due to its special qualities, it has found use in a variety of applications, including through-wall radar (TWR) and ground-penetrating radar (GPR) systems.

The UWB radar system is mostly divided among the following three architecture types:

- **IR-UWB radar:** Here, a fixed pulse repetition period is used to generate a series of pulse signals, most of which are monocycle pulse in shape. An UWB antenna emits these pulses at the target, and the reflected wave is captured and processed.
- **Stepped-frequency continuous wave (SFCW) radar:** CW signals are used in this situation. The CW signals' frequency is raised in predetermined steps. Prior to sending the subsequent frequency-incremented CW, each CW of a certain frequency is sent and received. After receiving all of the frequency components, inverse discrete Fourier transform (IDFT) is performed to produce a synthetic pulse.

- **Frequency-modulated continuous wave (FMCW) radar:** Here, the frequency of the CW is continuously altered in accordance with various waveforms, including sinusoidal, sawtooth, and triangular. A beat frequency is produced by combining the transmitted and received waveforms. This beat frequency is then used to gather target information.

For the sake of locating trapped people buried under rubble, Impulse radar can be used for the following reasons:

- High-range resolution because of their relatively broad bandwidth.
- A lower likelihood of multipath interference and good non-stationary clutter separation from the target.
- Less possibility of interference with essential radio services, including emergency radio, due to reduced power requirements.

21.5.2 Detailed Description of Rubble

Any time a structure collapses, a select few materials make up the majority of the debris. These materials are described in terms of their electrical characteristics. Based on Table 21.2, the real part of relative permittivity is defined as $\epsilon_r' = af^b$, conductivity, $\sigma = cf^d$, and imaginary part of relative permittivity, $\epsilon_r'' = \dfrac{j\sigma}{2\pi f \epsilon_0}$. Here, f is frequency in GHz, σ is in S/m, and both ϵ_r' and ϵ_r'' are dimensionless. Thus, frequency-dependent complex permittivity is given by $\epsilon_r(f) = \epsilon_r' - j\,\epsilon_r''$. A list of commonly found materials in the rubble along with their electrical properties is provided in Table 21.2 [14].

The foundation of any artificial structure is made up of brick, concrete, and wood; hence, these materials are most appropriate. And these are mostly the foundation of any rubble formed by the destruction of most of the structures. The medium-dry ground will be covered using these materials.

Table 21.2 Electrical Properties of Different Types of Rubble

Material Class	Real Part of Relative Permittivity		Conductivity (S/m)		Frequency Range (GHz)
	a	b	c	d	
Concrete	5.31	0	0.0326	0.8095	1–100
Brick	3.75	0	0.038	0	1–10
Wood	1.99	0	0.0047	1.0718	0.001–100
Medium dry ground	15	-0.1	0.035	1.63	1–10

21.5.3 Human Target Model

The layers of human tissue are drawn from Duke's anatomy [15] that goes across the heart tissues. The dispersive character of human tissues, caused by the UWB radar signals, is explained by taking into consideration the Cole-Cole model. The parameters of this model were derived from Gabriel *et al.* [16]. Table 21.3 lists the tissues' characteristics.

21.6 HUMAN VITAL SIGNS MATHEMATICAL MODEL

When a radar faces a human target, the breathing motion of the human torso can be translated as harmonic vibration with sinusoidal variation in the distance between the radar and human target changing in a sinusoidal manner. Figure 21.2 describes the positioning of the radar with respect to a human target.

Table 21.3 Electrical Properties of Human Tissue Layers from Duke's Anatomy at 1.5 GHz

Tissue Layer	Thickness (mm)	Permittivity	Conductivity (S/m)
Skin	3	44.412	1.0889
Fat	8	5.3833	0.068028
Muscle	10	53.963	1.1881
Cartilage	6	40.932	1.098
Heart	84	57.2	1.5749
Lung deflated	65	49.922	1.121
Lung inflated	70	21.211	0.57058
Bone	6	19.757	0.49783

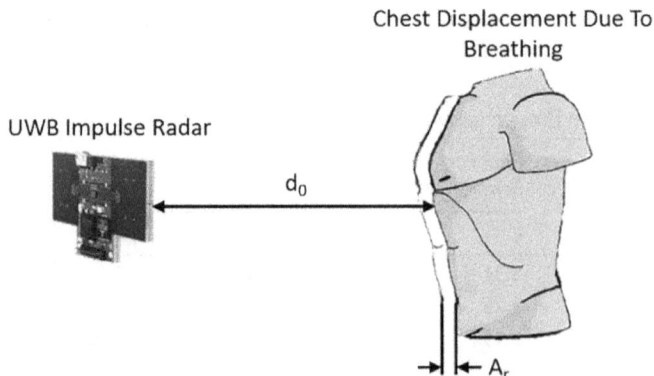

Figure 21.2 Human chest displacement due to breathing.

The radar to human subject range, upon incorporating the chest movement, is expressed as follows [17]:

$$d(t) = d_0 + r(t) = d_0 + A_r \sin(2\pi f_r t) + A_h \sin(2\pi f_h t) \tag{21.1}$$

where t represents a slow time, d_0 represents the range between the radar and human chest, A_r and A_h are the amplitude changes caused by chest and heart wall, respectively, and f_f is the frequency of chest displacement, whereas f_h is the frequency of heart wall motion. $r(t)$ is the total collective motion generated by the torso.

The impulse response of the radar can be represented by:

$$h(\tau, t) = a_v \delta(\tau - \tau_v(t)) + \sum_i a_i \delta(\tau - \tau_i) \tag{21.2}$$

where $\delta(\tau)$ is a delta function. $\sum_i a_i \delta(\tau - \tau_i)$ represents the reflections from stationary targets, where a_i denotes the amplitude and τ_i denotes the time delay of i reflection from a stationary target. $a_v \delta(\tau - \tau_v(t))$ represents the reflection from a vibrating target with a_v amplitude and $\tau_v(t)$ is the time delay in fast time. $\tau_v(t)$ can be calculated as follows:

$$\tau_v(t) = \frac{2d(t)}{v} = \tau_0 + \tau_r \sin(2\pi f_r t) + \tau_h \sin(2\pi f_h t) \tag{21.3}$$

where $\tau_0 = 2d_0 / v$, $\tau_h = 2A_h / v$, $\tau_r = 2A_r / v$. Considering, $s(\tau)$ is the transmitted signal, the received signal is represented as:

$$R(\tau, t) = s(\tau) * h(t, \tau) = a_v s(\tau - \tau_v(t)) + \sum_i a_i s(\tau - \tau_i) + n(t) \tag{21.4}$$

where n(t) denotes environmental noise. The received signal is subsequently down-translated to baseband components that are in phase (I) and quadrature phase (Q). The frames for each range bin are contained in the received echo matrix, which is created by merging the I and Q data, as illustrated next:

$$R = \{r^{(m)}[n] : m = 1, 2, .., M; n = 1, 2.., N\} \tag{21.5}$$

where the time index is represented by n and the range by m, respectively. The radar's detecting range is given to us in M, or range units. The overall time period of the data is determined by the time units, N. As a result, the echo data is M × N in size.

21.7 METHODOLOGY

Radar raw data pre-processing is utilized to eliminate the clutter present within.

21.7.1 Raw Data Preprocessing Stage

The entire preprocessing flow can be divided into three basic steps:

1. **Non-static clutter removal:** The input radar raw echo matrix is subjected to singular value decomposition (SVD) [18], which produces the singular values $S(m, n)$, $M \times M$ unitary matrix U, and $N \times N$ unitary matrix V as illustrated next:

$$SVD[R]_{(M \times N)} = U_{(M \times M)} \times S_{(M \times N)} \times V^T_{(N \times N)} \tag{21.6}$$

The reflections obtained from the target environment are only appropriately represented by a small number of dominating single values because there is insufficient target information. Using these dominating single values, the eigenvalue gradient method (EGM) [19] is used to obtain the desired target space. The modified radar echo matrix $R_1(m, n)$:

$$R_{1(M \times N)} = U_{(M \times o)} \times S_{(o \times o)} \times V^T_{(N \times o)} \tag{21.7}$$

The variable o displays the quantity of dominant eigenvalues in the received echo.

2. **Static clutter removal:** Static clutter includes both background clutter and direct current (DC) clutter. Every range bin in the slow-time direction contains DC clutter. Time mean subtraction (TMS) is processed upon estimating the DC component [20]:

$$DC = \frac{1}{M \times N} \sum_{m=1}^{M} \sum_{n=1}^{N} R_1(m, n) \tag{21.8}$$

The received echo matrix for the radar is altered as follows based on equation 21.8:

$$R_2(m, n) = R_1(m, n) - DC \tag{21.9}$$

Background clutter is also recognized as a secondary stationary clutter signal because of how diverse the background is. To eliminate this, a stable background estimate B(m, n) must be established for each range bin. The following weighting coefficient, lambda [21], is employed to achieve this:

$$B(m,n) = \lambda \times B(m,n-1) + (1-\lambda) \times R_2(m,n) \qquad (21.10)$$

Reliance on earlier time samples is decreased via the recursive equation 21.10. It is clear that finding a fair value for B(m, n) by utilizing the formula lambda = 0.95 is successful. In order to obtain the adjusted radar echo data, the input radar received echo matrix is subtracted using equation 21.11 after the acquisition of the full background estimate.

$$R_3(m,n) = R_2(m,n) - B(m,n) \qquad (21.11)$$

The bulk of the clutter is being filtered at this point.

3. **Linear trend removal:** The received data is affected by malfunctions in the radar's triggering mechanism. The slow-time data exhibits a linear trend as a result of amplitude instability. The following equation explains how to account for this linear trend using a linear trend suppression (LTS) technique based on the linear least square fit [22]:

$$X^T = R_3^T - z(z^T z)^{-1} z^T R_3^T \qquad (21.12)$$

Here z = [n, 1_N], n = [0, 1, ..., N − 1]^T, 1_N is a N × 1 vector with unit values. After detrending, X is obtained, where X = {$x^{(m)}$ [n]: m = 1, 2, ..., M;n = 1, 2, ..., N}.

21.8 BURIED HUMAN LOCATION AND VITAL SIGNS ESTIMATION STAGE

The entire process showcasing the steps in obtaining the exact target location and vital signs is described in detail in [23] and in Figure 21.3.

21.9 EXPERIMENTAL SETUP

Considering that the components of the experimental setup include a radar system, target, and surrounding environment, there can be a number of possible combinations in which the setup can be formulated. Figure 21.4 showcases how the experimental setup can be classified into different categories.

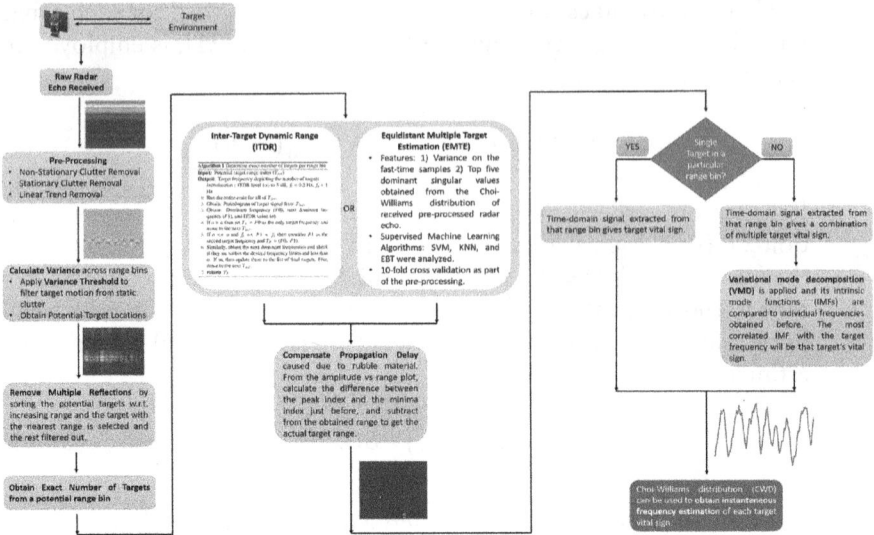

Figure 21.3 Overall signal processing steps to obtain the target location and vital signs.

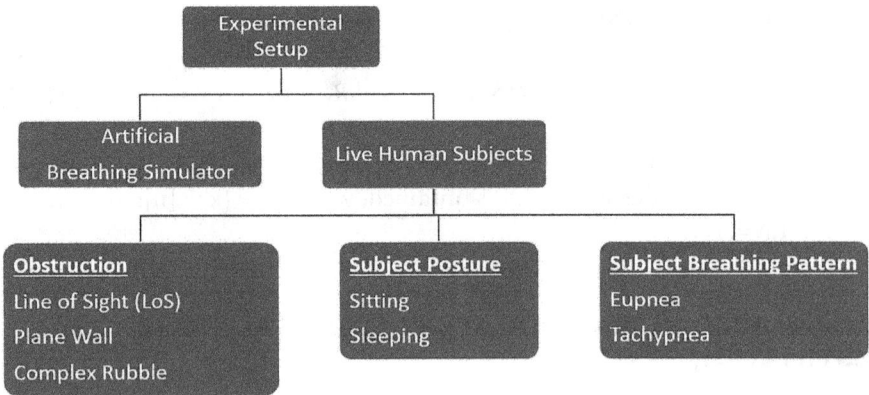

Figure 21.4 Experimental setup classification.

In this experiment, the following specifications of IR-UWB radar is used: center frequency of the NOVELDA X4M200 [24] = 7.29 GHz, unambiguous range of 2.5 m, sampling rate of 23.328 GS/s, frames per second of 17, and pulse repetition frequency (PRF) of 15.1875 MHz. The benchmark for breathing signals has been the Vernier Go Direct respiration belt [25]. For the purpose of gathering and processing data, a PC is used to communicate with the radar.

21.10 ARTIFICIAL BREATHING SIMULATOR

Humans are more likely to be engulfed completely by debris in real-world situations. It is not feasible to pile the debris in direct contact with the test individuals in a lab setting. Thus, Figure 21.5a mimics the back-and-forth movement of the torso using a breathing simulator. A plastic box containing water is used to resemble a human torso. A setup comprising a stepper motor and lead screw is built, whose movement is programmed by a microcontroller and controlled via a PC. The moving shaft is connected to the box that represents the human torso in such a way that the overall device mimics a human subject's breathing action. The entire setup is housed within a box, which as a whole represents a working breathing simulator.

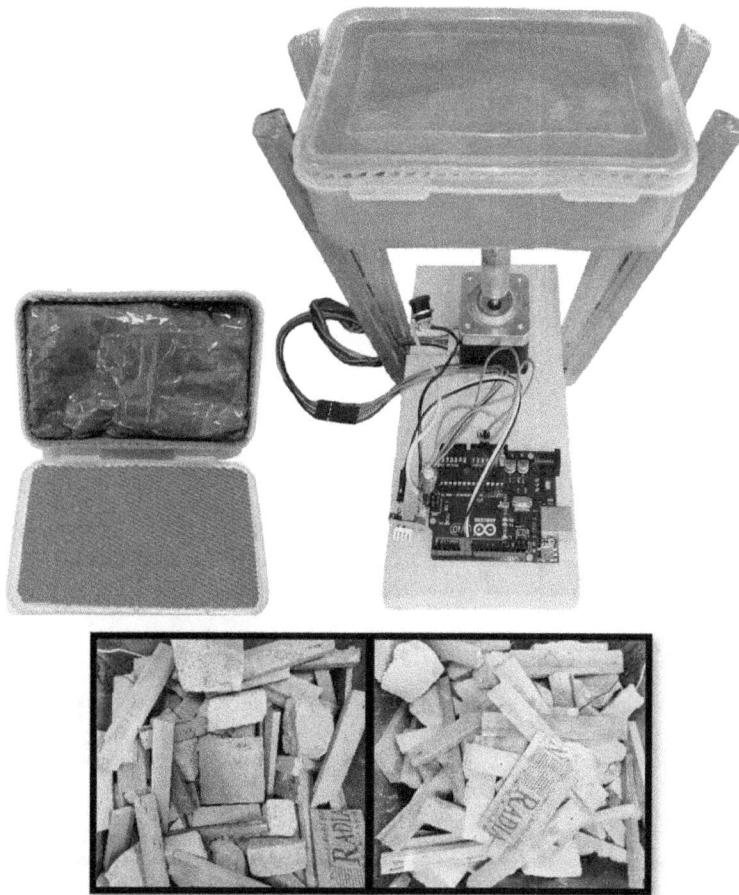

Figure 21.5 (a) Complete breathing simulator. (b) Breathing simulator housed inside a plywood box and placed inside rubble [23].

The breathing simulator is positioned within a $1\,m \times 1\,m \times 1\,m$ box containing bricks, concrete, and wood debris for the experiment. The radar is located 1 m above the simulator. A PC was used to control the radar and simulator at the same time. The simulator was covered with 0.25 m of debris. Figure 21.5b displays the before and after pictures of the simulator covered in debris. In this experiment, the simulator oscillates for the first 30 s with 0.2 Hz frequency and the next 30 s at 0.4 Hz frequency. The authors' design of a human torso mimic performs better than models that have been put forth in the literature [26–28], where metal sheets are frequently used as targets.

21.11 LIVE HUMAN SUBJECTS

Table 21.4 lists the physical details of the three people who were involved in all of the experiments.

21.11.1 No Obstruction between Radar and Target

For the single target, the radar is 1 m from the subject and a pulse sensor is connected to the PC to record heart rate information. For multiple targets, the radar is 2 m from the subjects. Breathing pattern analyzed: Eupnea. Recording time: 20 s. See Figure 21.6.

Table 21.4 Physical Characteristics of the Volunteers

Subject	Height (m)	Torso Dimension (L × W × H)
1	1.75	0.5 m × 0.35 m × 0.23 m
2	1.62	0.45 m × 0.32 m × 0.25 m
3	1.68	0.42 m × 0.33 m × 0.23 m

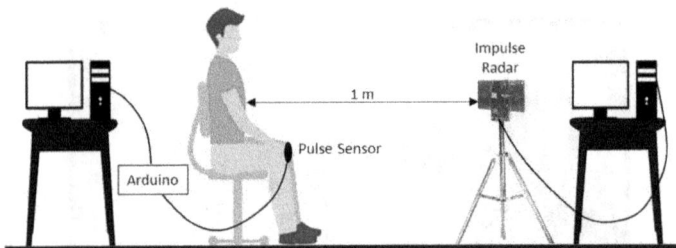

Figure 21.6 (a) Single target sitting at LoS from radar. (b) Multiple targets sitting at same range in LoS from radar [29].

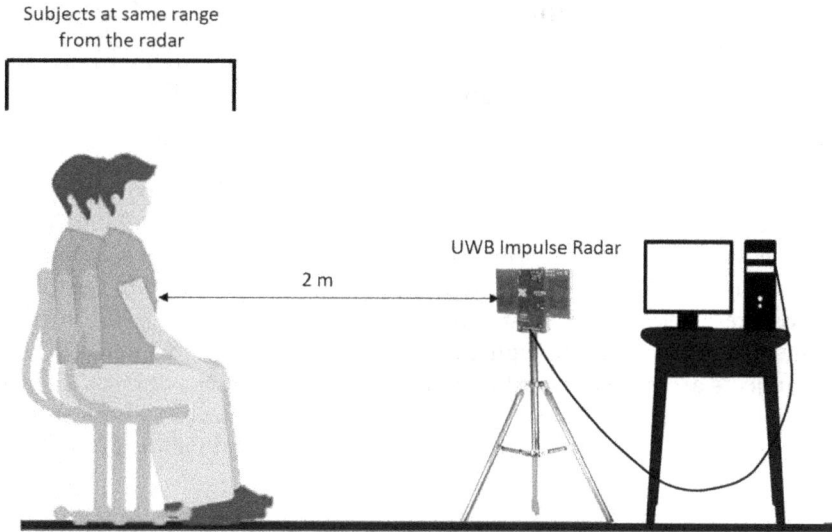

Figure 21.6 (Continued)

21.11.2 Planar-Wall Obstruction between Radar and Target

The radar is 1 m from the wall and the subject is sitting at 1 m on the other side of the wall. The wall is 30 cm thick and composed of concrete and bricks. Breathing pattern analyzed: Eupnea. Recording time: 20 s. See Figure 21.7.

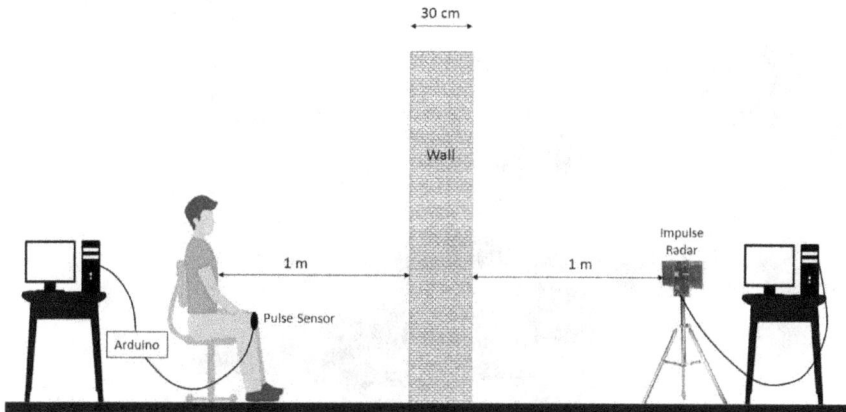

Figure 21.7 Planar wall between a single sitting target and radar [30].

21.11.3 Complex Rubble Obstruction between Radar and Target

Subjects are requested to lie down underneath a table. The rubble at the top of the table is roughly 2.1 m(L) × 1.3 m(W) × 0.25 m(H) in size. The rubble is made up of bricks, wood blocks, and concrete. The experimental specifications are as follows: Distance from radar to rubble top is 0.5 m. Two breathing patterns analyzed: Eupnea (normal resting breathing) and Tachypnea (heavy fast breathing). Time taken for each recording: 20 s. **Scenario 1:** Here, a single target is present with a eupnea breathing pattern in the following postures: Supine, prone, side and a tachypnea breathing pattern in the supine posture as shown in Figure 21.8a and Figure 21.8b. **Scenario 2:** Here, two subjects are present simultaneously within one case; a subject is sitting and another subject is in the supine position; in the other case, both subjects are in the supine position as shown in Figure 21.8c and Figure 21.8d.

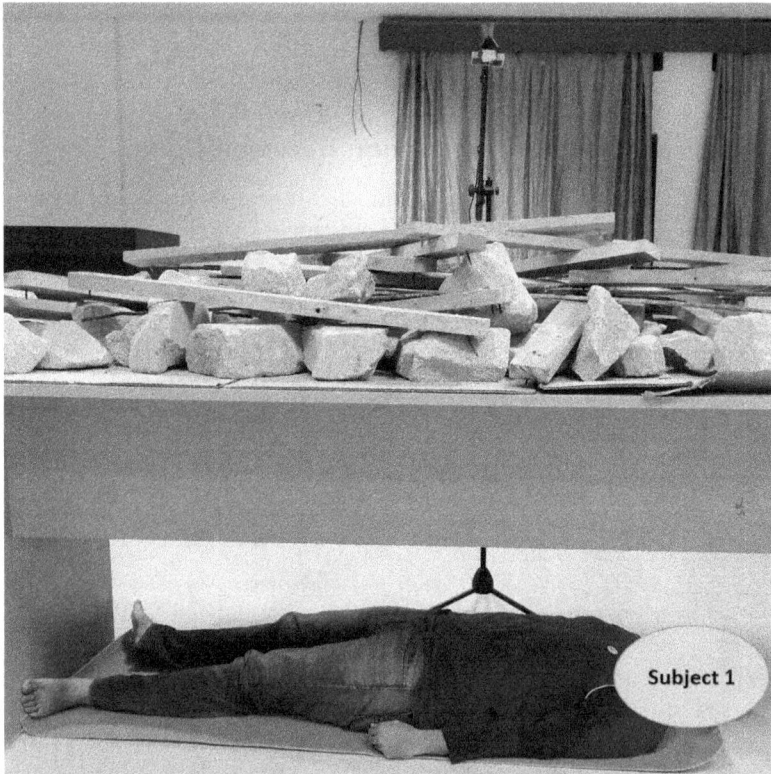

Figure 21.8 (a) One human subject sleeping beneath the rubble. (b) Three types of postures explored. (c) One person is in a sitting posture and the other is in a supine posture. (d) Both subjects are sleeping [23].

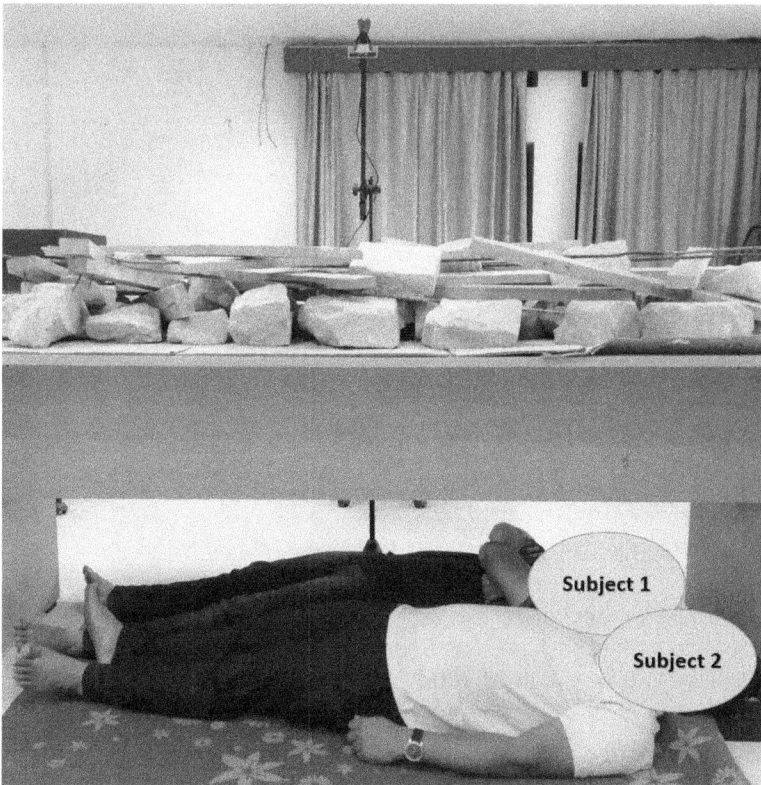

Figure 21.8 (Continued)

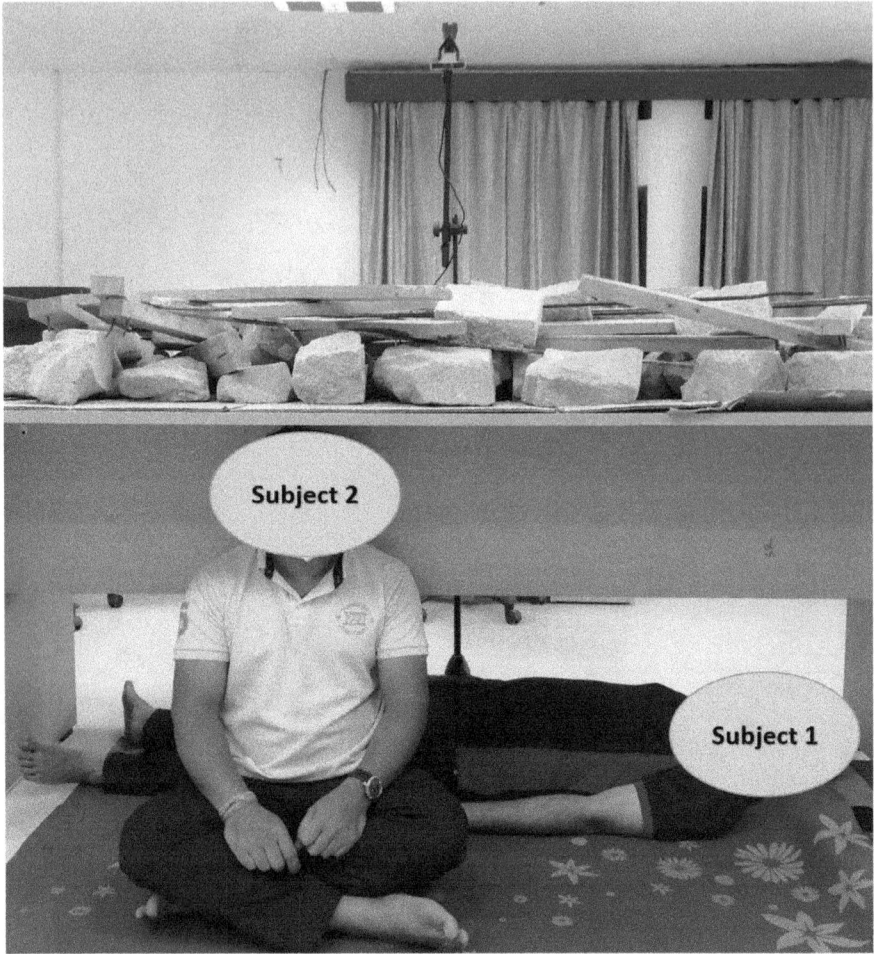

Figure 21.8 (Continued)

21.12 RESULTS AND DISCUSSION

As mentioned earlier, there can be a number of radar-target-environment scenarios. To satisfy the most realistic conditions possible, the environment with realistic rubble is considered for further analysis.

21.12.1 Estimation of Breathing Simulator Frequency

The raw image in Figure 21.9 shows that there is a lag in the measured range when the simulator is filled with debris.

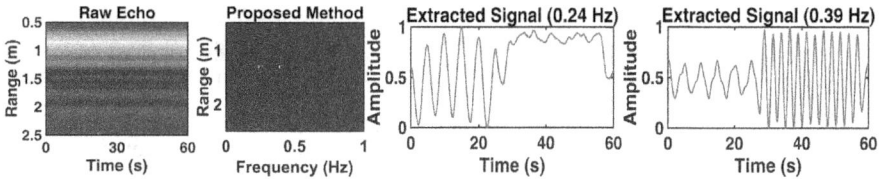

Figure 21.9 Estimation of breathing simulator movement when covered in rubble [23].

The resultant signal obtained from the movement of the simulator is collected from the finalized range bin. In the first 30 s, the estimated frequency obtained is 0.24 Hz, which is close to the actual frequencies of 0.2 Hz. And for the later 30 s, the 0.39 Hz frequency is obtained compared to 0.4 Hz. After range correction, the target range is determined to be 1 m.

21.13 ESTIMATION OF HUMAN TARGET LOCATION AND VITAL SIGNS

21.13.1 Single Human Target Per Range

Here, the situation of a single target buried by debris is considered. The target's respiration and range information were determined by processing the received data collected from each of the target postures. The scenario where an individual is in a supine position having eupneic and tachypneic breathing patterns is shown in Figure 21.10 to illustrate the systematic procedure for obtaining the attributes of the actual target.

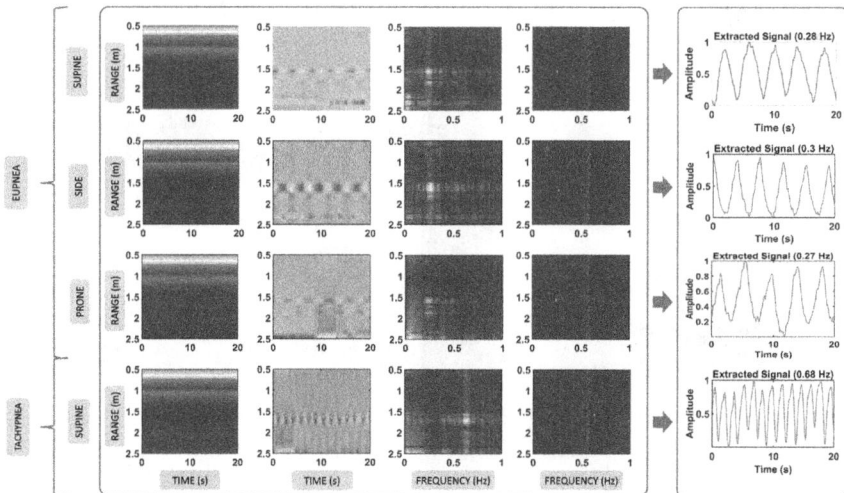

Figure 21.10 Single target vital signs and location estimation in different postures.

Figure 21.10 demonstrates how raw data obtained from different scenarios is filtered in a step-by-step fashion. This data is collected from one subject who is located under the debris. Hence, after removing the clutter and false targets, the true target location and its vital signs are obtained. In this experiment, variation of a single target's breathing frequency is also verified.

21.14 MULTIPLE TARGETS PER RANGE

This scenario can be handled in two ways as discussed next:

1. **Intertarget dynamic range (ITDR) processing:** In this case, there were two persons buried beneath the debris. In one instance, one participant was supine while the other was seated; in another, both subjects were supine. For the first case, both of the volunteers were advised to breathe in a eupneic pattern. In the second case, two volunteers were advised to breathe in different patterns: one in eupnea and another with tachypnea. Figure 21.11 shows the progression of determining the final target attributes.

   Figure 21.11 shows how the exact location and vital signs of two targets at different postures can be obtained uniquely. Even the different breathing patterns of the targets can be identified. Here, the two targets were at different ranges as well as at the same range.

   The target information at each range bin is determined by computing the ITDR values during the application of the proposed methodology. For one dataset of participants in various positions, for instance, the ITDR values estimated are between 14 dB for the seated subject and 7 dB for the supine subject. In comparison, the ITDR values for the dataset of both participants lying supine ranged from 0.8 dB to 2.8 dB.

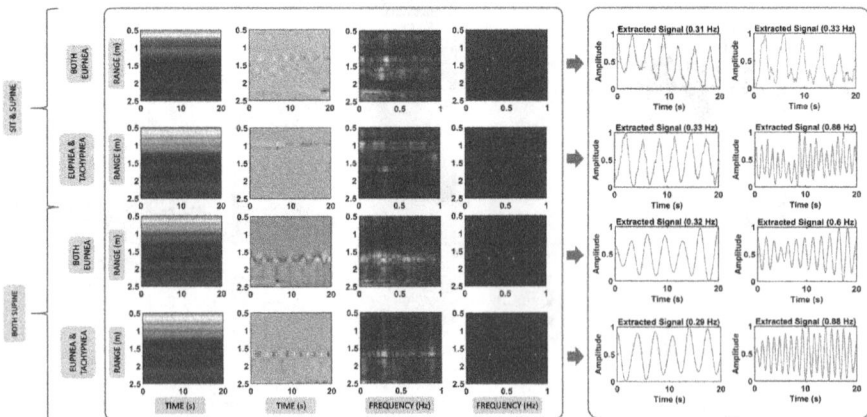

Figure 21.11 Two targets' vital signs and location estimation in different postures.

2. **Equidistant multiple target estimation (EMTE) processing:** Another methodology to differentiate the exact target count present equidistant from the radar can be accomplished by using the EMTE technique. Table 21.5 shows how different supervised machine learning algorithms performed when the dominant singular values and variance of the received echo were used as features.

The EBT classifier performed better than other classifiers. Along with accuracy, the EBT classifier also demonstrated high precision and recall, which points to the model's extreme dependability and sturdiness. The EBT classifier was also used to compute the confusion matrix, as seen in Figure 21.12.

Table 21.5 Comparison of Performance Metrics of Different Classifiers [29]

Performance Metrics	Classifiers		
	SVM	KNN	EBT
Accuracy	0.9688	0.9375	0.9750
Precision	0.9695	0.9383	0.9753
Recall	0.9688	0.9375	0.9750
F1 Score	0.9687	0.9377	0.9749

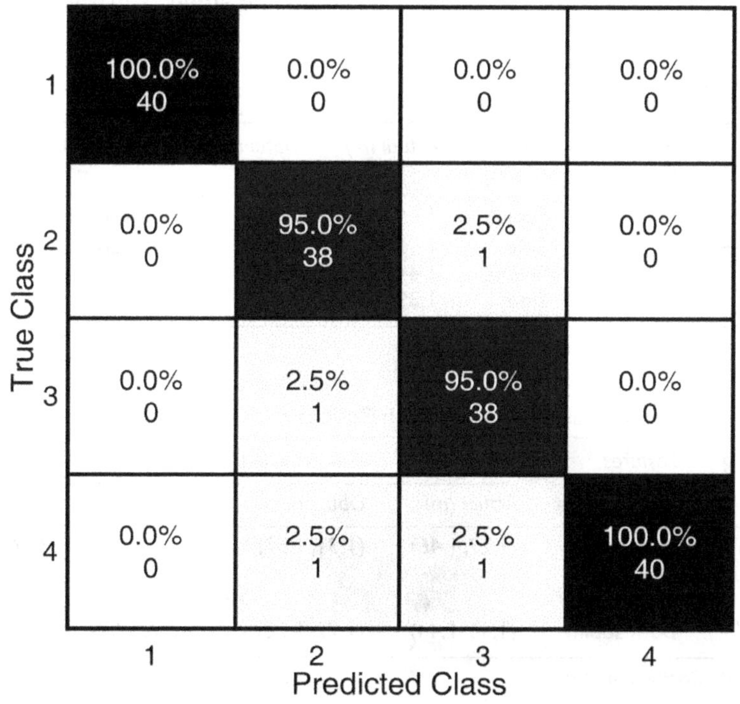

Figure 21.12 Confusion matrix obtained using EBT classifier [29].

According to the confusion matrix shown in Figure 21.12, classes 1 (no subjects) and 4 (three subjects) have been correctly identified. Classes 3 (two subjects) and 2 (one subject) were both classified with an accuracy of 95 %, although there was some overlap.

21.15 CORRELATION ANALYSIS

21.15.1 Single-Target Range Error

The value of the obtained range has grown due to the effect of the debris. For the purpose of obtaining the real range, the range compensation algorithm is used. Table 21.6 provides a summary of the discrepancy between the estimated range and the actual range.

The error values obtained are very low for the supine position and a bit higher when the target is on its side. One of the reasons for this discrepancy happened due to inconsistency in measuring the actual range when the subject is on its side.

21.15.2 Multitarget Range Error

Table 21.7 displays the range error for both the subjects at their different postures and breathing patterns.

Table 21.6 Error Analysis for Range of Individual Subjects [23]

Target	Breathing Pattern	Postures	Actual (m)	Obtained (m)	C.R. (m)	Error (%)
Subject 1	Eupnea	Supine	1.46	1.57	1.41	3.42
Subject 2		Side	1.36	1.67	1.52	11.76
		Supine	1.44	1.57	1.41	2.08
		Side	1.35	1.67	1.52	12.59

Note: C.R.: Compensated range.

Table 21.7 Error Analysis for Range of Both Co-located Subjects [23]

Breathing Pattern	Postures	Subject 1 (S1), Subject 2 (S2)			
		Actual (m)	Obtained (m)	C.R. (m)	Error (%)
Eupnea	Sitting (S2) and supine (S1)	(1.27, 1.46)	(1.31, 1.57)	(1.16, 1.42)	(8.66, 2.74)
	Both supine	(1.46, 1.44)	(1.72, 1.72)	(1.57, 1.57)	(7.53, 9.03)

Note: C.R.: Compensated range.

Furthermore, it should be observed that the targets in the first instance are not closely involved and seemed to be located at various ranges as a result of their varied positions. Additionally, compared to the situation of a single subject, the error in range is larger. It should be observed that several reflections occur between the subjects as a result of the targets' close proximity. And it's not possible to account for the extra delay brought on by the multiple reflections via the propagation delay compensation method.

21.15.3 Target Heartbeat Signal Accuracy

The instantaneous frequencies (IFs) obtained from the radar and pulse sensor data are used as the inputs for the Bland-Altman plots. The mean difference and the bounds of agreement are marked with horizontal lines, respectively. The 95 % confidence interval (CI) is the name given to this upper boundary. For subjects 1 and 2, the Hilbert method is used to generate two Bland-Altman plots, one with the presence of an obstruction in Figure 21.13b and the other without an obstruction in Figure 21.13a, to see whether their IFs agreed.

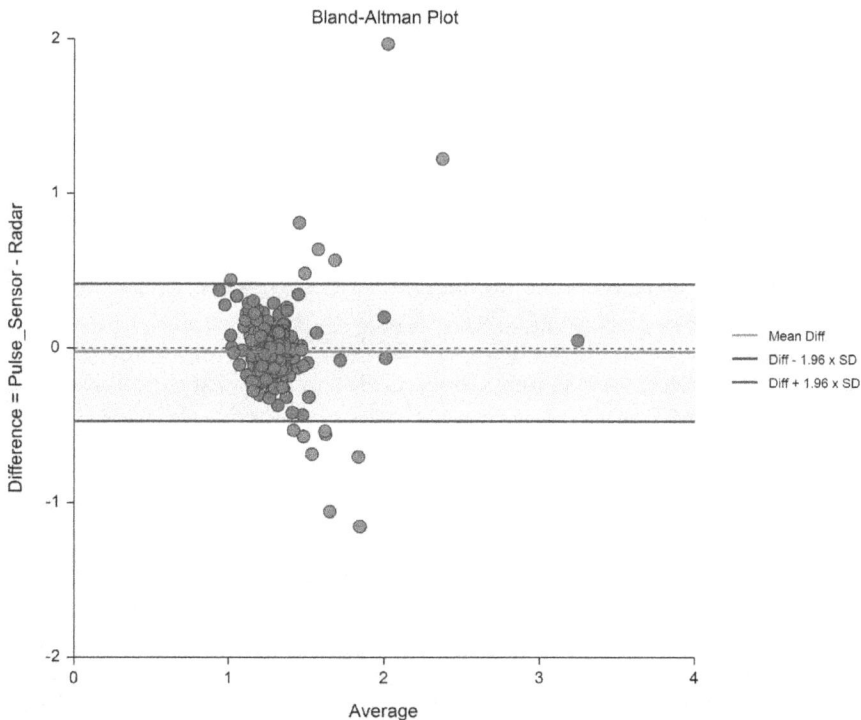

Figure 21.13 Bland-Altman plot (a) under no obstruction and (b) under an obstruction.

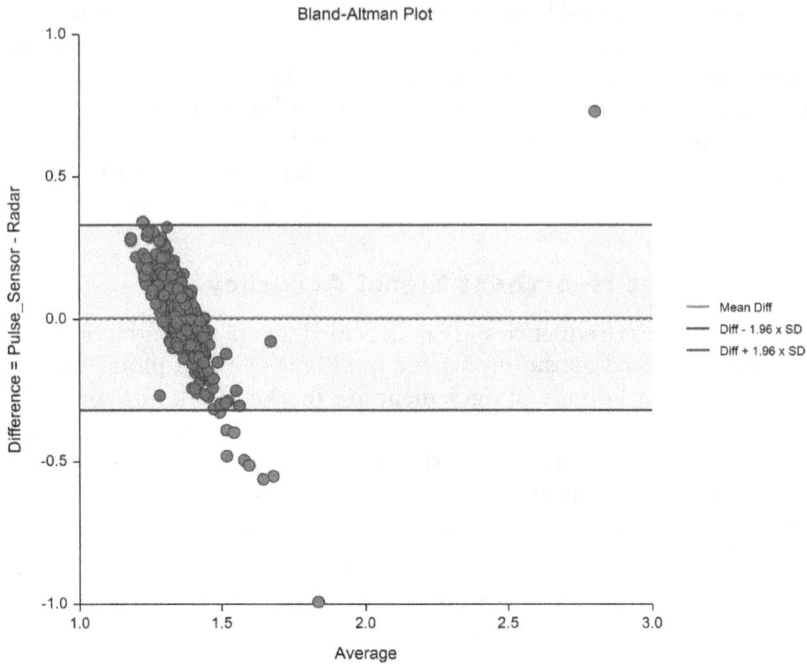

Figure 21.13 (Continued)

21.15.4 Target Breathing Signal Accuracy

To validate the chest displacement signal obtained from the received radar echo, a respiration belt was worn by the volunteers around their torso. Both of these signals were recorded simultaneously, and Figure 21.14 describes their comparison.

The following correlation coefficient is obtained the for supine posture: 86.1 %, when the subject is on its side: 83.4 %, and when the subject is in prone posture: 94.2 %.

21.16 CONCLUSION

This work highlights the problems faced by the search and rescue operators during the collapse of any artificial structures. UWB radar is shown to be very effective in locating people buried under rubble. Detailed signal processing analysis is performed on the raw received radar echo data. To replicate the true rubble conditions, extra care is taken to compose the rubble. Also, a breathing simulator is developed from scratch to be used as a human substitute covered with rubble. Human subjects are used in different experimental

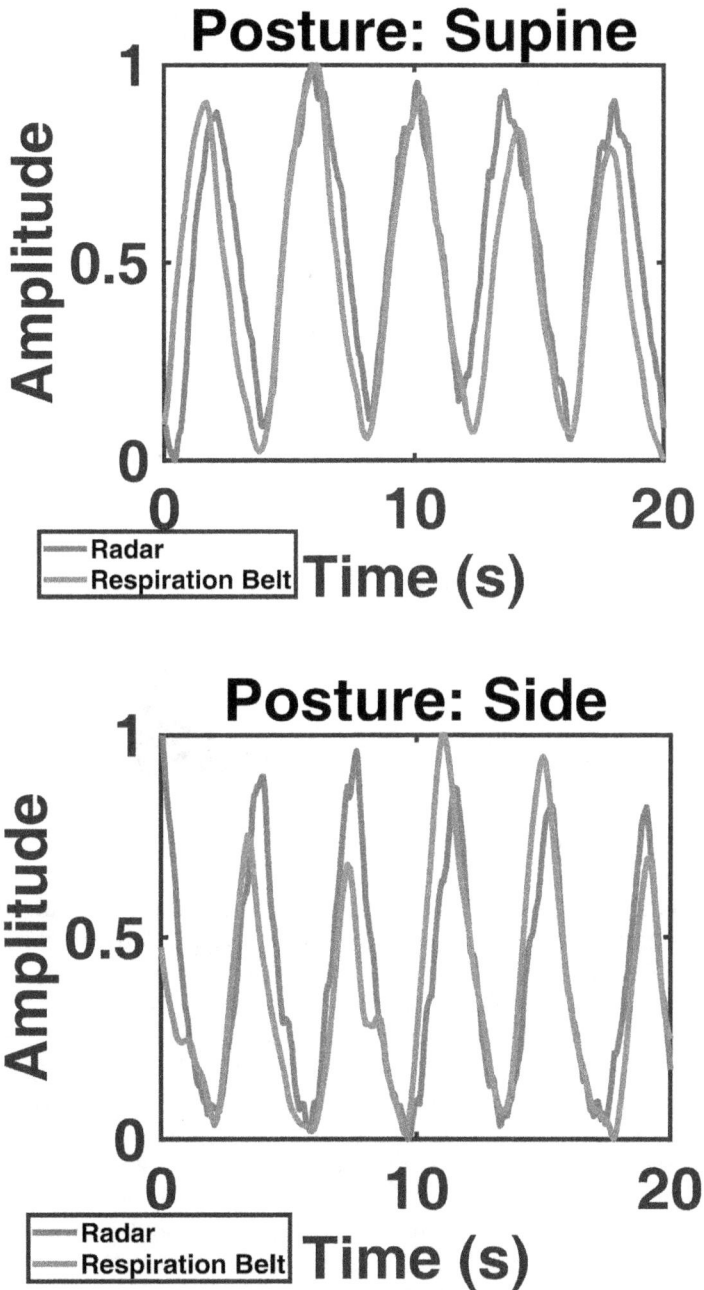

Figure 21.14 Signal similarity between the signals received from the radar and respiration belt from subject 1 in a eupnea breathing pattern. (a) Target in supine posture. (b) Target in side posture. (c) Target in prone posture [23].

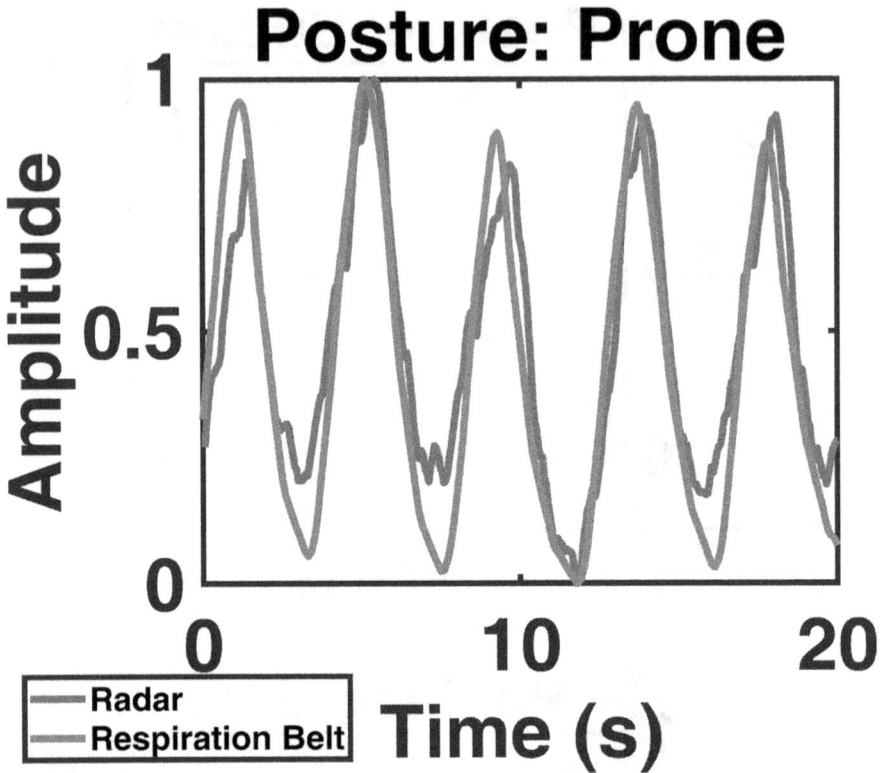

Figure 21.14 (Continued)

environments replicating different types of obstructions. The subject's posture and breathing pattern are also taken into account. Novel techniques are
applied to obtain the exact number of human subjects at each range from the
radar and consequently obtain their vital signs. The error analysis of subject
range, heart rate, and breathing rate are all documented. Overall, this work
shows extensively that a UWB radar is well equipped in accurately detecting
the location and vital signs of humans under any type of obstruction.

   There is a scope for improvement in the current work. Currently, a bistatic
radar is used for the entire experimentation. With the addition of multiple
antennas in a multiple-input, multiple-output (MIMO) configuration, beamforming as well as beam scanning can be implemented. A narrow beam will
be very helpful in locating nearby targets much more effectively without
depending too much on the signal processing algorithms. Also, a flexible
pulse repetition frequency (PRF) would provide more control in deciding the
maximum unambiguous range. To efficiently utilize the increased unambiguous range, the center frequency of the transmitted pulse needs to be lowered

more than it currently is in this work. In short, with the implementation of the previous modifications, the target detection efficiency of the radar system will increase greatly.

REFERENCES

[1] Hanifi, K., and Karsligil, M.E., 2021. Elderly fall detection with vital signs monitoring using CW doppler radar. *IEEE Sens. J.*, 21(15), pp. 16969–16978.

[2] Zhao, M., Adib, F., and Katabi, D., 2018. Emotion recognition using wireless signals. *Commun. ACM.*, 61(9), pp. 91–100.

[3] Ahmed, S., Kallu, K.D., Ahmed, S., and Cho, S.H., 2021. Hand gestures recognition using radar sensors for human-computer-interaction: a review. *Rem. Sens.*, 13(3), p. 527.

[4] Xu, H., Li, Y., Li, Y., Li, J., Wang, B., and Liu, L., 2022. Through-wall human motion recognition using random code radar sensor with multi-domain feature fusion. *IEEE Sens. J.*, 22(15), pp. 15123–15132.

[5] GSSI Geophysical Survey Systems, Inc (www.geophysical.com/products/lifelocator-trx)

[6] Cambridge Radar-Tech International Ltd (http://cambrate.com/product.php?id=17)

[7] Bresnahan, D.G., and Li, Y., 2022. Measurement of multiple simultaneous human vital signs using a millimeter-wave FMCW radar. *IEEE Radar Conference (RadarConf22)*, March 21–25, New York City, NY, pp. 1–5.

[8] Yan, J., Hong, H., Zhao, H., Li, Y., Gu, C., and Zhu, X., 2016. Through-wall multiple targets vital signs tracking based on VMD algorithm. *Sensors*, 16(8), p. 1293.

[9] Liang, X., Zhang, H., Ye, S., Fang, G., and Gulliver, T.A., 2018. Improved denoising method for through-wall vital sign detection using UWB impulse radar. *Dig. Sig. Proc.*, 74, pp. 72–93.

[10] Yang, Z., Ma, C., Qi, Q., Li, X., and Li, Y., 2021. Applications of TVF-EMD in vital signal detection for UWB radar. *J. Sens.*, p. 18.

[11] Yang, D., Zhu, Z., and Liang, B., 2019. Vital sign signal extraction method based on permutation entropy and EEMD algorithm for ultra-wideband radar. *IEEE Access*, 10(7), pp. 178879–178890.

[12] Rohman, B.P.A., Andra, M.B., and Nishimoto, M., 2021. Through-the-wall human respiration detection using UWB impulse radar on hovering drone. *IEEE J. Sel. Top. Appl. Earth Obs. Remote Sens.*, 14, pp. 6572–6584.

[13] Accidental Deaths & Suicides in India (ADSI) (https://ncrb.gov.in/en/accidental-deaths-suicides-india-adsi)

[14] Effects of Building Materials and Structures on Radiowave Propagation Above about 100 MHz, 2015. International Telecommunications Union Recommendation ITU-RP. 2040-1.

[15] Cavagnaro, M., Pittella, E., and Pisa, S., 2013. UWB pulse propagation into human tissues. *Phys. Med. Biol.*, 58(24), pp. 8689–8707.

[16] Andreuccetti, D., Fossi, R., and Petrucci, C., 1996. An internet resource for the calculation of the dielectric properties of body tissues in the frequency range

10 Hz–100 GHz. *IFAC-CNR*, Florence (Based on data published by C. Gabriel et al. in 1996). Available at: http://niremf.ifac.cnr.it/tissprop/.

[17] Yang, Z., Ma, C., Qi, Q., Li, X., and Li, Y., **2021**. Applications of TVF-EMD in vital signal detection for UWB radar. *J. Sens.*, p. 18.

[18] Klema, V.C., and Laub, A.J., **1980**. The singular value decomposition: its computation and some applications. *IEEE Trans. Autom. Cont.*, AC-25(2), pp. 164–176.

[19] Moghaddam, S.S., and Jalaei, S., **2013**. Determining the number of coherent/correlated sources using FBSS-based methods. *Front. Sci.*, 2(6), pp. 203–208.

[20] Yang, Z., Ma, C., Qi, Q., Li, X., and Li, Y., **2021**. Applications of TVF-EMD in vital signal detection for UWB radar. *J. Sens.*, p. 18.

[21] Liang, X., Zhang, H., Ye, S., Fang, G., and Gulliver, T.A., **2018**. Improved denoising method for through-wall vital sign detection using UWB impulse radar. *Dig. Sig. Proc.*, 74, pp. 72–93.

[22] Wu, S., Yao, S., Liu, W., Tan, K., Xia, Z., Meng, S., et al., **2016**. Study on a novel UWB linear array human respiration model and detection method. *IEEE J. Sel. Top. Appl. Earth Obs. Remote Sens.*, 9(1), pp. 125–140.

[23] Sarkar, A., and Ghosh, D., **2022**. Accurate sensing of multiple humans buried under rubble using IR-UWB SISO radar during search and rescue. *Sens. Actuators A Phys.*, 348, Part A, no. 113975.

[24] Laonuri Corp (www.laonuri.com/en/product/x4m200/)

[25] Vernier (www.vernier.com/product/go-direct-respiration-belt)

[26] Liang, X., Zhang, H., Fang, G., Ye, S., and Gulliver, T.A., **2017**. An improved algorithm for through-wall target detection using ultra-wideband impulse radar. *IEEE Access*, 5, pp. 22101–22118.

[27] Ren, L., Wang, H., Naishadham, K., Kilic, O., and Fathy, A.E., **2016**. Phase-based methods for heart rate detection using UWB impulse doppler radar. *IEEE Trans. Microw. Theory Tech.*, 64(10), pp. 3319–3331.

[28] Mishra, A., and Li, C., **2019**. A low power 5.8-GHz ISM-band intermodulation radar system for target motion discrimination. *IEEE Sens. J.*, 19(20), pp. 9206–9214.

[29] Sarkar, A., and Ghosh, D., **2020**. Detection of multiple humans equidistant from IR-UWB SISO radar using machine learning. *IEEE Sens. Lett.*, 4(1), pp. 1–4.

[30] Sarkar, A., and Ghosh, D., **2019**. Through-wall heartbeat frequency detection using ultra-wideband impulse radar. *2019 International Conference on Range Technology (ICORT)*, February 15–17, Balasore, pp. 1–5.

Chapter 22

Employment of Antennas in Biomedical Applications

A Review

Krishanu Kundu and Narendra Nath Pathak

22.1 INTRODUCTION

The growth of the healthcare sector has brought a lot of interest to bio-medical telemetry. Without regular hospital check-ups and follow-up routine check-ups, it is now possible to remotely monitor a patient's physiological indications. Implantable medical devices (IMDs) are crucial for wireless telemetry patient monitoring. IMDs are made up of nodes and implanted sensors, which heavily include antenna. The implanted sensors have several drawbacks. For the implanted sensors to operate reliably and continuously, a number of variables need to be taken into account, including miniaturisation, patient safety, biocompatibility, less power consumption, reduced frequency band along with dual-band operation. In the design of implanted sensors, choosing the antenna is a difficult problem since it affects how well the entire implant performs. The antenna, battery, and sensors are only a few of the many parts that make up implantable devices. The implanted antenna plays a significant role in the communication link's dependability and durability between the inner and outer equipment. An implanted device's antenna is a key component since it is crucial to the basic operating need of signal receiving and transmission. Also, it has an impact on the implanted device's total size and weight. The difficulty of the design is greatly increased by the extremely hostile electromagnetic atmosphere found in the body. Although research in this field is expanding, a thorough overview is still needed to familiarise antenna designers with the most recent and cutting-edge advancements. A few examples of such types of implantable antennas are planar antennas [1–5], wire antennas [6–9], conformal antennas [10–13], spiral antennas [14–17], slot antennas [18–20], and planar inverted F antennas (PIFAs) [21, 22]. Planar antennas are inexpensive ultra-wideband antennas, particularly when they are mass-produced using printed circuit board (PCB) technology. They are perfect for wireless applications since they may be quite tiny. The aperture of planar arrays is substantial. By altering the phase of every constituent, directional beam regulation can be achieved. Microstrip and PCB antennas are examples of planar antennas. The "patches" of the

DOI: 10.1201/9781003459880-22

antennas might be round, square, or triangular. A long wire hung over the ground makes up a wire antenna, a form of radio antenna. The antenna's wire gathers the signals and radiates them further. The length of the wire antenna is independent of its wavelength. To send or receive signals, the cable is only linked to the transmitter or receiver through the tuner of an antenna. These antennas are widely recognised for being portable and easy to install. Phased-array antennas come in the form of conformal antennas. They consist of a surface-covering array of several identical tiny, flat antenna components, such as dipole, horn, or patch antennas. Every location where an analytical function has a nonzero derivative is a conformal point. On the other hand, analytic conformal mappings of complex variables are those that have continuous partial derivatives. Complex analysis, as well as many physics and engineering fields, depend greatly on conformal mapping. A spiral antenna is a kind of radio frequency (RF) antenna used in microwave systems. It has a two-arm spiral form; however, more arms might be added. The first spiral antenna was described in 1956. The family of frequency-independent antennas that logarithmic spiral antennas belong to maintain its driving point impedance, radiation pattern, and polarisation throughout a wide bandwidth range. Circularly polarised spiral antennas have a poor gain by nature. To boost the gain, antenna arrays might be employed. Reduced in size, spiral antennas have a very tiny construction because of their windings. While a unidirectional pattern is typically sought in such antennas, lossy holes are typically added at the rear to eliminate back lobes. A metal surface, often a flat plate, has one or more slots or holes carved out of it to form a slot antenna. The slot emits electromagnetic waves in a manner akin to a dipole antenna when the plate is driven as an antenna by an RF current. Slot antennas are applied in the 300 MHz to 30 GHz frequency band. The ultra high frequency (UHF) and super high frequency (SHF) bands are where it operates. The PIFA is primarily suited for mobile devices. Due to its versatility and ability to support a variety of wireless services, PIFA is primarily developed for dual-band frequencies. As a result, it becomes more acceptable for use in mobile device design. The antenna spans six distinct frequency bands and resonates between 0.841 GHz and 2.12 GHz while varying the statuses of the diodes. Because of its ease of modification to fit the desired resonant frequency, impedance, polarisation, and radiation pattern, patch antennas are more common. Due to the fact that the human body functions as a dielectric lossy medium, the antenna can often lose performance while operating close to the human body. The directivity gain and efficiency of an antenna can be reduced as a result of the discontinuity in the varied dielectric characteristics of different human tissues. To increase the antenna's performance, a variety of strategies have been used, including slots on the patch, various dielectric materials, feeding techniques, and defective ground structures (DGSs).

22.2 DESIGN REQUIREMENTS OF BIO-IMPLANTABLE ANTENNAS

When compared to conventional antennas functioning in free space, implanted antennas operate entirely differently within the human body. Although some of their design specifications vary on the application, several important design clauses are universal and are included in this section.

A. Miniaturisation: The downsizing of the implanted antenna is a highly important element since it is a significant component of an implantable device. Here, several miniaturisation methods [23] are being used to make the antenna smaller. These are a few popular techniques:
  1. Usage of a high permittivity dielectric substrate: By employing a substrate with a higher dielectric permittivity, the size of the bio-implantable antenna can be lowered. The substrate with a high permittivity would have a shorter wavelength, which would reduce the resonance frequency. The antenna's input power may be transformed into surface waves if a substrate or superstrate with a very high permittivity is used, which will reduce the antenna's radiation efficiency.
  2. PIFAs: Creating implantable antennas with smaller dimensions is made simple using PIFA architecture. For this reason, microstrip patch antennas are also employed. The PIFA's wavelength is one-quarter of its resonant length, compared to the microstrip patch antenna's wavelength of half. Hence, in applications where size reduction is required, PIFAs are favoured in place of microstrip patch antennas.
  3. Radiator current route broadening: Lengthening the radiator's current path is a further technique for shrinking implantable antennas [24]. The implanted antenna's resonance frequency is lowered by a longer current route. Two radiators are placed on top of one another, either vertically or horizontally, in a method known as "radiator stacking," which effectively lengthens the current track with respect to the radiator. By using this strategy, it is possible to reduce the size by 33%.
B. Patient safety: Internally placed medical devices are known as implants. Thus, careful evaluation is required to rule out electromagnetic exposure–related tissue injury. As stated next, several measures are used to guarantee patient safety:
  1. SAR: The quantity of RF energy absorbed by human body tissues is known as the specific absorption rate (SAR) [25]. It serves as a gauge to guarantee the protection of biological tissues from electromagnetic radiation. There are two globally accepted SAR

standards. The SAR should be less than 1.6 W/kg on average across 1 g of tissue volume, as required by the IEEE C95.1–1999 standard [26]. According to the IEEE C95.1–2005 standard, the SAR must be less than 2 W/kg on average across a tissue's 10 g cubic volume [27]. The implanted devices need to use modest output power to keep typical SAR values.

2. Effective isotropic radiated power (EIRP): An implanted antenna with a high EIRP [28] can damage the body and cause interference with neighbouring radio equipment. For an implanted antenna operating in the MedRadio band, the conventional limit of EIRP is –16 dBm, while for the ISM band, it is –20 dB. The input power should be restricted if the implanted antenna is utilised for data telemetry to prevent tissue harm. If the implanted antenna serves as a receiver, these requirements should be followed by the external source power.

C. Biocompatibility: To guarantee the safety of the human subject, biocompatibility is a crucial consideration for implanted antennas. Because human body tissues are electrically conductive, if the antenna made direct contact with them, it would short-circuit. To ensure biocompatibility, two methods are frequently employed. Both constructing the antenna from biocompatible material and encasing it in a biocompatible superstructure are viable options.

D. Techniques for fabricating implantable antennas: Due to their intended use, implantable antennas require the highest level of care during fabrication. Several various manufacturing techniques are being suggested in the literature, such as 3D printed antennas, microfluidic antennas, photolithography, and antennas embroidered on fabric [29–31]. Polydimethylsiloxane (PDMS) was incorporated in conductive fabric in [32]. As compared to traditional fabrication techniques, utilising conductive fabric and PDMS to fabricate the antenna (which simultaneously serves as the substrate and a protective encapsulation) results in a more durable and flexible antenna construction. To accomplish the flexibility and robustness in [33], an ultra-wideband antenna is constructed using the same method. Similar antennas for biomedical applications have been discussed in [34].

22.3 CONCEPT BEHIND USAGE OF BIO-IMPLANTABLE ANTENNAS FOR TUMOUR DETECTION

We can characterise any biological tissue in terms of electrical characteristics like permittivity, tensor losses, etc., if we sandwich the biological tissue between two antennas that serve as sensors. Blood, fat, and several other biological components with varying permittivity and tensor losses make up biological tissue, which is fundamentally a very nonhomogeneous material. Nevertheless, we may efficiently represent this tissue using a single permittivity and tensor loss, which is essentially homogenous modelling at a certain frequency. Due to the existence of cancer cells, the homogeneity of such tissue will be compromised, and any changes to the tissue's (Figure 22.1) electric properties will be reflected in the transmission coefficient.

The signal propagation from bio-implantable antennas (Figure 22.2), namely from antenna 1 to antenna 2, will be affected by presence of a tumour cell inside the breast tissue. This is basically a two-port network. This entire system can be modelled using a scattering matrix. For the N = 2 port network the scattering matrix will contain $N^2 = 4$ elements. We will find a different transmission coefficient vs. frequency plots (responses) depending upon tumour size. By having multiple simulation plots (via CST/HFSS) (Figures 22.3 and 22.4) we can launch a correlation between the transmission coefficient along with tumour cell size and location.

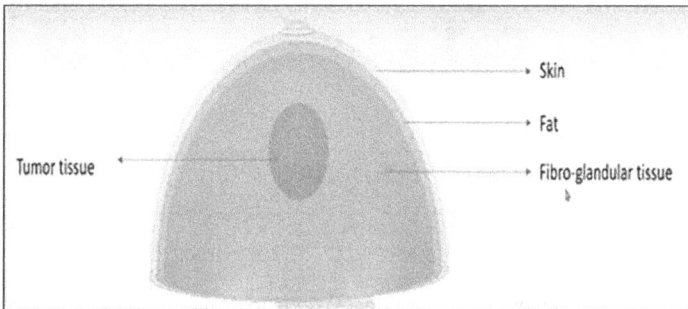

Figure 22.1 Tumour tissue inside breast tissue.

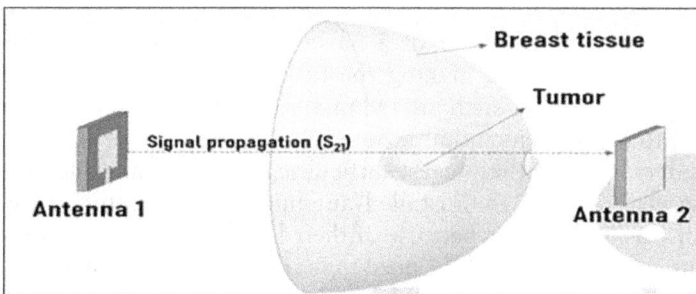

Figure 22.2 Usage of bio-implantable antennas for tumour detection.

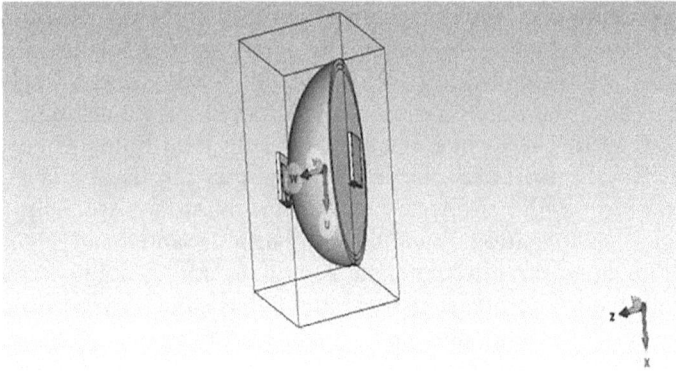

Figure 22.3 Simulation of bio-implantable antennas for tumour detection using CST.

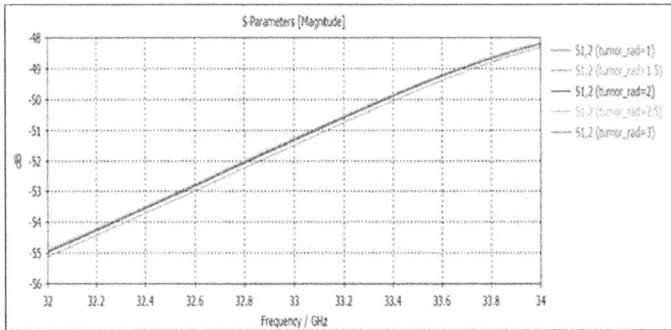

Figure 22.4 Gain vs. frequency plots (responses) depending upon tumour size (with respect to transmission coefficient).

22.4 CONCLUSION

The antenna is a crucial part of a biosensor that operates in the hostile environment of the body, making the task of designing it extremely difficult. While developing such an antenna, a number of parameters need to be taken into account, including size, gain, efficiency, radiation pattern, and patient safety. One of the biggest difficulties with implantable antennas is providing electricity to the installed antenna system. Batteries are a poor choice for this application because of their limited lifespan, the presence of dangerous components, and the need for surgery to replace them. The future of bio-implantable antennas will be smooth if proper attention is given to these challenges.

REFERENCES

[1] Komal Jaiswal, Ankit Kumar Patel, Shekhar Yadav, Sweta Singh, Ram Suchit Yadav, and Rajeev Singh. Probe-fed wideband implantable microstrip patch antenna for biomedical and telemetry applications. In Soft Computing and Signal Processing, pp. 635–642. Springer, 2019.

[2] Yusheng Fu, Jianmin Lei, Xiao Zou, and Jiuchuan Guo. Flexible antenna design on PDMS substrate for implantable bioelectronics applications. Electrophoresis, 40(8):1186–1194, 2019.

[3] Syed Ahson Ali Shah and Hyoungsuk Yoo. Scalp-implantable antenna systems for intracranial pressure monitoring. IEEE Transactions on Antennas and Propagation, 66(4):2170–2173, 2018.

[4] N.H. Sulaiman, N.A. Samsuri, M.K.A. Rahim, F.C. Seman, and M. Inam. Compact meander line telemetry antenna for implantable pacemaker applications. Indonesian Journal of Electrical Engineering and Computer Science, 10(3):883–889, 2018.

[5] Santanu Maity, Kuheli Roy Barman, and Shankar Bhattacharjee. Silicon-based technology: Circularly polarized microstrip patch antenna at ISM band with miniature structure using fractal geometry for biomedical application. Microwave and Optical Technology Letters, 60(1):93–101, 2018.

[6] Saber Helmy Zainud-Deen, Hend Abd El-Azem Malhat, and Asmaa Adel Balabel. Octafilar helical antenna for circular polarization wireless capsule endoscopy applications. Wireless Personal Communications, 1–11, 2019.

[7] Xin Wang, Jingjing Shi, Lisheng Xu, and Jianqing Wang. A wideband miniaturized implantable antenna for biomedical application at HBC band. In 2018 Cross Strait Quad-Regional Radio Science and Wireless Technology Conference (CSQRWC), pp. 1–3. IEEE, 2018.

[8] N.A. Malik, P. Sant, T. Ajmal, and M. Ur-Rehman. Implantable antennas for bio-medical applications. IEEE Journal of Electromagnetics, RF and Microwaves in Medicine and Biology, 5(1):84–96, 2020.

[9] Sang Heun Lee, Kihun Chang, Ki Joon Kim, and Young Joong Yoon. A conical spiral antenna for wideband capsule endoscope system. In 2008 IEEE Antennas and Propagation Society International Symposium, pp. 1–4. IEEE, 2008.

[10] Rongqiang Li, Yong-Xin Guo, and Guohong Du. A conformal circularly polarized antenna for wireless capsule endoscope systems. IEEE Transactions on Antennas and Propagation, 66(4):2119–2124, 2018.

[11] Kumar Naik Ketavath, Dattatreya Gopi, and Sriram Sandhya Rani. In vitro test of miniaturized CPW-fed implantable conformal patch antenna at ISM band for biomedical applications. IEEE Access, 7:43547–43554, 2019.

[12] Ke Zhang, Changrong Liu, Xueguan Liu, Honglong Cao, Yudi Zhang, Xinmi Yang, and Huiping Guo. A conformal differentially fed antenna for ingestible capsule system. IEEE Transactions on Antennas and Propagation, 66(4):1695–1703, 2018.

[13] N. Mahalakshmi and A. Thenmozhi. Design of hexagon shape bowtie patch antenna for implantable bio-medical applications. Alexandria Engineering Journal, 56(2):235–239, 2017.

[14] Li-Jie Xu, Yaming Bo, Wen-Jun Lu, Lei Zhu, and Cheng-Fei Guo. Circularly polarized annular ring antenna with wide axial-ratio bandwidth for biomedical applications. IEEE Access, 7:59999–60009, 2019.

[15] Izaz Ali Shah, Muhammad Zada, and Hyoungsuk Yoo. Design and analysis of a compact-sized multiband spiral-shaped implantable antenna for scalp implantable and leadless pacemaker systems. IEEE Transactions on Antennas and Propagation, 67(6):4230–4234, 2019.

[16] Gopinath Samanta and Debasis Mitra. Dual-band circular polarized flexible implantable antenna using reactive impedance substrate. IEEE Transactions on Antennas and Propagation, 67(6):4218–4223, 2019.

[17] Li-Jie Xu, Yong-Xin Guo, and Wen Wu. Miniaturized circularly polarized loop antenna for biomedical applications. IEEE Transactions on Antennas and Propagation, 63(3):922–930, 2015.

[18] Srinivasan Ashok Kumar and Thangavelu Shanmuganantham. Coplanar waveguide-fed ISM band implantable crossed-type triangular slot antenna for biomedical applications. International Journal of Microwave and Wireless Technologies, 6(2):167–172, 2014.

[19] Soumyadeep Das and Debasis Mitra. A compact wideband flexible implantable slot antenna design with enhanced gain. IEEE Transactions on Antennas and Propagation, 66(8):4309–4314, 2018.

[20] Huiying Zhang, Long Li, Changrong Liu, Yong-Xin Guo, and Sirao Wu. Miniaturized implantable antenna integrated with split resonate rings for wireless power transfer and data telemetry. Microwave and Optical Technology Letters, 59(3):710–714, 2017.

[21] Rongqiang Li, Bo Li, Guohong Du, Xiaofeng Sun, and Haoran Sun. A compact broadband antenna with dual-resonance for implantable devices. Micromachines, 10(1):59, 2019.

[22] Lingzhi Luo, Bing Hu, Jiahui Wu, Tianwen Yan, and Li-Jie Xu. Compact dual-band antenna with slotted ground for implantable applications. Microwave and Optical Technology Letters, 61(5):1314–1319, 2019.

[23] Muhammad Zada and Hyoungsuk Yoo. A miniaturized triple-band implantable antenna system for bio-telemetry applications. IEEE Transactions on Antennas and Propagation 66(12):7378–7382, 2018.

[24] Nabeel Ahmed Malik, Tahmina Ajmal, Paul Sant, and Masood Ur-Rehman. A compact size implantable antenna for bio-medical applications. In 2020 International Conference on UK-China Emerging Technologies (UCET), pp. 1–4. IEEE, 2020.

[25] K.C. Perumalla and P. Muthusamy. Opportunistic control of vertical stub compact pentagonal slot circular polarized implantable antenna with SAR analysis for bio-medical applications. Optik, 258:168882, 2022.

[26] Federal Communications Commission. Establishment of a medical implant communications service in the 402–405 MHz band. Federal Register, 64(240):69926–69934, 1999. https://www.govinfo.gov/app/details/FR-1999-12-15/99-32454.

[27] James C. Lin. Safety standards for human exposure to radio frequency radiation and their biological rationale. IEEE Microwave Magazine, 4(4):22–26, 2003.

[28] Tianlin Wang, Christopher S. Ruf, Scott Gleason, Andrew J. O'Brien, Darren S. McKague, Bruce P. Block, and Anthony Russel. Dynamic calibration of GPS effective isotropic radiated power for GNSS-reflectometry earth remote sensing. IEEE Transactions on Geoscience and Remote Sensing, 60: 1–12, 2021.

[29] Bahare Mohamadzade, Raheel M. Hashmi, Roy B. V. B. Simorangkir, Reza Gharaei, Sabih Ur Rehman, and Qammer H. Abbasi. Recent advances in fabrication methods for flexible antennas in wearable devices: State of the art. Sensors, 19(10):2312, 2019.

[30] Karu P. Esselle, Basit A. Zeb, Raheel Hashmi, and Roy B. V. B. Simorangkir. Antennas for wireless medical devices. In Antennas for Small Mobile Terminals, pp. 183–202. Artech House, 2018.

[31] Anindya Nag, Roy B. V. B. Simorangkir, Elizabeth Valentin, Toni Björninen, Leena Ukkonen, Raheel M. Hashmi, and Subhas Chandra Mukhopadhyay. A transparent strain sensor based on PDMS-embedded conductive fabric for wearable sensing applications. IEEE Access, 6:71020–71027, 2018.

[32] Roy B. V. B. Simorangkir, Yang Yang, Raheel M. Hashmi, Toni Björninen, Karu P. Esselle, and Leena Ukkonen. Polydimethylsiloxane-embedded conductive fabric: Characterization and application for realization of robust passive and active flexible wearable antennas. IEEE Access, 6:48102–48112, 2018.

[33] B. Mohamadzade, R. B. V. B. Simorangkir, R. M. Hashmi, and A. Lalbakhsh. A conformal ultrawideband antenna with monopole-like radiation patterns. IEEE Transactions on Antennas and Propagation, 68(8):6383–6388, Aug. 2020. doi: 10.1109/TAP.2020.2969744.

[34] Ankan Bhattacharya and Souvik Pal. An extremely compact and low-cost antenna sensor designed for IoT-Integrated biomedical applications. In Internet of Things and Data Mining for Modern Engineering and Healthcare Applications. Taylor & Francis. https://www.taylorfrancis.com/chapters/edit/10.1201/9781003217398-17/extremely-compact-low-cost-antenna-sensor-designed-iot-integrated-biomedical-applications-ankan-bhattacharya-souvik-pal?context=ubx&refId=d1883476-1f52-42cc-941e-5ad5e6673051.

Index

For Product Safety Concerns and Information please contact our EU
representative GPSR@taylorandfrancis.com
Taylor & Francis Verlag GmbH, Kaufingerstraße 24, 80331 München, Germany